Additive Manufacturing for Bio-Composites and Synthetic Composites

Additive Manufacturing for Bio-Composites and Synthetic Composites focuses on processes, engineering, and product design applications of bio-composites and synthetic composites in additive manufacturing (AM). It discusses the preparation and material characterization and selection, as well as future opportunities and challenges.

- Reviews the latest research on the development of composites for AM and the preparation of composite feedstocks.
- Offers an analytical and statistical approach for the selection of composites for AM, including characterization of material properties.
- Emphasizes the use of environmentally friendly composites.
- Analyzes the lifecycle including costs.
- Considers potential new fibers, their selection, and future applications.

This book provides a comprehensive overview of the application of advanced composite materials in AM and is aimed at researchers, engineers, and advanced students in materials and manufacturing engineering and related disciplines.

Additive Manufacturing for Bio-Composites and Synthetic Composites

Edited by
M. T. Mastura, S. M. Sapuan,
and R. A. Ilyas

CRC Press
Taylor & Francis Group
Boca Raton London New York

CRC Press is an imprint of the
Taylor & Francis Group, an **informa** business

First edition published 2024
by CRC Press
6000 Broken Sound Parkway NW, Suite 300, Boca Raton, FL 33487-2742

and by CRC Press
4 Park Square, Milton Park, Abingdon, Oxon, OX14 4RN

CRC Press is an imprint of Taylor & Francis Group, LLC

ISBN: 978-1-032-42293-0 (hbk)
ISBN: 978-1-032-42294-7 (pbk)
ISBN: 978-1-003-36212-8 (ebk)

DOI: 10.1201/9781003362128

Typeset in Times
by codeMantra

Contents

Preface..xi
Editors...xiii
List of Contributors..xvii

Chapter 1 Sustainable Biocomposites: Characterization and Applications
in 3D and 4D Printing..1

Vasi Uddin Siddiqui, Sapuan S.M., Sulaiman Ahmad,
Luqman Hakim Noorazlan, Neassan Muthamil Selvan,
Shudharson Ilengo, Pavithren Mailvaganan, and
Vairravell Khumar Ravindran

1.1   Introduction 1
1.2   Materials Used for 3D/4D Printing 4
     1.2.1   Poly Lactic Acid 6
     1.2.2   Poly Butylene Succinate (PBS) 7
     1.2.3   Acrylonitrile Butadiene Styrene................................7
     1.2.4   PBS Blends with PLA 9
     1.2.5   ABS Blend..9
     1.2.6   Smart Material 10
1.3   Application of Biocomposites for 3D and 4D Printing 13
1.4   Conclusion 16
Acknowledgments 17
References 17

Chapter 2 Fabrication Process of Bio-Composites Filament for Fused
Deposition Modeling: A Review 20

Hazliza Aida C.H., Mastura M.T., Abdul Kudus S.I.,
Ilyas R.A., and Loh Y.F.

2.1   Introduction 20
2.2   Bio-Composite Filament Production 24
     2.2.1   Composite Filament Preparation..............................24
     2.2.2   Compounding Process...29
     2.2.3   Extrusion Process...31
2.3   Mechanical Attributes of FDM
Bio-Composites Filaments 34
2.4   Future Trends in FDM Bio-Composite 36
2.5   Conclusion 37
Acknowledgment...37
References 37

Chapter 3   Biodegradable Natural Fiber Polymer Composite as Future 3D
               Printing Feedstock: A Review..42

               Hamat S., Ishak M.R., Sapuan S.M., and Yidris N.

               3.1   Introduction 42
               3.2   Biodegradable Polymers, Properties, and Applications 44
               3.3   Natural Fibers, Properties, and Applications 48
               3.4   Polymer Composite of FFF 50
               3.5   Biodegradable Polymer and Natural Fiber for FFF...............52
               3.6   Challenges and Future Opportunities.......................................52
               3.7   Conclusion 53
               Acknowledgment...53
               References 53

Chapter 4   Development of 3D Printing Filament Material Using Recycled
               Polypropylene (rPP) Reinforced with Coconut Fiber 59

               *Yusliza Yusuf, Nuzaimah Mustafa, Mastura M.T.,
               Muhamad Ariffadzilah Mohd Latip, and Dwi Hadi S.*

               4.1   Introduction 59
               4.2   rPP and Its Application 60
               4.3   Biocomposites Materials 61
               4.4   Composites with Coconuts Fiber Reinforcement...................62
               4.5   Advantages and Disadvantages of Existing Filament
                     Materials for Commercial 3D Printing 65
               4.6   Producing Filament for 3D Printing from rPP Reinforced
                     with Coconut Fiber 66
               4.7   Conclusion 74
               Acknowledgment...74
               References 74

Chapter 5   Advances in Polylactic Acid Composites with Biofiller as 3D
               Printing Filaments in Biomedical Applications 76

               *Azmah Hanim Mohamed Ariff, Mohammad Akram Syazwie
               Mohd Areff, Che Nor Aiza Jaafar, Zulkiflle Leman,
               and Recep Calin*

               5.1   Biodegradable Polymer 76
                     5.1.1   Biodegradable Polymers Derived from Petroleum
                             Resources 77
                     5.1.2   Biodegradable Polymers Derived from
                             Renewable Resources...78
               5.2   Polylactic Acid (PLA)..79
               5.3   Application of PLA...80
                     5.3.1   PLA in Textile Industry..81
                     5.3.2   PLA in the Packaging Industry 81

　　　　　5.3.3　PLA in the Automotive Industry 82
　　　　　5.3.4　PLA in Building 83
　　　5.4　Application of PLA in Biomedicine 83
　　　5.5　PLA with Filler 85
　　　5.6　Conclusions 91
　　　Acknowledgments 91
　　　References 91

Chapter 6　An Overview of the Compression and Flexural Behaviours of Sandwich Composite Structure with *3D-Printed Core* 96

Nur Ainin F., Azaman M.D., Abdul Majid M.S., and Ridzuan M.J.M.

　　　6.1　Introduction 96
　　　6.2　AM Technology 97
　　　6.3　Materials for 3D-Printed Core Structures 99
　　　6.4　3D-Printed Honeycomb Core Structure 99
　　　6.5　Sandwich Composite Structure 102
　　　6.6　Types of Quasi-Static Loadings 103
　　　　　6.6.1　Flatwise Compression Test 104
　　　　　6.6.2　In-Plane Compression Test 105
　　　　　6.6.3　Flexural Test 105
　　　6.7　Significant Data Obtained under Quasi-Static Loadings 107
　　　6.8　Energy Absorption 109
　　　6.9　Failure Modes of the 3D-Printed Core Structure 110
　　　6.10　Conclusion 112
　　　Acknowledgment 113
　　　References 113

Chapter 7　A Brief Review of the Structure Designed Using Metallic 3D Printing for Biomechanics Applications 120

Rayappa Shrinivas Mahale, Shika Shaygan, Ali Attaeyan, Shamanth Vasanth, Hemanth Krishna, Sharath P.C., Shashanka R., Malekipour M.H., Atefeh Ghorbani, Harsha Prasad, and Ghaffari Y.

　　　Nomenclature 120
　　　7.1　Introduction 121
　　　7.2　Selective Laser Sintering (SLS) 122
　　　　　7.2.1　Selective Laser Sintering 122
　　　　　7.2.2　Sintering Mechanisms 123
　　　　　7.2.2　Difference between SLM and Other AM Methods 124
　　　7.3　Advantages and Disadvantages of the SLM Method 124
　　　　　7.3.1　Advantages 124
　　　　　7.3.2　Disadvantages 124

7.4   Applications.. 124
       7.4.1   Industrial Applications... 125
       7.4.2   Applications of SLS in Biomedicine....................... 125
       7.4.3   SLS in Dentistry... 127
       7.4.4   Use of 3D Printing Models in Liver Surgeries......... 127
       7.4.5   SLS in Pharmacy.. 128
7.5   Conclusions... 129
Acknowledgment.. 129
References 129

Chapter 8   Investigation on the Effect of Different Joint-Based Topology of
                 PLA Core Structure Using 3D Printing Technology 138

*Zichen W., Zuhri M.Y.M., Hang Tuah B.T.,
As'arry A., and Lalegani Dezaki M.*

8.1   Introduction 138
8.2   Methodology.. 140
       8.2.1   Design.. 140
       8.2.2   Fabrication of Sample... 140
       8.2.3   Compression Testing 141
8.3   Results and Discussion 142
       8.3.1   Failure Deformation 142
       8.3.2   Compression Strength 143
       8.3.3   Energy Absorption Capability................................... 146
8.4   Conclusions... 148
Acknowledgment.. 149
References 149

Chapter 9   Application of Artificial Intelligence (Machine Learning) in
                 Additive Manufacturing, Bio-Systems, Bio-Medicine, and
                 Composites 152

Vahid Monfared

9.1   Introduction 152
9.2   Some Questions and Challenges 154
       9.2.1   Generalities, Ambiguities, and Questions................ 154
       9.2.2   Short Answers to Ambiguities and Questions 155
       9.2.3   Where Did the Problem (Difficulty/Need) Start? 156
       9.2.4   Why Do We Need It (ML/AI)? 156
       9.2.5   Where ML Don't Work?.. 157
       9.2.6   The Future of Some Labs (Automation)................... 157
9.3   A Review on ML/AI in Healthcare and Medicine/Bio 158
9.4   ML/AI in Bio-Systems 164
9.5   Brief Illustrations about the AI/ML Process........................ 166
       9.5.1   General Information... 166

9.5.2   Introduction to Some Models 169
9.5.3   Accuracy 178
9.5.4   Preprocessing the Data 181
9.6   ML and AI in AM 181
9.7   Solved and Analyzed Practices 185
9.7.1   Example I 185
9.7.2   Example II 189
9.7.3   Example III 192
9.7.4   Example IV 195
9.7.5   Example V 195
9.8   Conclusion 196
References 198

Chapter 10   Additive Manufacturing of Bio-Inspired Ceramic and
Ceramic-Polymer Composite Lattice Structures 204

He R., and Zhang X.

10.1   Introduction 204
10.2   Mechanical Properties of CCSs 205
10.2.1   Effect of Relative Density 205
10.2.2   Effect of Structural Configuration 208
10.3   Mechanical Properties of Bio-Composites 210
10.3.1   Biological Structures in Nature Materials 211
10.3.2   Mechanical Properties of Bio-Inspired CCS/
Metal Composite 212
10.4   Conclusions 214
Acknowledgment 215
References 216

Index 221

Preface

This book will present the current research on the development of AM technology that has been done extensively and reviewed comprehensively in all areas by the researchers. The studies that are presented in this book will showcase more potential of bio-composites in AM and guide researchers toward more ideas and solutions for future studies. Studies on AM technology include fused deposition modeling (FDM), stereolithography, 3D printing, selective laser sintering (SLS), laminated object manufacturing (LOM), PolyJet, electron beam manufacturing (EBM), and laser engineered net shaping (LENS). Various materials used as feedstock where they are of different types and shapes have been discussed. Using natural or synthetic fibers could reduce the consumption of primary materials and improve the environmental impact of AM. Therefore, the studies gathered in this book provide recent information and bring cutting-edge developments to the attention of young researchers to encourage further advances in the field of bio-composites and synthetic composites for AM. Generally, the chapters will discuss advanced techniques for the development of bio-composites and synthetic fiber composites. By introducing these topics, the book highlights a totally new and recent research theme in composites for AM. Moreover, composite filaments for FDM may use composites of various types for engineering applications. Besides, this book covers the mechanical behavior, thermal, flammability, and functional properties of the composites for AM. This book also discusses the latest developments in the field of composites beyond the state of the art. The latest results in material evaluation for targeted applications are also presented. In particular, the book highlights the latest advances and future challenges in bio-composites and synthetic composites for AM in various types of engineering applications. As such, it offers a valuable reference guide for scholars interested in bio-composites as well as an evaluation of nano/microstructure-based materials. It also provides essential insights for graduate students and scientists pursuing research in the broad fields of composite materials and AM.

Editors

Mastura M.T. started her tertiary educational journey at the Language Institute of Seoul National University, Seoul, South Korea, from 2003 to 2004 where she studied the Korean language (Level 1 until Level 6). She graduated with a Diploma in System Design from Dongyang Mirae University, Seoul, South Korea, formerly known as Dongyang Technical College in 2007, and afterward, in 2009, she graduated with a Bachelor of Science in Engineering from Korea University, Seoul, South Korea. In 2011, she completed her Master of Engineering in the field of Mechanics at Universiti Kebangsaan Malaysia, Selangor, Malaysia. In 2017, she graduated with her Ph.D. degree in the field of Mechanical Engineering at the Faculty of Engineering, Universiti Putra Malaysia, Malaysia. She started her work as an academician in 2009 and was appointed as senior lecturer in 2018. As an academician in an engineering school, she was awarded the Most Cited Article Award in the *International Journal of Precision Engineering and Manufacturing – Green Technology* (*IJPEM-GT*), Springer, in 2020. She was also a recipient of the Best Manuscript Award in the fifth Postgraduate Seminar on Natural Fiber-Reinforced Polymer Composites, 2016. Her main research interests are: (1) Concurrent Engineering, (2) Natural Fiber Composites and Design and (3) Fused Deposition Modeling. To date, she has authored or co-authored more than 40 research articles including journal papers, chapters in books, edited books, proceedings, etc., and has 480 citations. She has been involved in research and education for more than ten years and has been awarded more than 50,000 USD research grants in total.

Sapuan S.M. is Professor (A grade) of Composite Materials at the Department of Mechanical and Manufacturing, Universiti Putra Malaysia (UPM), Head of Advanced Engineering Materials and Composite Research Centre (AEMC), UPM, and Chief Executive Editor of Pertanika journals under the office of Deputy Vice Chancellor, UPM. He earned a B.Eng. in Mechanical Engineering from the University of Newcastle, Australia, an M.Sc. in Engineering Design, Loughborough University, UK, and a Ph.D. in Material Engineering, De Montfort University, UK. He is a professional engineer, and Fellow of Society of Automotive Engineers, Academy of Science Malaysia, International Society for Development and Sustainability, Plastic and Rubber and Institute, Malaysia, Malaysian Scientific Association, International Biographical Association, and Institute of Materials, Malaysia. He is a Honorary Member of the Asian Polymer Association and the Founding Chairman and Honorary Member of Society of Sugar Palm Development and Industry, Malaysia. He is the Editor-in-Chief of the *Journal of Natural Fibre Polymer Composites*, co-editor-in-chief of *Functional Composites and Structures*, Associate Editor-in-Chief of *Defence Technology*, Elsevier, and editorial board member of 30 journals. He has produced over 2,000 publications including over 900 journal papers, 55 books, and 190 chapters in books. He has delivered over 60 plenary and keynote lectures, and over 150 invited lectures. He organized 30 journal special issues as guest editor, reviewed over 1,500 journal papers, and has 8 patents. He successfully supervised 95 Ph.D. and 70 M.Sc. students and 15 postdoctoral researchers. His h-index is 102 with 38,309 citations (Google Scholar). He received the ISESCO Science Award, the Khwarizimi International Award, the Kuala Lumpur Royal Rotary Gold Medal Research Award, two National Book Awards, the Endeavour Research Promotion Award, TMU/IEEE India, the Citation of Excellence Award, Emerald, UK, Malaysia's Research Star Award, the Publons Peer Review Award, USA, the Professor of Eminence Award, Aligarh Muslim University, India, the Top Research Scientists' Malaysia Award, and the PERINTIS Publication Award. He was listed in World's Top 2% Scientists, Stanford University, USA. He received the IET Achievement Award, the SAE Subir Chowdhury Medal of Quality Leadership, Anugerah Tokoh Pekerja, UPM, Anugerah Khas Akademia Putra, UPM, the International Society of Bionic Engineering Outstanding Contribution Award, China, Ikon Akademia 2022 from the Ministry of Higher Education Malaysia, the TWAS Award in Engineering Sciences by the World Academy of Sciences, and the IET Malaysia Leadership Award.

Ilyas R.A. is a senior lecturer in the Faculty of Chemical and Energy Engineering, Universiti Teknologi Malaysia, Malaysia. He is also a Fellow of the International Association of Advanced Materials (IAAM), Sweden, a Fellow of the International Society for Development and Sustainability (ISDS), Japan, a member of the Royal Society of Chemistry, UK, and Institute of Chemical Engineers (IChemE), UK. He received his Diploma in Forestry at Universiti Putra Malaysia, Bintulu Campus (UPMKB), Sarawak, Malaysia, from May 2009 to April 2012. In 2012, he was awarded the Public Service Department (JPA) scholarship to pursue his Bachelor's Degree (BSc) in Chemical Engineering at Universiti Putra Malaysia (UPM). Upon completing his B.Sc. program in 2016, he was again awarded the Graduate Research Fellowship (GRF) by the Universiti Putra Malaysia (UPM) to undertake a Ph.D. degree in the field of Biocomposite Technology & Design at the Institute of Tropical Forestry and Forest Products (INTROP) UPM. Ilyas R.A. was the recipient of the MVP Doctor of Philosophy Gold Medal Award UPM 2019, for Best Ph.D. Thesis and Top Student Award, INTROP, UPM. He was awarded with Outstanding Reviewer by Carbohydrate Polymers, Elsevier United Kingdom, Top Cited Article, 2020–2021, *Journal Polymer Composite*, Wiley, 2022, and Best Paper Award at various international conferences. Ilyas R.A. was also listed and awarded the World's Top 2% Scientist (Subject-Wise) Citation Impact during the Single Calendar Year 2019, 2020, and 2021 by Stanford University, US, the PERINTIS Publication Award 2021 and 2022 by Persatuan Saintis Muslim Malaysia, the Emerging Scholar Award by Automotive and Autonomous Systems 2021, Belgium, Young Scientists Network – Academy of Sciences Malaysia (YSN-ASM) 2021, the UTM Young Research Award 2021, the UTM Publication Award, 2021, and the UTM Highly Cited Researcher Award, 2021. In 2021, he won the

Gold Award and Special Award (Kreso Glavac (The Republic of Croatia)) at the Malaysia Technology Expo (MTE2022), the Gold Award and the Special Award at the International Borneo Innovation, Exhibition & Competition 2022 (IBIEC2022), and a Gold Award at New Academia Learning Innovation (NALI2022). His main research interests are: (1) Polymer Engineering (Biodegradable Polymers, Biopolymers, Polymer Composites, Polymer Gels) and (2) Material Engineering (Natural Fiber Reinforced Polymer Composites, Biocomposites, Cellulose Materials, Nano-Composites). To date he has authored or co-authored more than 431 publications (published/accepted): 188 journals indexed in JCR/Scopus, 3 non-index journals, 17 books, 104 book chapters, 78 conference proceedings/seminars, 4 research bulletins, 10 conference papers (abstract published in the book of abstract), 17 Guest Editor of Journal special issues and 10 editor/co-editor of conference/seminar proceedings on green materials-related subjects.

Contributors

Abdul Kudus S.I.
Fakulti Teknologi dan Kejuruteraan
  Industri dan Pembuatan, Universiti
  Teknikal Malaysia Melaka, Hang
  Tuah Jaya, Durian Tunggal, Melaka,
  Malaysia

Abdul Majid M.S.
Faculty of Mechanical Engineering
  Technology, Universiti Malaysia
  Perlis, Arau, Perlis, Malaysia; Centre
  of Excellence Frontier Materials
  Research, Universiti Malaysia Perlis,
  Arau, Perlis, Malaysia

Ali Attaeyan
Department of Biomedical Engineering,
  Biomechanics Faculty, Najafabad
  Branch, Islamic Azad University,
  Najafabad, Iran

As'arry A.
Department of Mechanical and
  Manufacturing Engineering, Faculty
  of Engineering, Universiti Putra
  Malaysia, UPM Serdang, Selangor,
  Malaysia

Atefeh Ghorbani
Biotechnology Department, Falavarjan
  Branch, Islamic Azad University,
  Isfahan, Iran

Azaman M.D.
Faculty of Mechanical Engineering
  Technology, Universiti Malaysia
  Perlis, Arau, Perlis, Malaysia; Centre
  of Excellence Frontier Materials
  Research, Universiti Malaysia Perlis,
  Arau, Perlis, Malaysia

Azmah Hanim Mohamed Ariff
Department of Mechanical and
  Manufacturing Engineering, Faculty
  of Engineering, Universiti Putra
  Malaysia (UPM), Serdang, Malaysia;
  Advance Engineering Materials
  and Composites Research Center,
  (AEMC), Faculty of Engineering,
  Universiti Putra Malaysia, Serdang,
  Selangor, Malaysia

Che Nor Aiza Jaafar
Department of Mechanical and
  Manufacturing Engineering, Faculty
  of Engineering, Universiti Putra
  Malaysia (UPM), Serdang, Malaysia;
  Advance Engineering Materials
  and Composites Research Center,
  (AEMC), Faculty of Engineering,
  Universiti Putra Malaysia, Serdang,
  Selangor, Malaysia

Dwi Hadi S.
Faculty of Industrial Engineering,
  Brawijaya University, Malang, Jawa
  Timur, Indonesia

Ghaffari Y.
Department of Biomedical, Science
  and Research Branch, Islamic Azad
  University, Tehran, Iran

Hamat S.
Faculty of Mechanical Engineering &
  Technology, Universiti Malaysia
  Perlis, Ulu Pauh, Perlis, Malaysia

Hang Tuah B.T.
Department of Mechanical and
   Manufacturing Engineering, Faculty
   of Engineering, Universiti Putra
   Malaysia, UPM Serdang, Selangor,
   Malaysia

Harsha Prasad
Department of Materials Science and
   Engineering, Indian Institute of
   Technology, Kanpur, India

Hazliza Aida C.H.
Fakulti Teknologi Kejuruteraan
   Mekanikal, Universiti Teknikal
   Malaysia Melaka, Hang Tuah Jaya,
   Durian Tunggal, Melaka, Malaysia

He R.
State Key Laboratory of Explosion
   Science and Technology, Beijing
   Institute of Technology, Beijing,
   China

Hemanth Krishna
School of Mechanical Engineering,
   REVA University, Bengaluru,
   Karnataka, India

Ilyas R.A.
Faculty of Chemical and Energy
   Engineering, Universiti Teknologi
   Malaysia, UTM, Centre for
   Advanced Composite Materials,
   Universiti Teknologi Malaysia
   (UTM), Johor Bahru, Malaysia

Ishak M.R.
Department of Aerospace Engineering,
   Faculty of Engineering, Universiti
   Putra Malaysia, UPM Serdang,
   Selangor, Malaysia; Aerospace
   Malaysia Research Centre (AMRC),
   Universiti Putra Malaysia, Serdang
   Selangor, Malaysia; Laboratory of
   Biocomposite Technology, Institute
   of Tropical Forestry and Forest
   Products (INTROP), Universiti
   Putra Malaysia, Serdang, Selangor,
   Malaysia

Lalegani Dezaki M.
Department of Engineering, School of
   Science and Technology, Nottingham
   Trent University, Nottingham, UK

Loh Y.F.
Lembaga Perindustrian Kayu Malaysia,
   Pusat Pembangunan Fibre and
   Biokomposit, Kompleks Perabot
   Olak Lempit, Banting, Selangor,
   Malaysia

Luqman Hakim Noorazlan
Advanced Engineering Materials
   and Composites Research Centre
   (AEMC), Faculty of Engineering,
   Universiti Putra Malaysia, UPM
   Serdang, Malaysia

Malekipour M.H.
School of Dentistry, Isfahan University
   of Medical Sciences, Isfahan, Iran

Mastura M.T.
Fakulti Teknologi dan Kejuruteraan
    Industri dan Pembuatan, Universiti
    Teknikal Malaysia Melaka, Hang
    Tuah Jaya, Durian Tunggal, Melaka,
    Malaysia

**Mohammad Akram Syazwie Mohd
Areff**
Department of Mechanical and
    Manufacturing Engineering, Faculty
    of Engineering, Universiti Putra
    Malaysia (UPM), Serdang, Malaysia

Muhamad Ariffadzilah Mohd Latip
Fakulti Teknologi dan Kejuruteraan
    Industri dan Pembuatan, Universiti
    Teknikal Malaysia Melaka, Hang
    Tuah Jaya, Durian Tunggal, Melaka,
    Malaysia

Neassan Muthamil Selvan
Advanced Engineering Materials
    and Composites Research Centre
    (AEMC), Faculty of Engineering,
    Universiti Putra Malaysia, UPM
    Serdang, Malaysia

Nur Ainin F.
Faculty of Mechanical Engineering
    Technology, Universiti Malaysia
    Perlis, Arau, Perlis, Malaysia

Nuzaimah Mustafa
Fakulti Teknologi dan Kejuruteraan
    Industri dan Pembuatan, Universiti
    Teknikal Malaysia Melaka, Hang
    Tuah Jaya, Durian Tunggal, Melaka,
    Malaysia

Pavithren Mailvaganan
Advanced Engineering Materials
    and Composites Research Centre
    (AEMC), Faculty of Engineering,
    Universiti Putra Malaysia, UPM
    Serdang, Malaysia

Rayappa Shrinivas Mahale
School of Mechanical Engineering,
    REVA University, Bengaluru,
    Karnataka, India

Recep Calin
Material Engineering Department,
    Engineering Faculty, Kirikkale
    University, Kirikkale, Turkey

Ridzuan M.J.M.
Faculty of Mechanical Engineering
    Technology, Universiti Malaysia
    Perlis, Arau, Perlis, Malaysia; Centre
    of Excellence Frontier Materials
    Research, Universiti Malaysia Perlis,
    Arau, Perlis, Malaysia

Sapuan S.M.
Advanced Engineering Materials
    and Composites Research Centre
    (AEMC), Faculty of Engineering,
    Universiti Putra Malaysia, UPM
    Serdang, Malaysia

Shamanth Vasanth
School of Mechanical Engineering,
    REVA University, Bengaluru,
    Karnataka, India

Sharath P.C.
Department of Metallurgical and
    Materials Engineering, JAIN
    Deemed to be University, Bengaluru,
    Karnataka, India

Shashanka R.
Department of Metallurgical and
    Materials Engineering, BARTIN
    University, Bartin, Turkey

Shika Shaygan
Department of Pharmacy, Cyprus
    Health and Social Science,
    Guzelyurt, TRNC via Mersin,
    Turkey

Shudharson Ilengo
Advanced Engineering Materials
    and Composites Research Centre
    (AEMC), Faculty of Engineering,
    Universiti Putra Malaysia, UPM
    Serdang, Malaysia

Sulaiman Ahmad
Advanced Engineering Materials
    and Composites Research Centre
    (AEMC), Faculty of Engineering,
    Universiti Putra Malaysia, UPM
    Serdang, Malaysia

Vahid Monfared
Department of Mechanical Engineering,
    Zanjan Branch, Islamic Azad
    University, Zanjan, Iran

Vairravell Khumar Ravindran
Advanced Engineering Materials
    and Composites Research Centre
    (AEMC), Faculty of Engineering,
    Universiti Putra Malaysia, UPM
    Serdang, Malaysia

Vasi Uddin Siddiqui
Advanced Engineering Materials
    and Composites Research Centre
    (AEMC), Faculty of Engineering,
    Universiti Putra Malaysia, UPM
    Serdang, Malaysia

Yidris N.
Department of Aerospace Engineering,
    Faculty of Engineering, Universiti
    Putra Malaysia, UPM Serdang,
    Selangor, Malaysia;

Yusliza Yusuf
Fakulti Teknologi dan Kejuruteraan
    Industri dan Pembuatan, Universiti
    Teknikal Malaysia Melaka, Hang
    Tuah Jaya, 76100 Durian Tunggal,
    Melaka, Malaysia

Zhang X.
State Key Laboratory of Explosion
    Science and Technology, Beijing
    Institute of Technology, 100081
    Beijing, China

Zichen W.
Advanced Engineering Materials
    and Composites Research Centre
    (AEMC), Department of Mechanical
    and Manufacturing Engineering,
    Faculty of Engineering, Universiti
    Putra Malaysia, 43400 UPM
    Serdang, Selangor, Malaysia

Zuhri M.Y.M.
Advanced Engineering Materials
    and Composites Research Centre
    (AEMC), Department of Mechanical
    and Manufacturing Engineering,
    Faculty of Engineering, Universiti
    Putra Malaysia, 43400 UPM
    Serdang, Selangor, Malaysia.
    Laboratory of Biocomposite
    Technology, Institute of Tropical
    Forestry and Forest Products
    (INTROP), Universiti Putra
    Malaysia, 43400 UPM Serdang,
    Selangor, Malaysia

Zulkiflle Leman
Department of Mechanical and
    Manufacturing Engineering, Faculty
    of Engineering, Universiti Putra
    Malaysia (UPM), 43400 Serdang,
    Malaysia; Advance Engineering
    Materials and Composites Research
    Center, (AEMC), Faculty of
    Engineering, Universiti Putra
    Malaysia, 43400 Serdang, Selangor,
    Malaysia

1 Sustainable Biocomposites

Characterization and Applications in 3D and 4D Printing

Vasi Uddin Siddiqui, Sapuan S.M., Sulaiman Ahmad, Luqman Hakim Noorazlan, Neassan Muthamil Selvan, Shudharson Ilengo, Pavithren Mailvaganan, and Vairravell Khumar Ravindran

1.1 INTRODUCTION

Biocomposites are a class of multi-phase materials that exhibit the reinforcement of one or more fillers composed of distinct constituent materials with a polymer matrix derived from biological sources. As per the findings of Ilyas and Sapuan (2020), the mechanical properties of biocomposites are observed to be superior to those of their individual constituent materials. The identification of suitable fibers for composite materials is contingent upon a number of significant attributes, including torsional stiffness, rigidity, dilatation following failure, fiber-matrix adhesion, chemical inertness, kinetic and long-term performance, operational costs, and processing expenses (Ilyas & Sapuan, 2020). Fibers that are obtained from plants or animals are classified as natural fibers, which are not synthetic or man-made in nature. A diverse range of materials, including agricultural by-products and cellulosic fibers such as flax, sisal, jute, banana, kenaf, hemp, and oil palm fruit bunch cellulose fiber, have been identified as potential sources (Keya et al., 2019). The utilization of natural fibers in the manufacturing of biocomposites, also referred to as organic nutrient polymer natural fiber-reinforced polymer (FRP) composites, has been on the rise. This is attributed to their cost-effective substitution of synthetic fibers and their eco-friendly properties, as reported by Oyeniran and Ismail (2021). Hybrid biocomposites, which are a combination of natural fibers and other materials, have been identified as having significant potential for use in various industries including transportation, building, sports, and medicine. The potential of utilizing alternatives to glass and fiber-reinforced composites in various products such as boats, kayaks, pipes, and storage tanks is currently under investigation (Guna et al., 2018). Different types of biocomposites

DOI: 10.1201/9781003362128-1

1

are illustrated in Figure 1.1. The fabrication of biocomposites encompasses a different technique such as extrusion, injection molding, filament winding, compression, infusion, and autoclaving.

In contrast, additive manufacturing, also known as 3D printing, enables the production of tangible 3D entities through the utilization of volumetric information. As per the findings of Zarna et al. (2022), 3D printers differ from conventional printers in that they create 3D objects by adding layers of various materials. The utilization of 3D printing technology has been extended beyond the aerospace and automobile sectors to encompass medical applications, as reported by Ganapathy et al. (2022). According to Li et al. (2018), the 3D printing process comprises several essential components, including a feeder, filament, frame, print bed, among others. The filament, which acts as the primary raw material for printing, is a crucial component of the process. The 3D printer and its parts are illustrated in Figure 1.2.

The categorization of 3D printing into seven subcategories is outlined by the ISO/ASTM 52900 classification. Each subcategory is characterized by unique approaches

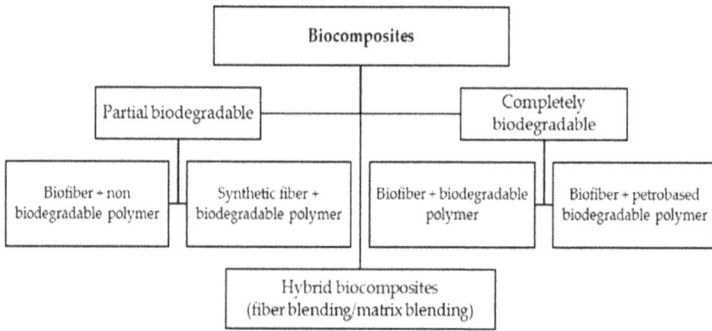

FIGURE 1.1   Type of biocomposites (Bahrami et al., 2020).

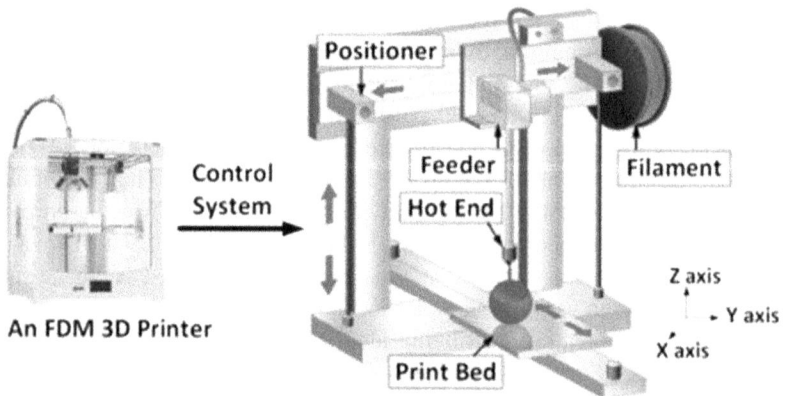

FIGURE 1.2   Three-dimensional printer components (Li et al., 2018).

and benefits. Each of these approaches is detailed in Table 1.1 and depicted in Figure 1.3. The various additive manufacturing techniques that have been identified in the literature include powder bed fusion, vat polymerization, binder jetting, direct energy deposition, material extrusion, sheet lamination, and material jetting (Hodgdon et al., 2018). A thorough comprehension of these methodologies is essential in order to choose the suitable approach that aligns with particular prerequisites.

Four-dimensional printing is a cutting-edge iteration of 3D printing technology that enables the production of objects with the ability to undergo transformational changes over a period of time. The achievement of this objective is facilitated through the utilization of shape-memory materials or multi-material structures that exhibit diverse deformation tendencies in response to external input, as noted by Ahmed et al. (2021). In a recent study, Goo and colleagues (2020) have introduced a novel 4D printing technique that leverages the anisotropic thermal deformation of 3D-printed objects.

TABLE 1.1

Extensive and comprehensive explanations of the seven 3D printing methods (Carew & Errickson, 2020)

Method	Description
Material extrusion	When using an extruder, a spool of material is fed into a heated extrusion head, and the material is then extruded out, where it is deposited in layers onto a construction platform. Many other kinds of materials can be used, each having its own set of advantages and limitations, as well as its own palette of colors.
Vat polymerization	Vat polymerization is the process of selectively curing a liquid photopolymer material using a light source within the build chamber. After each successive layer of the print has cured, the build platform is lowered to expose the next layer for printing. After the part is finished, it undergoes post-production steps including cleaning, post-exposure to UV radiation to harden the build, and support material removal.
Binder jetting	To selectively bind a bed of powder, a liquid bonding agent is applied to the bed. Printed objects have unfused powder removed from them after they have been printed, with successive layers being added and fused. The building can't be finished till the postprocessing is done. Prototypes in full color can be made with binder jetting without the need for any support structures.
Material jetting	Material jetting is a process in which a liquid photopolymer material is jetted onto a build tray and cured in a fashion analogous to that of an inkjet printer. There is no strict rule about what colors or materials can be utilized to construct a structure. It also makes use of individual, detachable structural parts.
Powder bed fusion	A powder bed is fused selectively using a high-energy source contained within a chamber. The build is supported by the un-sintered powder material, so details like overhanging edges can be made without any additional framework.
Sheet lamination	As a substance is placed in layers, they are adhered to one another with an adhesive. Each successive layer is sculpted to get the final form.
Direct energy deposition	As the material is put onto the build platform in layers, it is fused together as the energy is transferred from the source to the platform. Direct energy deposition (DED) utilizes a multi-axis arm equipped with a deposition nozzle for controlled depositing.

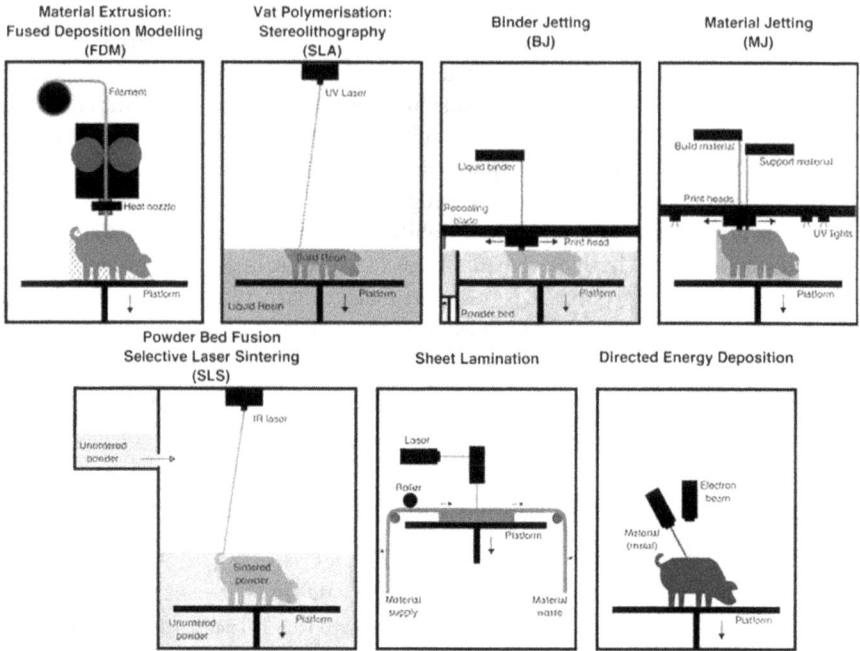

FIGURE 1.3 Diagrams showing the seven different types of 3D printing technology (Carew & Errickson, 2020).

The method is deemed practical and holds promising potential for various applications. Through the consideration of thermal anisotropy in the design of printing routes, directional size alterations were successfully attained in the printed objects. Residual tension caused a reduction in the longitudinal and transverse dimensions, whereas the lateral and laminating dimensions experienced an increase following heat treatment. The manipulation of thermal deformation parameters in a controlled manner facilitates the investigation of fundamental elements of 4D printing.

1.2 MATERIALS USED FOR 3D/4D PRINTING

The characteristics of composite materials produced through 3D and 4D printing techniques are subject to the influence of multiple factors. These factors comprise the nature of the fibers utilized, and their rigidity, durability, and capacity to adhere to the polymer matrix. Ideally, the composite material should possess a microstructure that is homogeneous in nature, featuring well-dispersed fibers with a high aspect ratio. In order to achieve optimal load transfer, it is imperative to maximize the interfacial surface area. The primary objective is to minimize imperfections and voids within the matrix and interfacial area. Natural fibers, on the other hand, have anisotropic mechanical properties that are determined by their hierarchical microstructure and biochemical makeup, which consists of cellulose, hemicelluloses, pectins, lignin, and water. Table 1.2 contains data on the tensile qualities, biological content, and microstructure of several natural fibers, which are required for the analysis

TABLE 1.2

Mechanical properties (tensile modulus and stress at break) and biochemical composition (cellulose, hemicelluloses, lignin, pectin concentration, and microfibrillar angle (MFA)) of several natural fibers used in 3D and 4D printing (Le Duigou et al., 2020)

Properties/fibers	Flax	Jute	Kenaf	Coir	Cotton	Wood	Bamboo	Harakeke	Hemp
E (GPa)	46–85	24.7–26.5	11–60	3.44–4.16	3.5–8	15.4–27.5	10–40	14–33	14.4–44.5
σ (MPa)	600–2,000	393–773	223–930	120–304	287,597	553–1,500	340–510	440–990	270–889
Cellulose (%)	64–85	61–75	45–57	32–46	82–99	38–45	34.5–50		55–90
Hemicellulose (%)	11–17	13.6–20.4	21.5	0.15–0.3	4	19–39	20.5		04–16
Lignin (%)	02–03	12–13	8–13	40–45	0.75	22–34	26		02–05
Pectines (%)	1.8–2.0	0.6±0.6	03–05	03–04	6	0.4–5	<1		0.8–8

outlined in the 3D and 4D printing sections. Acrylonitrile butadiene styrene (ABS) and poly lactic acid (PLA) are now the most popular polymers used in 3D printing, and they may be purchased in spools for $15–30 per pound in either 1.75- or 3.00-mm diameters. However, there are a variety of substitute polymer-based filaments on the market (Tymrak et al., 2014).

1.2.1 POLY LACTIC ACID

PLA is a type of polymer that is known for its biodegradability. It is derived from renewable resources such as sugarcane or maize starch. PLA polymers are widely recognized for their potential to exhibit compostable and biodegradable characteristics. PLA, a thermoplastic polymer, exhibits notable strength and modulus properties. The utilization of continuously renewable resources for the production of industrial packaging materials and biocompatible/bioabsorbable medical devices is a viable option. After the water has been removed, the lactide dimer can be polymerized without any further water being created. Recently, polymer scientists have become interested in PLA as a viable biopolymer to replace plastics derived from fossil fuels. PLA, a biodegradable polymer, has found extensive applications in the biomedical industry owing to its biocompatibility properties. The lactic acid utilized in the production of PLA is derived from the process of fermenting corn or sugar beets. PLA presents a noteworthy advantage of being capable of undergoing biodegradation, which allows it to return to biomass and participate in the natural cycle. Recent research and findings on PLA have shown that the biopolymer is easily produced and has strong mechanical properties (Balakrishnan et al., 2012). The "rediscovery" of various polymers that may have been made from renewable resources (some biodegradable) resulted from rising plastic waste and oil prices. The polymer industry can handle the critiques directed at petroleum-based plastics if the greenness of bio-based polymers is evaluated. "Old" polyesters, which are essentially aliphatic ones, are becoming more and more popular for this purpose. The latter type of bio-based polyester is more widely used in commerce. New bio-based monomers can be created thanks to recent advancements in fermentation technologies (Jacquel et al., 2011).

Biodegradable polymers like polylactide have numerous lucrative markets. Examples of common uses for plastic bags include barriers for sanitary goods and diapers, planting cups, disposable cups, and plates. There is currently no large-scale commercial production of polylactide, although this is about to change. Lactic acid, the starting component, will also require a new capability. In the upcoming years, commercial markets for biodegradable polymers are anticipated to grow significantly. Despite PLA having a broad range of uses, there are certain drawbacks to its use, including a sluggish rate of degradation, hydrophobicity, and low-impact toughness. Combining PLA with other polymers provides handy ways to enhance related features or create brand-new PLA polymers or mixes for application-specific uses. Applications in the biomedical field include stents, dialysis fluid, and medication delivery systems. Additionally, it is being considered as a material for disposable tableware, loose-fill packaging, compostable bags, and tissue engineering. PLA can be used in upholstery, disposable clothing, awnings, and feminine hygiene items. PLA can also be used to make fibers and nonwoven fabrics (Table 1.3).

TABLE 1.3

Mechanical properties of PLA (Leksakul & Phuendee, 2018)

Properties	Values obtained
Young's modulus	3,600 MPa
Break elongation	2.4%
Impact strength	16.5 kJ/m^2
Moisture absorption	0.3%
Tensile strength	70 MPa
Flexural strength	98 N/mm^2
Notched impact strength	3.3 kJ/m^2
Density	1.25 g/cm^2

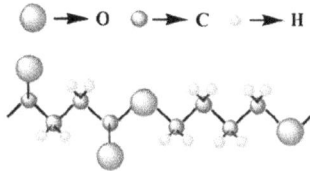

FIGURE 1.4 Chemical structure of PBS (Rafiqah et al., 2021).

1.2.2 POLY BUTYLENE SUCCINATE (PBS)

PBS is an aliphatic polyester- a succinate derived polymer and can be produced by renewable sources like sugarcane, cassava, and corn, and by bacterial fermentation route (Aliotta et al. 2022). PBS is a potential biopolymer because of its mechanical resemblance to commonly used high-density polyethene. Its melting temperature is lower than PLA's (115 vs. 160°C), but it is far more flexible and does not require plasticizers. PBS is a relatively new polymer with intriguing properties that make it appealing for a wide range of applications. PBS is currently manufactured (largely) from petrochemical sources. It is mostly used in fields like packaging. Production of 50% bio-based PBS is already possible thanks to the growth and extension of bio-based succinic acid production capacity, and production of 100% bio-based PBS is possible with a further switch to bio-based butanediol. Additionally, the cost of PBS will rapidly decline over the ensuing years as technology advances and production scales rise. PBS has the potential to be more economical than biopolymers like PLA, polybutylene adipate terephthalate (PBAT), and polyhydroxybutyrate (PHB) among the available biopolymers. It can be used alone or in conjunction with other biopolymers. A few examples are flowerpots, personal hygiene products, fishing nets, and fishing lines (Figure 1.4).

1.2.3 ACRYLONITRILE BUTADIENE STYRENE

ABS is a particularly attractive material for structural applications due to its exceptional mechanical properties, including high stiffness, superior impact resistance even at low temperatures, good insulating qualities, good abrasion and strain resistance, and high dimensional stability. These properties contribute to the material's mechanical robustness and stability over time, making it an excellent choice for a wide range of structural applications. ABS exhibits superior electrical insulation properties and a notable level of surface condition. The susceptibility of pure ABS polymers to stress corrosion cracking is well-documented in the presence of certain organic molecules. Additionally, these polymers are known to be vulnerable to degradation from oxidizing agents and strong acids. While exhibiting reactivity toward aromatic or chlorinated hydrocarbons, the material shows notable resistance to aliphatic hydrocarbons. ABS exhibits high strength, resilience, and chemical resistance; however, its susceptibility to rapid degradation in the presence of polar solvents is noteworthy. When compared to high impact polystyrene (HIPS), the material exhibits superior impact properties and a marginally elevated heat distortion temperature. A diverse array of conventional equipment can be employed for the purpose of processing.

ABS plastic is commonly used for the housings of various appliances, including vacuum cleaners, food processors, and refrigerator liners. The principal applications of ABS are observed in articles intended for household and consumer use. It is a commonly utilized material in the production of various industrial components such as keyboard keycaps, blowers, gears, pump impellers, chemical tanks, and camera housings. At present, this material is widely acknowledged as one of the most diverse materials for 3D printing. The ABS material is available in the form of a lengthy filament that has been coiled around a spool. The design is fabricated using the FDM (fusion deposition modeling) technique with ABS material, wherein the material is heated and extruded through a small nozzle in 250-m layers (Figure 1.5 and Table 1.4).

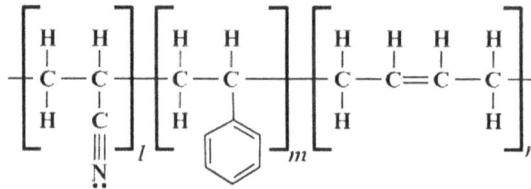

FIGURE 1.5   Formula of ABS (Okada et al., 2016).

TABLE 1.4

Mechanical properties of the ABS matrix

Properties	Values obtained
Young's modulus	1.79–3.2 GPa
Elongation at break	10–50%
Elongation at yield	1.7–6%
Toughness	200–215 J/m
Stiffness	1.6–2.4 GPa
Hardness shore D	100
Toughness at low temperature	20–160 J/m

1.2.4 PBS Blends with PLA

The modification is broken down into the following subsections in this review: copolymerization, simple blending, plasticization, and reactive compatibility. A straightforward and affordable technique to combine the desired qualities of each component and create a possible matrix for creating composites without losing their biodegradable behavior is to simply blend PLA and PBS. The solution blending process typically entails mechanical mixing, solvent evaporation, and the dissolution of the components of a polymer blend, which is achieved by using a suitable solvent. Another method for increasing the ductility of PLA-based polymers is plasticization. Utilizing a plasticizer makes materials more flexible and processable by lowering intermolecular tensions and enhancing polymer chain mobility. To avoid having poor miscibility with PLA or PBS, plasticizers must be chosen taking into consideration the solubility idea.

Immiscible polymer blends are considered more compatible and have less phase separation when they undergo reactive compatibility. Reactive compatibilizers are capable of reacting with both components of a blend during the mixing process to couple the phases quickly, improving the blends' compatibility and interfacial interactions. Including block copolymers is the most typical non-reactive compatibility technique for polymer blends. A copolymer becomes entangled with mixed components, promoting strong interfacial adhesion that enhances the physical qualities. First, an evaluation of PBS's impact on the crystallization of blends based on PLA has revealed that PBS speeds up PLA's crystallization rate (Lin et al., 2022) (Figure 1.6).

1.2.5 ABS Blend

ABS containing 5% carbon fibers (CFs) has a higher porosity (up to 19%), which reduces its mechanical and structural stability when compared to ABS with carbon nanotube reinforcement (Vakharia et al., 2022). In a study conducted by Tekinalp et al. (2014), 3D-printed ABS samples containing up to 40% chopped CFs exhibited a 115% increase in tensile strength (TS) and a 700% increase in modulus. The samples under investigation displayed a notable difference in fiber orientation between the 3D printing process and compression molding. The results indicate that the first method produced a notably elevated fiber orientation along the printing axis, with measurements peaking at 91.5%. Conversely, the second method yielded a reduced fiber orientation. The study found that 3D-printed composites demonstrated similar TS and modulus to those manufactured through conventional compression molding, despite their relatively high porosity, which can be attributed to differences in fiber orientation, dispersion, and void formation. Ning et al. (2015) conducted a study to examine the effects of varying CFs contents and lengths on the characteristics of ABS plastic. The results suggest that the inclusion of CFs led to an increase in both TS and Young's modulus, while simultaneously resulting in a decrease in toughness, yield strength, and ductility. The experimental findings suggest that the specimens that were incorporated with 150-μm CFs demonstrated greater TS and Young's modulus values when compared to those that contained 100-μm CFs. The findings suggest that the specimens with a thickness of 150-μm exhibited decreased toughness and ductility when compared to those with a thickness of 100 μm. The results

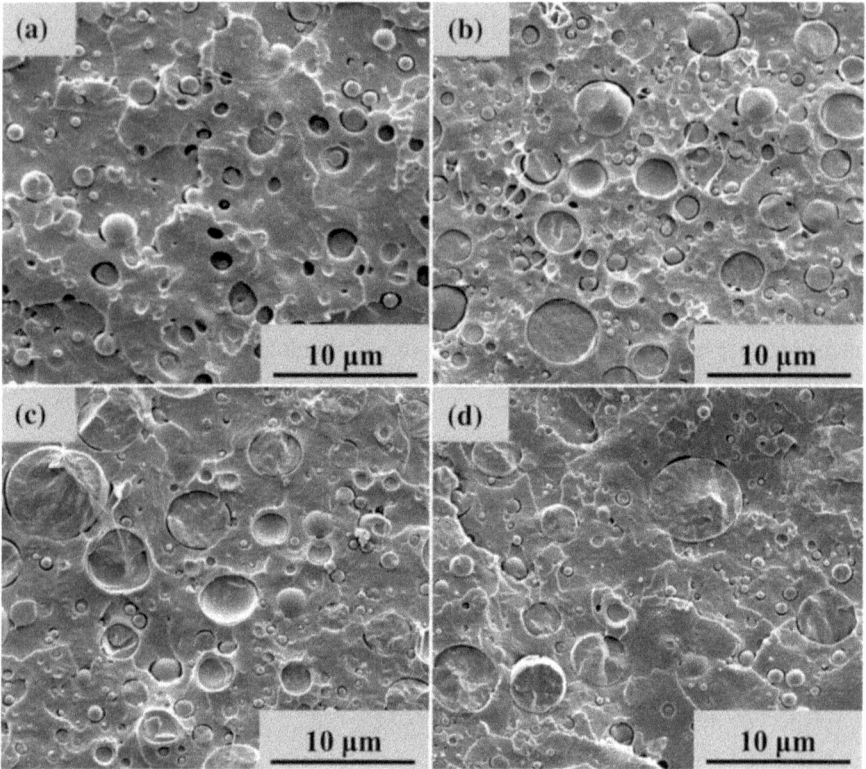

FIGURE 1.6 SEM images of PLA/PBS blends at different PBS contents: (a) 10 wt%, (b) 20 wt%, (c) 30 wt%, and (d) 40 wt% with particle diameters from 0.2 to 6.5 μm (Su et al., 2019).

indicate that there was no significant difference in yield strength between the two groups, as determined by statistical analysis. In a study conducted by Love and colleagues (2014), it was observed that the addition of CFs to ABS led to enhancements in thermal conductivity, reductions in the coefficient of thermal expansion (CTE), and mitigations in part distortion (Table 1.5).

1.2.6 Smart Material

The common materials utilized in 3D printing include nylon, ABS plastic, resin, wax, and polycarbonate. The accessibility of these conventional materials in the market renders their utilization for printing purposes facile. Four-dimensional printing technology employs smart materials. Smart materials are a class of materials that possess one or more properties that can be controlled in a precise manner through the application of external energy. These materials are typically composed of multiple components and are capable of undergoing transformation in a controlled fashion. Smart materials typically encompass piezoelectric, electrostrictive, magnetostrictive, thermoelectric, and shape-memory alloys. Shape-memory alloys exhibit high strength, hardness, toughness, and excellent electrical conductivity, albeit at a relatively high

TABLE 1.5

Summary of mechanical, thermal, and morphological properties of petroleum-based polymers with biofillers (Wasti & Adhikari, 2020)

Matrix	Filler	Filler content (wt%)	Mechanical properties	Thermal properties	Morphology properties
ABS	Lignin-coated cellulose nanocrystal (L-CNC)	2, 4, 6, 8, 10	Up to 4wt% L-CNC, tensile strength (TS) increases, then rapidly declines to 10wt%. The E increases to 6wt% L-CNC, then declines.	L-CNC lowers degradation temperature. At high temperature thermal stability increases with L-CNC.	Uniform distribution of L-CNC in the ABS matrix. Pores in ABS/L-CNC composite filament 3D-printed object. Increases pore diameter to >30 µm with above 6wt% L-CNC.
ABS	Rice straw (RS)	5, 10, 15, 20	Decreases in tensile strength (TS) and elastic modulus (E) with the addition of RS. With increases in RS content, flexure modulus and strength decreases; however, both increase at 15wt% RS.		With RS, printed pieces became darker, with high porosity, and wood-like.
ABS	Lignin	40	Incorporation of 40wt%.	Tg of composite decreases on adding lignin, NBR42, and CFs.	Lignin was well-dispersed. Composite CFs percolated. 10wt% CFs in ABS/lignin/NBR41 composite increased interlayer adhesion between printing layers.
		10	Lignin in ABS decreases TS of composite, which was improved on adding NBR41 and CFs.		
	NBR41 carbon fibers (CFs) (1/8 inch)	10	Lignin and NBR41 decreased E of ABS while CFs increased it.		

(*Continued*)

TABLE 1.5 (*Continued*)

Summary of mechanical, thermal, and morphological properties of petroleum-based polymers with biofillers (Wasti & Adhikari, 2020)

Matrix	Filler	Filler content (wt%)	Mechanical properties	Thermal properties	Morphology properties
ABS	Lignin Polyethylene oxide (PEO) CFs	10, 20, 30 10 (relative to lignin amount) 20 vol%	PEO did not affect E. PEO lengthens failure. CFs increase TS in ABS/lignin/PEO (70/27/3).	PEO delayed lignin degradation. PEO raised degradation peak temperature. PEO decreased Tg of ABS.	PEO improves ABS matrix lignin particle adhesion. PEO inclusion reduces ABS matrix lignin domain size from 300–1,000 nm to 200–500 nm.
Nylon 12	HW lignin	40–60	40wt% lignin in nylon 12 matrix increased E but TS was about the same as plain nylon 12.	CFs and lignin lower recrystallization and Tm.	CFs were well-dispersed in the polymer matrix.
	CFs	4–16	12wt% CFs in nylon 12/lignin (6:4) composite increases TS and E.	CFs increase nylon 12/lignin (6:4) composite thermal conductivity.	Polymer matrix lignin phases were spherical.
Poly(ε-caprolactone) (PCL)	Cocoa shell waste (CSW) (50 μm)	10, 20, 30, 40, 50	Filament E increases and then drops with 30 wt% CSW. Tensile strain at break decreases with addition of CSW.	CSW blending little affects PCL matrix thermal characteristics.	CSW was evenly distributed in the PCL matrix. The 3D-printed specimen demonstrated high interlayer adhesion and no gaps.

cost. Smart metal alloys such as CuAl-Ni alloy, Ni-Ti alloy, and Cu-Au-Zn alloy have been identified as potential candidates for various applications.

According to Momeni et al. (2017), the literature distinguishes two subclasses of shape-transformation materials, namely, shape-changing materials and shape-memory materials. The phenomenon of shape memory pertains to the capacity of materials to recollect and restore their initial shape. This observation suggests that the initial configuration of the object is malleable and can be altered to assume a different shape through the application of an external force. Materials that can change their shape are characterized by their ability to undergo a reversible transformation in response to an external stimulus. Specifically, the transformation occurs from the material's original shape to a new shape when the stimulus is applied, and then reverts to the original shape when the stimulus is removed. The phenomenon of oscillation between two equilibrium states in the absence of an external force can be assumed to occur in the material, leading to multiple cycles of transformation. According to the research carried out by Zhou and Sheiko (2016), it has been observed that the material structure possesses a pre-determined direction and amplitude of movement. The exploration of humidity as a plausible trigger for shape-shifting materials is an expanding field of study in the domain of 4D printing technologies. Several thermoplastic polymers have been utilized in FDM 4D printing, alongside PLA and PU. The materials that have been previously mentioned, specifically ABS, styrene ethylbutylene styrene (SEBS)/PCL, polyvinylidene fluoride (PVDF), and polyamide (PA), are listed and organized in Table 1.6.

1.3 APPLICATION OF BIOCOMPOSITES FOR 3D AND 4D PRINTING

The 4D printing technique is a type of additive manufacturing that integrates stimuli-responsive materials with 3D printing technology. Smart polymeric materials and polymer-based composites are used in this procedure to amplify the characteristics of the printed components. Figure 1.7 provides examples of representative applications of this technology.

The manipulation of surface wettability is highly dependent on the surface microstructure. In the majority of cases, once the surface microstructure has been established, the surface wettability remains unchanged. The utilization of 4D printing technology presents a promising avenue for the facile creation of tunable surface wettability through reversible changes in surface microstructure. The observed phenomenon in Figure 1.8b indicates that the microsheets have the ability to fold upon evaporation of ethyl acetate (EA) and open upon application and evaporation of isopropyl alcohol. The observed behavior exhibits the ability of the surface to demonstrate both penetration and anti-penetration properties, correspondingly. The observed behavior can be attributed to the relative height difference between the folded sheets and the re-entrant pillars. In a previous study, Liu and colleagues (2021) utilized two-photon photoinitiator (TPP) technology to print a microstructure resembling a butterfly (as shown in Figure 1.8a) using acrylic monomers. The microsheets, resembling butterfly wings, were divided into two parts with varying cross-linking densities. This was done asymmetrically, as shown in Figure 1.8. As a result, when

TABLE 1.6

The following is a list of categories of recently reported polymeric materials, actuation mechanisms, and structure designs used in 4D printing by FDM (Fu et al., 2022)

Polymer	Fillers	Actuation method	Structure design
PLA	CFs, CNTs	Temperature	Multilayer anisotropic blocks with different printing path directions
	CNTs	Electric current	Layer-by-layer printing with diverse raster angles
	Ag nanoparticles	Electric current	A bilayer-based FDM printer with two extruder heads
Polyurethane (PU)		Temperature	Step-by-step strain-temperature programming of shape-memory PU
	Carbon black (CB)	Light	3D frame made of photo-responsive shape-memory CB/PU
ABS		Temperature	Laminating transverse and longitudinal layers sequentially in a bidirectional manner programmed printing pathways
SEBS/PCL blends		Temperature	Shape-memory effect (SME) was controlled by SEBS/PCL composition and improved by preference orientation along the printing direction
PVDF	Graphene and barium titanate (BTO)	Piezoelectric	A twin-screw extruder developed a piezoelectric BTO/graphene/PVDF filament that was used to FDM a cylindrical disk
PA11	BTO	Mechanical force	Piezoelectric part with a circular shape
PP/PA6 alloy		Temperature	Four different infill orientations were created by modifying the nozzle path. The experimental results indicate that the orientation of the sample tended to align with the direction of the applied force. This alignment led to a decrease in the shape fixity of the sample and an increase in its shape recovery

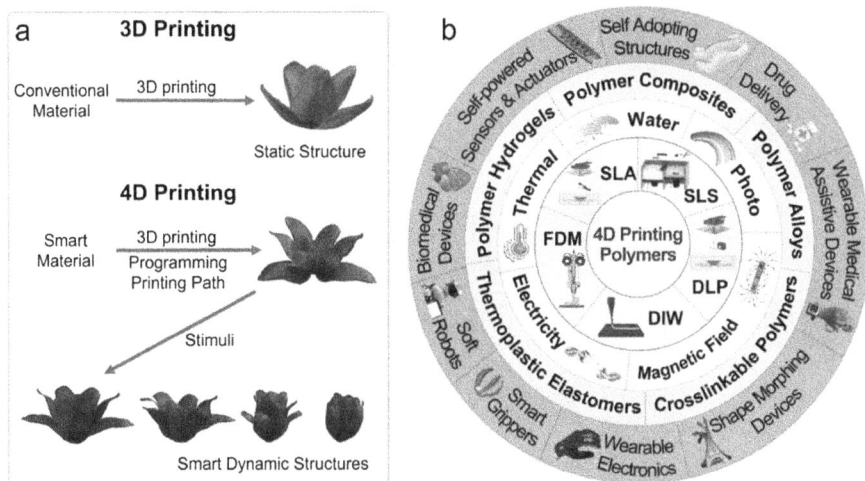

FIGURE 1.7 (a) Brief comparison of 3D and 4D printing. (b) Representative printing methods, external stimuli, printable polymeric materials, and applications of 4D printing (Fu et al., 2022).

the microsheets were transferred from one solvent to another, one part experienced a greater degree of shrinkage than the other, leading to the formation of a curved microstructure, as depicted in Figure 1.8a. The initial microstructure prior to any movement is depicted in the upper left-hand corner of Figure 1.8a. Subsequently, the microstructure was subjected to various solvents and subsequently subjected to evaporation to induce motion. The phenomenon of wing folding upon the evaporation of EA can be explained by the higher capillary force acting between the wings as compared to the force between the wing and the substrate. This can be attributed to the surface tension of EA, which is measured to be 23.5 mN m^{-1}. Upon the evaporation of a solvent with lower activity, such as n-pentane ($\gamma = 18.7$ mN m^{-1}) or isopropyl alcohol (IPA, $\gamma = 22.6$ mN m^{-1}), a partially or fully open structure was observed to form. The observed phenomenon can be attributed to the decrease in capillary force resulting from solvent evaporation. The study's findings indicate that all states are reversible, as demonstrated by the authors' observation of a rapid recovery to the vertical state following re-immersion in EA. The printed multi-material grippers, depicted in Figure 1.8b, exhibit various designs that enable them to effectively grasp and release objects. Upon programming, the closed (open) grippers that were printed were observed to have opened (closed). The ability to grasp (release) objects was activated through a heating process.

Natural fiber reinforced PLA composites are preferred for usage in the automotive industry because of their superior mechanical strength. According to a study, PLA reinforced with natural fibers exhibited a mechanical strength that was on par with glass fiber. The results of the study indicate that the specific strength of the composites was found to be three times greater than that of mild steel. In addition to this, it was found that the specific strength of the composites was comparable to that of conventional laminated glass-fiber-reinforced plastics. The automobile

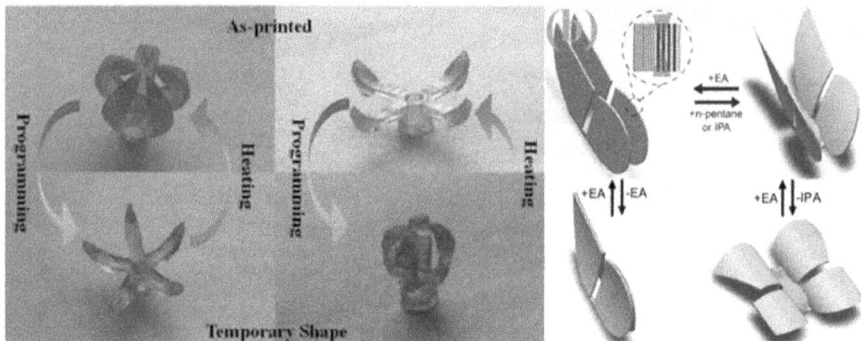

FIGURE 1.8 (a) Demonstration of the transition between the as printed shape and the temporary shape of the multi-material; (b) TPP printed responsive microstructure grippers (Fu et al., 2022).

industry employs coconut fiber for the production of car furnishings, and cotton is utilized for noise reduction purposes. The utilization of wood fiber as furnishings and accessories in vehicles, along with the use of sisal, hemp, and flax in the refinement of seatback linings and floor panels, has been observed. Furthermore, the construction sector offers a diverse range of natural fiber composite alternatives. Research is being done on how renewable resources, like natural fibers, can be used to evaluate building materials and reinforce cement-based materials. In the United States, bio-based composite materials for structural beams and panels were developed and produced (Ilyas et al., 2021). Bioprinting is a promising medical technology. Four-dimensional bioprinting can precisely manage medication delivery by producing foldable or unfoldable bi-layered structures that respond to temperature (Figure 1.9).

The evolution of 3D-printed models has enhanced both doctor-patient interaction and surgical success. Digital techniques, such as the printing of occlusal splints, have replaced many of the traditional procedures. Osteotomy and genioplasty surgical guides printed on a 3D printer shorten the duration of the procedure and lessen the risk of damage to sensitive structures like the inferior alveolar nerve (Khorsandi et al., 2021).

1.4 CONCLUSION

The fields of materials science and manufacturing are currently witnessing the emergence of biocomposites and 3D/4D printing. The amalgamation of natural fibers and a polymer matrix presents enhanced mechanical properties and ecological advantages. Diverse 3D printing methodologies offer a range of possibilities for fabricating intricate 3D structures, surpassing conventional industrial applications and encompassing medical implementations. Furthermore, the implementation of 4D printing technology introduces the notion of dynamic objects that have the ability to alter their shape over a period of time, thereby expanding the potential for innovative design and enhanced functionality. The selection of materials, including but not limited to

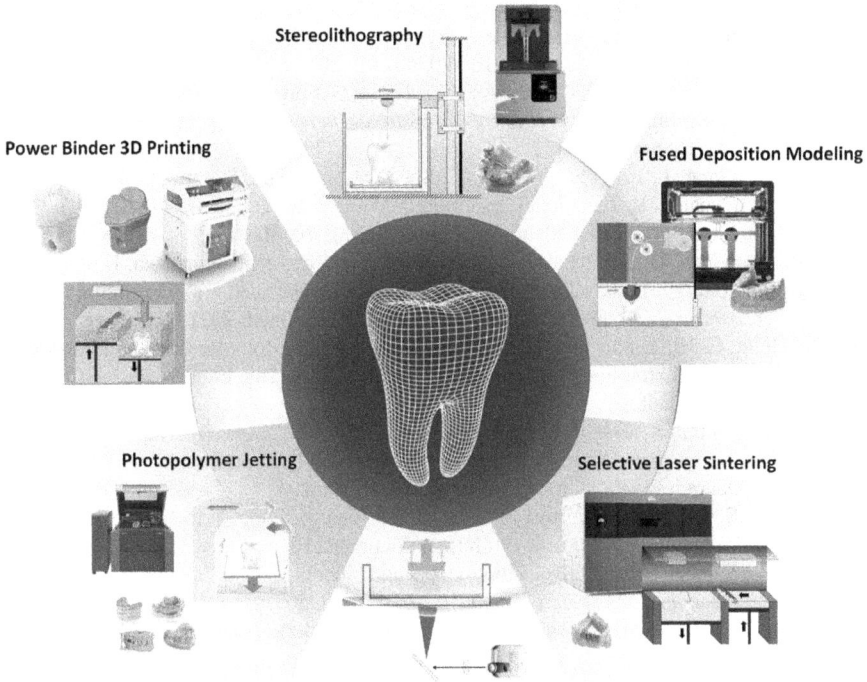

FIGURE 1.9   Current 3D and/or 4D printing techniques used in dentistry (Khorsandi et al., 2021).

PLA, PBS, and ABS, is a critical factor in determining the characteristics and potential uses of 3D/4D-printed items. Continued investigation and advancement in these domains will result in pioneering and enduring production methodologies that have a diverse array of pragmatic uses.

ACKNOWLEDGMENTS

The authors are thankful to Universiti Putra Malaysia for providing the platform and facilities.

REFERENCES

Ahmed, A., Arya, S., Gupta, V., Furukawa, H., & Khosla, A. (2021). 4D printing: Fundamentals, materials, applications and challenges. *Polymer*, 228. https://doi.org/10.1016/j.polymer.2021.123926

Aliotta, L., Seggiani, M., Lazzeri, A., Gigante, V., & Cinelli, P. (2022). A brief review of poly (butylene succinate) (PBS) and its main copolymers: synthesis, blends, composites, biodegradability, and applications. *Polymers*, 14(4), 844. https://doi.org/10.3390/polym14040844

Bahrami, M., Abenojar, J., & Martínez, M. Á. (2020). Recent progress in hybrid biocomposites: Mechanical properties, water absorption, and flame retardancy. *Materials*, 13(22), 5145. https://doi.org/10.3390/ma13225145

Balakrishnan, H., Hassan, A., Imran, M., & Wahit, M. U. (2012). Toughening of polylactic acid nanocomposites: A short review. *Polymer – Plastics Technology and Engineering*, 51(2), 175–192. https://doi.org/10.1080/03602559.2011.618329

Carew, R. M., & Errickson, D. (2020). An overview of 3D printing in forensic science: The tangible third-dimension. *Journal of Forensic Sciences*, 65(5), 1752–1760. https://doi.org/10.1111/1556-4029.14442

Fu, P., Li, H., Gong, J., Fan, Z., Smith, A. T., Shen, K., ... & Sun, L. (2022). 4D printing of polymeric materials: Techniques, materials, and prospects. *Progress in Polymer Science*, 101506. https://doi.org/10.1016/j.progpolymsci.2022.101506

Ganapathy, A., Chen, D., Elumalai, A., Albers, B., Tappa, K., Jammalamadaka, U., Hoegger, M. J., & Ballard, D. H. (2022). Guide for starting or optimizing a 3D printing clinical service. *Methods*, 206, 41–52. https://doi.org/10.1016/j.ymeth.2022.08.003

Goo, B., Hong, C. H., & Park, K. (2020). 4D printing using anisotropic thermal deformation of 3D-printed thermoplastic parts. *Materials and Design*, 188. https://doi.org/10.1016/j.matdes.2020.108485

Guna, V., Ilangovan, M., Ananthaprasad, M. G., & Reddy, N. (2018). Hybrid biocomposites. *Polymer Composites*, 39, E30–E54). https://doi.org/10.1002/pc.24641

Hodgdon, T., Danrad, R., Patel, M. J., Smith, S. E., Richardson, M. L., Ballard, D. H., ... & Decker, S. J. (2018). Logistics of three-dimensional printing: Primer for radiologists. *Academic Radiology*, 25(1), 40–51. https://doi.org/10.1016%2Fj.acra.2017.08.003

Ilyas, R. A., & Sapuan, S. M. (2020). Biopolymers and biocomposites: Chemistry and technology. *Perspectives in Analytical Chemistry*, 16(5), 500–503.

Ilyas, R. A., Sapuan, S. M., Harussani, M. M., Hakimi, M. Y. A. Y., Haziq, M. Z. M., Atikah, M. S. N., Asyraf, M. R. M., Ishak, M. R., Razman, M. R., Nurazzi, N. M., Norrrahim, M. N. F., Abral, H., & Asrofi, M. (2021). Polylactic acid (PLA) biocomposite: Processing, additive manufacturing and advanced applications. *Polymers*, 13(8). https://doi.org/10.3390/polym13081326

Jacquel, N., Freyermouth, F., Fenouillot, F., Rousseau, A., Pascault, J. P., Fuertes, P., & Saint-Loup, R. (2011). Synthesis and properties of poly(butylene succinate): Efficiency of different transesterification catalysts. *Journal of Polymer Science, Part A: Polymer Chemistry*, 49(24), 5301–5312. https://doi.org/10.1002/pola.25009

Keya, K. N., Kona, N. A., Koly, F. A., Maraz, K. M., Islam, Md. N., & Khan, R. A. (2019). Natural fiber reinforced polymer composites: History, types, advantages, and applications. *Materials Engineering Research*, 1(2), 69–87. https://doi.org/10.25082/MER.2019.02.006

Khorsandi, D., Fahimipour, A., Abasian, P., Saber, S. S., Seyedi, M., Ghanavati, S., Ahmad, A., de Stephanis, A. A., Taghavinezhaddilami, F., Leonova, A., Mohammadinejad, R., Shabani, M., Mazzolai, B., Mattoli, V., Tay, F. R., & Makvandi, P. (2021). 3D and 4D printing in dentistry and maxillofacial surgery: Printing techniques, materials, and applications. *Acta Biomaterialia*, 122, 26–49. https://doi.org/10.1016/j.actbio.2020.12.044

Le Duigou, A., Correa, D., Ueda, M., Matsuzaki, R., & Castro, M. (2020). A review of 3D and 4D printing of natural fibre biocomposites. *Materials & Design*, 194, 108911. https://doi.org/10.1016/j.matdes.2020.108911

Leksakul, K., & Phuendee, M. (2018). Development of hydroxyapatite-polylactic acid composite bone fixation plate. *Science and Engineering of Composite Materials*, 25(5), 903–914. https://doi.org/10.1515/secm-2016-0359

Li, Z., Rathore, A. S., Song, C., Wei, S., Wang, Y., & Xu, W. (2018). PrinTracker: Fingerprinting 3D printers using commodity scanners. In *Proceedings of the ACM Conference on Computer and Communications Security* (pp. 1306–1323). https://doi.org/10.1145/3243734.3243735

Lin, C., Liu, L., Liu, Y., & Leng, J. (2022). 4D printing of shape memory polybutylene succinate/polylactic acid (PBS/PLA) and its potential applications. *Composite Structures*, 279, 114729. https://doi.org/10.1016/j.compstruct.2021.114729

Liu, X., Wei, M., Wang, Q., Tian, Y., Han, J., Gu, H., et al. (2021). Capillary-force-driven self-assembly of 4D-printed microstructures. *Advanced Materials*, 33, 2100332. https://doi.org/10.1002/adma.202100332

Love, L. J., Kunc, V., Rios, O., Duty, C. E., Elliott, A. M., Post, B. K., Smith, R. J., Blue, C. A. (2014). The importance of carbon fiber to polymer additive manufacturing. *Journal of Materials Research*, 29, 1893–1898 https://doi.org/10.1557/jmr.2014.212

Momeni, F., Medhi Hassani, S., Liu, X., & Ni, J. (2017). A review of 4D printing, *Materials & Design*, 122, 42–79. https://doi.org/10.1016/j.matdes.2017.02.068.

Ning, F., Cong, W., Qiu, J., Wei, J., & Wang, S. (2015). Additive manufacturing of carbon fiber reinforced thermoplastic composites using fused deposition modeling. *Composites Part B: Engineering*, 80, 369–378 https://doi.org/10.1016/j.compositesb.2015.06.013

Okada, T., Faudree, M. C., Tsuchikura, N., & Nishi, Y. (2016). Improvement of low-temperature impact value of sandwich-structural (CFRP/ABS/CFRP) laminate plies by Homogeneous Low-Energy Electron Beam Irradiation (HLEBI). *Materials Transactions*, 57(3), 305–311. https://doi.org/10.2320/matertrans.M2015332

Oyeniran, A. A., & Ismail, S. O. (2021). Mechanical Behaviors of Natural Fiber-Reinforced Polymer Hybrid Composites. *Mechanical and Dynamic Properties of Biocomposites*, 1–26. https://doi.org/10.1002/9783527822331.ch1

Rafiqah, S. A., Khalina, A., Harmaen, A. S., Tawakkal, I. M., Zaman, K., Asim, M., Nurrazi, M. N., & Lee, C. H. (2021). A review on properties and application of bio-based poly(butylene succinate). *Polymers* 13(9), 1436. https://doi.org/10.3390/polym13091436

Su, S., Kopitzky, R., Tolga, S., & Kabasci, S. (2019). Polylactide (PLA) and its blends with poly (butylene succinate) (PBS): A brief review. *Polymers*, 11(7), 1193. https://doi.org/10.3390%2Fpolym11071193

Tekinalp, H. L., Kunc, V., Velez-Garcia, G. M., Duty, C. E., Love, L. J., Naskar, A. K., Blue, C. A., & Ozcan, S. (2014). Highly oriented carbon fiber–polymer composites via additive manufacturing. *Composites Science and Technology*, 105, 144–150 https://doi.org/10.1016/j.compscitech.2014.10.009

Tymrak, B. M., Kreiger, M., & Pearce, J. M. (2014). Mechanical properties of components fabricated with open-source 3D printers under realistic environmental conditions. *Materials & Design*, 58, 242–246. https://doi.org/10.1016/j.matdes.2014.02.038

Vakharia, V. S., Singh, M., Salem, A., Halbig, M. C., & Salem, J. A. (2022). Effect of reinforcements and 3-D printing parameters on the microstructure and mechanical properties of acrylonitrile butadiene styrene (ABS) polymer composites. *Polymers*, 14, 2105. https://doi.org/10.3390/polym14102105

Wasti, S., & Adhikari, S. (2020). Use of biomaterials for 3D printing by fused deposition modeling technique: A review. *Frontiers in Chemistry*, 8, 315. https://doi.org/10.3389/fchem.2020.00315

Zarna, C., Rodríguez-Fabià, S., Echtermeyer, A. T., & Chinga-Carrasco, G. (2022). Preparation and characterisation of biocomposites containing thermomechanical pulp fibres, poly(lactic acid) and poly(butylene-adipate-terephthalate) or poly(hydroxyalkanoates) for 3D and 4D printing. *Additive Manufacturing*, 59. https://doi.org/10.1016/j.addma.2022.103166

Zhou, J., & Sheiko, S. (2016). Reversible shapeshifting in polymeric materials. *Journal of Polymer Science Part B: Polymer Physics*, 54, 1365–1380. https://doi.org/10.1002/polb.24014.

2 Fabrication Process of Bio-Composites Filament for Fused Deposition Modeling
A Review

Hazliza Aida C.H., Mastura M.T.,
Abdul Kudus S.I., Ilyas R.A., and Loh Y.F.

2.1 INTRODUCTION

Natural fibers have great potential to be used as renewable resources because they are a cheaply available sustainable material that can act as a new source of reinforcement materials to substitute materials based on petroleum or manufactured fibers (Lotfi et al., 2019). In general, natural fibers can be classified based on the source of their origin, whether it is plant-based, animal-based, or from minerals (George et al., 2016), which is summarized in Figure 2.1. Plant fibers are the best recognized by the industry and the subject of most studies. Natural fibers have found new uses in biomedical gadgets and civil buildings (Khan et al., 2018). It is because of the product's quick growth time, reusability, and general availability (Cicala et al., 2010). Figure 2.2 shows the most often utilized plant fibers.

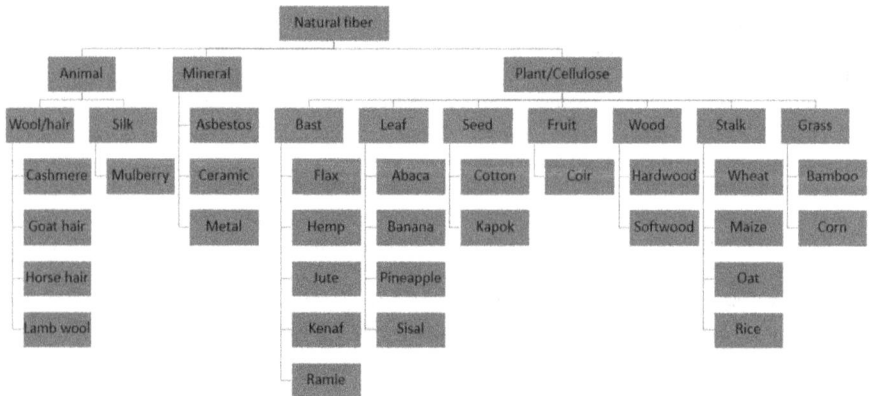

FIGURE 2.1 Classification of natural fibers.

DOI: 10.1201/9781003362128-2

FIGURE 2.2 An overview of some of the most common sources of natural fibers.

According to Le Duigou et al. (2020), cellulose, hemicellulose, and lignin are plant or cellulose fibers' primary components. In natural plant fibers, the lignin matrix binds cellulose micro-fibrils, forming cellulose fibers. Lignin serves as a biological barrier and a reinforcing agent in the fiber to withstand the effects of gravity and wind. According to Sharath Shekar and Ramachandra (2018), cellulose is the primary constituent responsible for natural fiber's intrinsic strength and stability, whereas hemicellulose aids the fiber's structure. Therefore, a fiber cell membrane is not uniform.

For example, the fiber cell wall is separated into main, secondary, and intermediate lamellae. The fiber's secondary layer is further subdivided into exterior (S1), middle (S2), and interior (S3) sections. Minerals and water are transported via the lumen in the secondary walls and the central one. According to Lotfi et al. (2019), various variables like the plant's age and type, weather effects, harvesting information, and details related to the processing of the fiber all have an important role in influencing the final product of the fibers. As Ho et al. (2012) suggested, the fiber cell shape and the degree of polymerization determine the qualities of each form of cellulose. Linear cellulosic macro-molecules are connected by hydrogen bonds and intimately coupled with hemicelluloses and lignin, which contribute to the fiber's rigidity, respectively. In addition to binding fibers together, cellulose is a structural component of the fiber cell wall (Chen, 2014). Figure 2.3 (Ho et al., 2012) shows how its elements influence fiber properties.

There is a crystalline structure to all plant fibers, with roughly 65%–70% cellulose $(C_6H_{10}O_5)_n$ made up of carbon (C), hydrogen (H), and oxygen (O). Lignin and other non-cellulosic components impact the final quality and characteristics of cellulosic fibers. Because of the huge number of hydroxyl groups and hygroscopic nature, plant-based fibers have enhanced moisture absorption capacity, which is often needed for manufactured composites (Witayakran et al., 2017). The stiffness and strength of natural fiber are anisotropic and very much dependent on their hierarchical microstructure

Strength	Lignin	Hemicellulose + lignin	Non-crystalline cellulose	Crystalline
Thermal degradation	Lignin	Cellulose	Hemicellulose	
Biological degradation	Lignin	Crystalline cellulose	Non-crystalline cellulose	
Moisture absorption	Crystalline cellulose	Lignin	Non-crystalline cellulose	Hemicellulose
UV degradation	Crystalline cellulose	Non-crystalline cellulose	Hemicellulose	Lignin

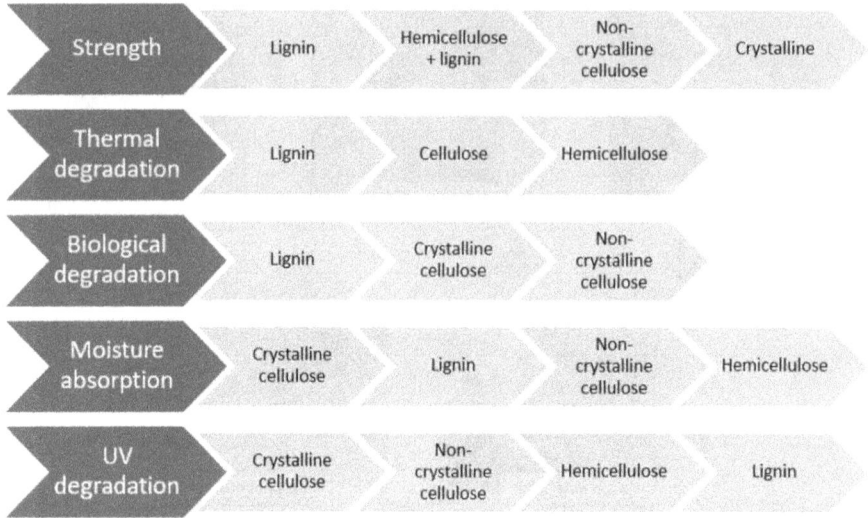

FIGURE 2.3 Plant constituents' influence. The rightward-pointing arrows indicate an increase.

and biochemical content, according to the research of Le Duigou et al. (2020). Several natural fibers may be analyzed using the tensile properties, biochemical compositions, and microstructures listed in Table 2.1. It is characteristic of naturally tuned biological materials that the link between microstructure, composition, and properties is so strong. For example, bast fibers that support the plant (e.g., hemp, flax) are stiffer and stronger than fruit fibers (e.g., coir fibers) that are used to dissipate energy (Defoirdt et al., 2010). The angle of the cellulose micro-fibrils (MFA) (Fratzl et al., 2008; Joffre et al., 2014), the interconnections between the components, and the overall structure will determine the natural fibers' overall hygro-mechanical properties. According to Zhan et al. (2020), cellulose increases the strength of the natural fiber and Young's modulus, and the micro-fibrillar angle influences fiber stiffness. Micro-fibrils that spiral around the fiber axis in natural fibers are more malleable. The microfiber cannot bend or deform when positioned perpendicularly to the fiber's axis.

Many other fields can attest to the superior qualities of natural fibers. High-quality mechanical capabilities and biodegradable features are drawing attention to natural fibers at the moment, particularly in the automotive sector and for general engineering applications (Nurazzi et al., 2021). In addition, a lot of scientists are concentrating their efforts on eco-friendly materials. The polymer composites sector, in particular, has responded enthusiastically to the rising need for engineering materials by doing extensive research and developing new and better materials. In the field of polymer composites, natural fibers have often been employed as reinforcement for the bio-composite's end result. Natural fibers are gaining popularity for several reasons, not the least of which is their potential to replace relatively inexpensive synthetic fiber-reinforced plastics with superior sustainability, environmental friendliness, and renewable energy. In addition, structural polymers that have fiber reinforcement have seen widespread usage in many fields (Izwan, Sapuan, and Ilyas, 2019).

TABLE 2.1

Biochemical composition (cellulose, hemicellulose, lignin, pectins concentration, and micro-fibrillar angle (MFA)) and mechanical properties of several natural fibers utilized in FDM (Le Duigou et al., 2020)

Fibers	E (Gpa)	σ (Mpa)	Cellulose [%]	Hemicellulose [%]	Lignin [%]	Pectins [%]	MFA (°)
Flax	46–85	600–2,000	64–85	11–17	2–3	1.8–2.0	10
Jute	24.7–26.5	393–773	61–75	13.6–20.4	12–13	0.6±0.6	7–12
Kenaf	11–60	223–930	45–57	21.5	8–13	3–5	7–12
Coir	3.44–4.16	120–304	32–46	0.15–0.3	40–45	3–4	30–49
Cotton	3.5–8	287–597	82–99	4	0.75	6	30–40
Wood	15.4–27.5	553–1,500	38–45	19–39	22–34	0.4–5	5–45
Bamboo	10–40	340–510	34.5–50	20.5	26	<1	2–10
Harakeke	14–33	440–990	55–90	4–16	2–5	0.8–8	6.2–11.2
Hemp	14.4–44.5	270–889					

The limited material characteristics of popular thermoplastics prompted the introduction of the composite idea to fused deposition modeling (FDM), which involves reinforcing the polymeric feedstock with a second phase (Sang et al., 2018; Sodeifian, Ghaseminejad, and Yousefi, 2019). The polymeric matrix is mixed with several fillers to boost its mechanical and thermal qualities. This study thus compiles publications about the fiber composite filament manufacturing technique, and proceeds to detail how changes in material parameters might affect the mechanical properties of FDM created components.

2.2 BIO-COMPOSITE FILAMENT PRODUCTION

Regarding 3D printing, the FDM technique stands out because it can manufacture intricate pieces while maintaining precise weight control via infill structures (Dey et al., 2019; Mohan et al., 2017). These features draw researchers and business people to the FDM process. Moreover, by expanding the availability of filament materials and a wide variety of qualities, the FDM technique may be used in a wider range of industries. Figure 2.4 depicts the categorization of many currently available filament materials (Dey et al., 2021), and this chapter focuses on bio-plastic composite materials.

2.2.1 COMPOSITE FILAMENT PREPARATION

According to Kerni et al. (2020), natural fiber-reinforced polymer composites may be made in various ways, comprising hand layup, injection molding, compression molding, and resin transfer molding. Three phases in the processing procedure are usually incorporated, including drying the natural fiber, combining the fiber with the matrix, and fabricating the component (Figure 2.5). Several investigations have revealed that natural fibers' moisture is removed at 80°C to achieve the ideal method (Sapuan

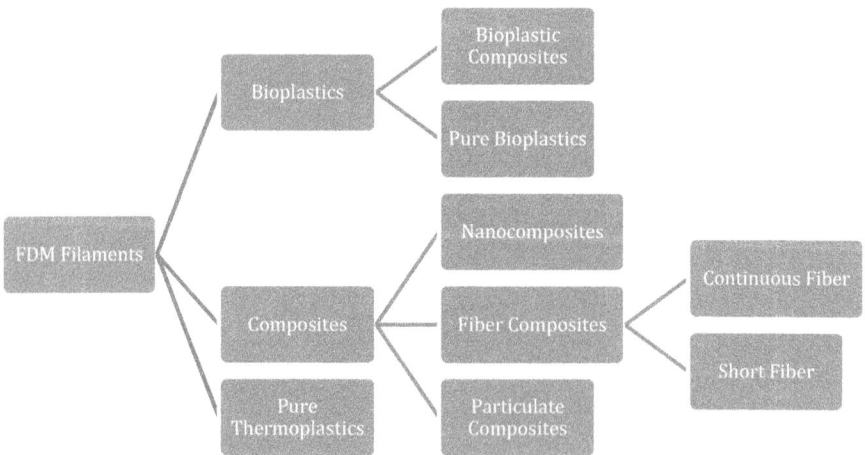

FIGURE 2.4   The categorizing of currently available filament materials.

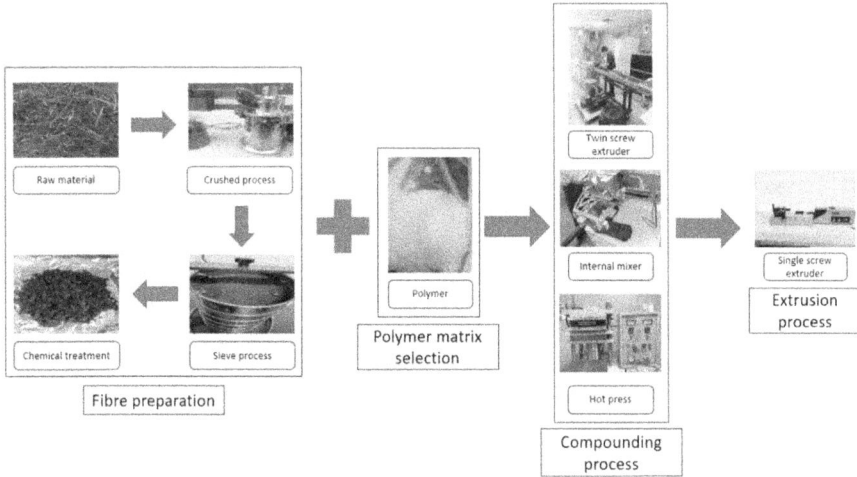

FIGURE 2.5 Fiber composite filament fabrication process.

et al., 2011; Siregar, 2011). In addition, extrusion and an internal mixing machine are the two most frequent approaches to combining fiber and matrix (Jaafar et al., 2019). Westman et al. (2010) shared that the two most often employed processes are compression molding and injection molding for natural fiber-reinforced polymer composites. More importantly, processing depends on several variables, such as the moisture level of natural fibers, the fiber type and content, and fiber length (Faruk et al., 2012). According to the type and length of fiber used, the mixing and manufacturing methods for composites can be seen in Table 2.2. The manual mixing method and hand layup techniques are the best for making composites with long natural fibers. In contrast, the automated mixing process and compression molding technique are better for making composites with short natural fibers.

2.2.1.1 Fiber Preparation

Retting and decortication methods are generally the best options (Sharath Shekar and Ramachandra, 2018). A decorticator machine removes fiber bundles from the leaf and the fiber plant. A revolving wheel with blunt blades is used for crushing and pulverizing leaves until only the fibers remain. Water removes the leaf's remaining sections. After being decorated, the fibers are washed and dried in the sun or hot air. Various grades of dry fibers are combed and separated.

Retting is traditionally divided into two categories: dew retting and water retting. Allowing the plant's stems to decompose naturally is called dew retting. With this method, the bast fibers are regularly examined to guarantee they do not lose their original nature. Dew retting is a well-known method across Europe, despite being very location-dependent and resulting in coarser, lower-quality fibers than those obtained from water retting. Furthermore, stems are submerged in water during water retting (ponds or tanks and slow-moving rivers). Consequently, water retting requires a considerable volume of clean water and is costly and undesirable to the environment.

TABLE 2.2

Various mixing and manufacturing methods composed from diverse natural fiber sources with differing lengths

No.	Natural fiber	Matrix	Mixing process	Manufacturing processes	Critical length (mm)
1	Hemp fiber	Polyester	Manual mixing	Vacuum bagging method hand layup technique	3.4
2	Kenaf	Unsaturated polyester	Kenaf fiber was put into the mold, into which resin was injected with a pressure of 1.3 bar	Resin transfer molding	6
3	Pineapple leaves fiber (PALF)	High-impact polystyrene (HIPS)	Internal mixing	Compression molding	0.5
4	Banana	Epoxy	Epoxy resin was put into the mold, into which fiber was manually laid	Hand layup technique	15
5	Kenaf	Soy-based resin	Extrusion	Compression molding	6
6	Bamboo	Starch	Manual mixing	Dried in an oven	15
7	Kenaf-PALF	High-density polyethylene (HDPE)	Internal mixing	Compression molding	1.0
8	Basalt	Polyester	Manual mixing	Compression molding	10
9	Agave	Epoxy	Stirring process	Compression molding	3

According to Zhan et al. (2020), natural fiber is found to have a profound impact on the mechanical performance of the composites. Fiber type, the orientation of the fiber, the length, fiber loading, and fiber hybrid all have an important role in the mechanical characteristics of natural fiber composites (Pickering et al., 2016). Table 2.3 displays the components of certain natural fibers and the most common form of glass fiber (E-glass). Flax, hemp, and ramie fiber are among the cellulose-based natural fibers with a high value of Young's modulus and tensile strengths. However, it should be noted that there is substantial diversity in the literature. For example, natural fibers have lower tensile strength and stiffness than glass fiber, but they may obtain stiffness equivalent to those produced with glass fiber. On the other hand, natural fibers may have greater specific Young's modulus and lower specific tensile strength than E-glass fibers.

The natural fiber's chemical makeup influences the qualities of natural fiber-reinforced composites. In addition, the cellulose content influences the natural fibers' mechanical characteristics, indirectly influencing the properties of natural fiber-reinforced composites (Jaafar et al., 2018; Sena Neto et al., 2015). The attributes

TABLE 2.3
Mechanical properties of natural and synthetic fiber

Fiber	Density (g/cm³)	Length (mm)	Failure strain (%)	Tensile strength (MPa)	Stiffness/Young's modulus (GPa)	Specific tensile strength (MPa/g cm⁻³)	Specific Young's modulus (GPa/g cm⁻³)
Ramie	1.5	900–1,200	2.0–3.8	400–938	44–128	270–620	29–85
Flax	1.5	5–900	1.2–3.2	345–1,830	27–80	230–1,220	18–53
Hemp	1.5	5–55	1.6	550–1,110	58–70	370–740	39–47
Jute	1.3–1.5	1.5–120	1.5–1.8	393–800	10–55	300–610	7.1–39
Sugar palm	1.292		7.98	156.96	4.96		
Kenaf	1.2		1.6–4.3	223–1,191	11–60	641	10–42.9
Harakeke	1.3	4–5	4.2–5.8	440–990	14–33	338–761	11–25
Pineapple	1.56		1.6–2.4	170–1,627	60–82.5		
Sisal	1.3–1.5	900	2.0–2.5	507–855	9.4–28	362–610	6.7–20
Alfa	1.4	350	1.5–2.4	188–308	18–25	134–220	13–18
Cotton	1.5–1.6	10–60	3.0–10	287–800	5.5–13	190–530	3.7–8.4
Coir	1.2	20–150	15–30	131–220	4–6	110–180	3.3–5
Silk	1.3	Continuous	15–60	100–1,500	5–25	100–1,500	4–20
Feather	0.9	10–30	6.9	100–203	3–10	112–226	3.3–11
Wool	1.3	38–152	13.2–35	50–315	2.3–5	38–242	1.8–3.8
E-glass	2.5	Continuous	2.5	2,000–3,000	70	800–1,400	29

of natural fibers depend on the amount of hemicellulose and lignin present in the fibers. However, it is undesirable because it promotes fiber breakdown and moisture absorption (Azwa et al., 2013). Therefore, natural fiber with high cellulose content and low hemicellulose and lignin content was recommended by Kerni et al. (2020) for generating a composite with excellent mechanical qualities.

On the other hand, the performance of composite filaments is strongly influenced by the fibers' length, diameter, and aspect ratio (length/diameter) (Le Duigou et al., 2020). To control the bio-composite's viscosity and minimize fiber aggregation, natural fibers that have a minimum length/diameter (L/d) ratio (i.e., in powder form) are often utilized to make filaments with a diameter of 1.75–3 mm (Tran et al., 2017; Zhao et al., 2019). Wood/polylactic acid (PLA) composites' complicated viscosity, for example, was shown to drop from 1.683 Pa.s to 1.549 Pa.s when the length of the poplar fiber was lowered from 850 to 236 μm to 180 μm and below, as stated by Zhao et al. (2019). The authors also suggested that low resistance to particle change when the shear rates are at an increased level is necessary for a simple extrusion process with minimal energy usage. On the contrary, high zero-shear viscosities also have an advantage by helping the extruded materials maintain their form after they come out of the nozzle and are deposited. According to the literature, extrusion-based hydrogel 3D printing systems have an ideal viscosity range of 100–10,000 Pa.s (Kyle et al., 2017).

Generally speaking, reinforced filaments have a greater tensile modulus than pure polymers. Still, their rupture characteristics are worse because of the inadequate fiber length/diameter ratio, and the dispersion rate is weak. As an example, Depuydt et al. (2018) found that by adding 15% flax or bamboo fibers, there is an increment in the tensile modulus to 2.4 from 1.3 GPa, but the strength initially was 31 MPa and decreased to 23–30 MPa. In Kelly-Tyson equations, an L/d ratio less than the crucial L/d prevents the stress from being transferred to the fiber effectively. A crucial L/d of 35.3 was established for PLA bio-composites reinforced with bamboo fibers, which was much higher than the actual aspect ratio. Fibers might be seen as flaws that cause stress concentrations from this viewpoint.

Additionally, the percentage of fibers in the material is also influenced by the composites' mechanical characteristics. An increase in the percentage of high-strength fibers leads to an improvement in mechanical qualities. However, if it is increased above the optimal amount, mechanical qualities may be degraded (Sastra et al., 2006; Siregar et al., 2009) because it has been found that increasing fiber concentration reduces matrix composition, weakening the interfacial connection between matrixes and composite (Shalwan, 2013).

2.2.1.2 Polymer Matrix Selection

Plastic is a common name for polymeric materials, consisting of several polymer molecules with identical chemical structures (Kaushik et al., 2017; Mallick, 2007). According to Nirmal et al. (2015), fossil fuels and natural gas are the two most commonly used synthetic materials as the basic material in the polymer. Interestingly, Rajeshkumar et al. (2021) highlighted that polymers are quickly replacing traditional materials because of their desired properties, such as ease of processing, unique mechanical properties, and being economically viable and lighter.

Polymer matrixes are usually divided into thermoplastic, thermoset, and elastomeric types. Secondary bonds or weak intermolecular forces hold thermoplastic polymers together but are not chemically bonded (Rajeshkumar et al., 2021). As a result, even after curing, thermoplastics could be converted into liquid form, thus able to be recycled several times. Cross-links chemically connect the particles in a thermoset polymer to produce a three-dimensional network. Polymers that have been cured or polymerized cannot be recycled, according to Pickering et al. (2016), since they cannot be treated by heat. Elastomeric polymers, on the other hand, have exceptional elasticity. They are found to be able to maintain their original form when the force applied, lower than the yield strength, is withdrawn.

Fiber composites made from natural fibers have high macroscopic stiffness and strength because the matrix holds them together while also transferring stress to the fibers through bonding or friction (Zhan et al., 2020). In addition, the matrix determines the composite's form, surface appearance, and environmental tolerance. In natural fiber matrixes, thermoplastic and thermosetting polymers have become widespread (Holbery and Houston, 2006). However, according to Summerscales et al. (2010), most natural fibers begin to deteriorate at above 200°C (Table 2.4), which limits the sorts of matrixes that may be used.

The thermoplastic polymers that are generally used are polypropylene (PP), polystyrene (PS), polyethylene (PE), PLA, and acrylonitrile butadiene styrene (ABS) (Zhan et al., 2020). Table 2.5 lists the physical (density ρ, glass transition temperature Tg, and melting point Tm) and mechanical (such as tensile strength S, tensile modulus E, and specific modulus E/ρ) properties of the standard thermoplastic polymers. According to Zhan et al. (2020), the mechanical characteristics of the compound in the same reinforced phase are often improved by polymers with good mechanical properties.

2.2.2 COMPOUNDING PROCESS

Understanding the correlation between fibers and matrix is critical, especially how it influences the composite's mechanical attributes. Exceptional interrelation between

TABLE 2.4
Printing temperature range of polymer filament materials

Material	Printing temperature range (°C)	References
Acrylonitrile butadiene styrene (ABS)	210–230	Kariz et al. (2017);
Polylactic acid (PLA)	230–275	Kariz et al. (2017); Stoof et al. (2017);
Polyhydroxyalkanoates (PHA)/PLA blend	210	Le Duigou et al. (2016)
Recycled polypropylene (rPP)	230	Stoof and Pickering (2018)
Biobased polyethylene (bioPe)	210	Filgueira et al. (2018)
Polycaprolactone (PCL)	120	Tran et al. (2017)
Thermoplastic polyurethane (TPU)	190	Bi et al. (2018)

TABLE 2.5

Mechanical characteristics of typical thermoplastics

Polymer	ρ (g/cm³)	Tg (°C)	Tm (°C)	S (MPa)	E (GPa)	E/ρ	References
PP	0.899–0.920	−23–10	2.0–3.8	26–41.4	0.95–1.77	1.03–1.97	Zhan et al.
PS	1.04–1.06		1.2–3.2	25–69	4–5	3.77–4.81	(2020)
LDPE	0.910–0.925	−125	1.6	40–78	0.055–0.38	0.059–0.418	
HDPE	0.94–0.96	−133–100	1.5–1.8	14.5–38	0.4–1.5	0.417–1.596	
PLA	0.9–1.27	56–65	4.2–5.8	60	1.5–2.7	1.18–3	
ABS	0.67–1.06		2.0–2.5	42.7	2.48		Kariz et al.
							(2018)

these two components is necessary to produce optimal reinforced fibers since stress is passed between the matrix and fibers through the interface. However, a too-strong interface might enable fracture propagation, causing the toughness and strength to degrade. On the other hand, fiber composites from plants tend to have poor interfacial bonding between hydrophilic fibers and hydrophobic matrixes, thus harming the mechanical performance and long-term properties. Fiber and matrix have to be close to each other for bonding; wettability might be considered a crucial requirement. When fiber wetting is insufficient, interfacial defects may serve as stress concentrators (Chen et al., 2006). There is evidence that wetting the fibers in composites affects their toughness, stiffness, and elasticity (Wu and Dzenis, 2006). A fiber's wettability and interfacial strength may be improved by physical and chemical modifications, according to Gholampour and Ozbakkaloglu (2019).

Chemical treatments are more prevalent in the literature than physical ones and have yielded the most significant advances in overall performance. The primary goal of this particular process is to improve fiber-matrix interrelation based on bonding and stress. Chemical treatments are employed to eliminate contaminants from the fiber surface and interrupt the moisture absorption process by coating the fibers with −OH groups to improve the fiber-matrix adhesion. By changing the surface and cleaning the fiber surface in this way, chemical treatment or pre-treatment may reduce moisture absorption and increase the unevenness of the surface. Table 2.6 summarizes the effects of a few of the chemical treatments now on the market.

Research on bio-composites is picking up steam across several fields. According to Zwawi (2021), the increased concerns about environmental pollution and heightened environmental laws by the respective agencies, even globally, have caused phenomenal growth in the market for bio-composites in the domestic and industrial sectors. Different customized chemical treatments and efficient process procedures are required to attain desired qualities and make bio-composites cost-effective. Conventional processes such as compression molding, hand layup, injection, extrusion, and pultrusion are used in manufacturing bio-composites (Bajpai et al., 2017). Fiber dispersion, orientation, and aspect ratio are all considered when deciding on the appropriate fabrication method (Le Duc et al., 2011). When choosing a method or technique, several factors, such as manufacturing costs, final design, form, and size, as well as the qualities of raw materials, are considered (Zwawi, 2021).

TABLE 2.6

How chemical treatments affect natural fiber's performance

Treatments	Effect	References
Alkali	Reduce the lignin content	Zin et al. (2018)
	Improve fiber-matrix adhesion, thermal stability and heat resistivity	Kabir et al. (2011)
Acetylation	Improve tensile and flexural strength	Azam et al. (2016)
Benzoylation	Improve hydrophobicity	
Enzyme	Reduce the lignin content	
Grafting	Improve UV-protective properties, hydrophobicity and mechanical properties	
Isocyanate	Surface modification	
Mercerization	Reduce the moisture regain	
	Improve the mechanical properties	
Methacrylate	Improve tensile and flexural strength	
Ozone	Affect surface energy and contact angle	
Peroxide	Reduce the moisture regain	
Plasma	Improve hydrophobicity	
Silane	Improve hydrophobicity and mechanical properties	
Sodium chlorite	Improve tensile strength, Young's modulus and elongation at break	

The mechanical characteristics of bio-composites are enhanced by fiber/matrix adhesion, homogeneous fiber dispersion, and a high aspect ratio. Fabrication methods and qualities are affected by various parameters, including the length of the fiber, content, type, orientation, and moisture content. Moisture may modify the qualities and process parameters of fibers. Thus, they must be properly dried (Zwawi, 2021). The mechanical characteristics of bio-composites are affected by moisture in fibers, which causes the void content and porosity to increase (Bledzki et al., 2008; Faruk et al., 2014). There are several ways to deal with the problem of moisture in fibers, according to Faruk et al. (2012). Bio-composites' production procedures and qualities are affected by silica and chemical structural variances across fibers (Bledzki et al., 2008; Chaudhary et al., 2018). According to Pickering et al., natural fibers deteriorate at high temperatures, resulting in a limited temperature range for composite fabrication (2016).

2.2.3 EXTRUSION PROCESS

The FDM technique may use various thermoplastic substances such as filaments. In order to create filaments from thermoplastics and composites, extruders like Filabot are employed. Figure 2.6 (Dey et al., 2021) shows a typical filament extruder. The unprocessed components are inserted into the barrel with the help of the hopper during the filament-creation process. The turning screw(s) is housed in the barrel, while the barrel is also used to heat the raw ingredients. The feed zone, transition zone, and metering zone are inside the barrel (Chaturvedi et al., 2017). This zone

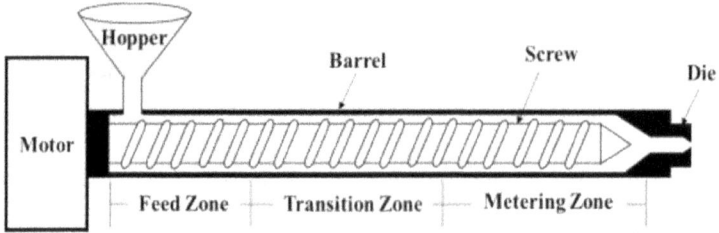

FIGURE 2.6   A layout of filament fabrication extruder (Dey et al., 2021).

is where the raw ingredients undergo three different softening and melting stages before being metered into the final product. The input materials determine how hot or cold each zone should be. First, material enters the metering zone from the feed zone and moves through the revolving screw's surface. The barrel might have one or two screws, referred to as a single-screw or a twin-screw extruder, by referring to the number of screws. Next, material from the metering zone is extruded via a die. The diameter of the die is chosen following the requisite diameters of the filaments to be formed. For 1.75-mm filaments, a brass die with a diameter of 2.5–3.5 mm is often used, according to Nassar et al. (2019). Materials that have been extruded travel through a cooling zone after passing through the die. The cooling rate has a considerable impact on filament quality. Filaments are collated using various methods, including a water bath and cold air cannon. The filament's cross-section may be further reduced using a drawing procedure. Typically, the drawing process uses a direct current (DC) motor.

Single-screw extruders and multi-screw extruders are two types of screw extruders. These components are popular because of their inexpensive cost, simple design, robustness, and dependability. A single-screw extruder's basic functioning is discussed in the preceding section. First, the screw's rotational action transports the content from the hopper into the feed throat. Next, an extruder pushes molten material out of a die with the help of the screw and barrel's thermal heat. Single-screw extruders provide several advantages, including:

  i. With the adoption of a double-order integral design, the single-screw extruder has a good plasticizing function, helping ensure stable, high-speed, and high-performance extrusion.
 ii. In addition, it also bears a large capacity and emits low noise since the whole process involves involute gear transmission.
iii. It likewise ensures the mixing effect of material because it encompasses the distinctive barrier of comprehensive mixing design.

The low plasticizing temperature and high shear provide the low-temperature, high-performance, and low-pressure metering extrusion of materials.

Specifically, the twin-screw extruders are said to have better performance mixing additives, fillers, and liquids uniformly compared to the single-screw extruders.

(a) Corotating screws

(b) Counterrotating screws

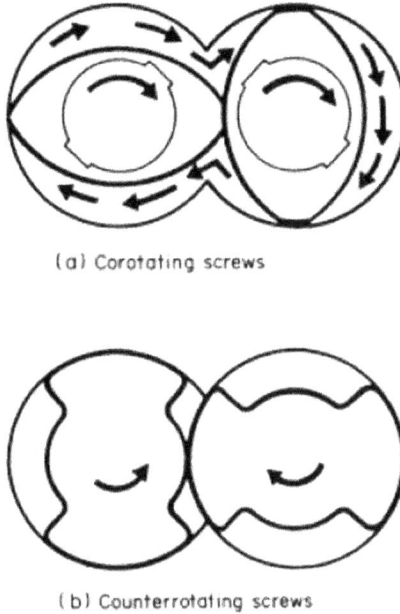

FIGURE 2.7   Twin-screw extruder cross-section: (a) co-rotating screws and (b) counter-rotating screws.

It is possible to increase the mixing performance of a single-screw extruder by utilizing mixing components; however, this is not as effective as for a twin-screw extruder. Twin-screw extruders are available in many constructions, with parallel or conical screws that may revolve in the same direction (co-rotating) or the opposite way (counter-rotating) and with varying degrees of intermeshing, as shown in Figure 2.7.

Using the extruders with twin screws as reaction vessels has certain distinct benefits because of their capacity to remove heat, devolatilize undesirable byproducts, and pump high-viscosity materials. In addition, twin-screw extruders provide several advantages, including:

i. Positive displacement capability, especially for counter-rotating screws – makes it easier to pump difficult materials, reduces the flow rate's susceptibility to pressure, and provides a more uniform residence time for the material.

ii. Heat-sensitive materials may be processed with reduced shear rates and improved temperature control.

iii. Superior blending without the use of high local heat.

The utilization of modular parts for specific purposes throughout the screw's length allows for more flexibility and adaptability.

2.3   MECHANICAL ATTRIBUTES OF FDM BIO-COMPOSITES FILAMENTS

Interestingly, the use of composites in areas like the automotive industry and residential construction is rising. As a result, several studies highlight and discuss the improved performance of bio-composites by focusing on the mechanical attributes of naturally fiber-reinforced composites. Moisture absorption, flaws, orientation, volume fraction, and physical qualities of natural fibers significantly impact the mechanical properties. As a result, the weak interfacial strength between natural fiber and matrix causes less efficient transmission of stress from matrix to fiber. On top of that, natural fiber compound alone has worse mechanical characteristics as it has a low fiber dispersion. Moreover, the mechanical attributes of natural fiber composites are affected by numerous aspects, including the type of fiber, interface parameters, and matrix selection.

For the FDM filament, Tao et al. (2017) employed PLA and 5 wt. % wood flour (WF) as reinforcement, as shown in Figure 2.8. Adding wood reinforcing enhances elastic modulus while lowering thermal degradation temperature. Compressive strength is superior to tensile strength in WF-PLA composite components. Unlike WF, PLA is hydrophobic, which means weak interfacial adhesion between the two materials. According to Kariz et al. (2018), tensile strength increases with wood content up to 10% but decreases as the percentage of wood increases. The WF-PLA components generated by the FDM technique have an improved surface texture or roughness and void in the reinforced polymers than pure PLA-built parts. Adding various toughening agents and graft copolymers (GC) may influence the tensile and flexural strength of WF-PLA composite components (Guo et al., 2018). TPU outperforms PCL and polyethene-co-octene (POE) as a toughening agent.

In contrast to other filament materials, WF-PLA's surface roughness rose as the layer thickness grew (Ayrilmis, 2018), as illustrated in Figure 2.9. His research also revealed that the thickness of the layer has an important role in determining the contact angle. On thicker layers, the wettability is better. Daver et al.'s (2018) study used combined FDM filaments made from PLA, cork, and tributyl citrate (TBC). The percentage of TBC in the composites was set at 5%, while the percentage of cork

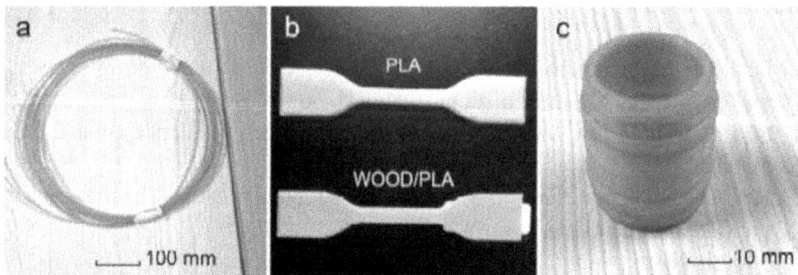

FIGURE 2.8   Filament, test specimen, and 3D product: (a) WF/PLA composite filament; (b) specimens for tensile properties measurement; (c) a barrel made by FDM 3D printer (Tao et al., 2017).

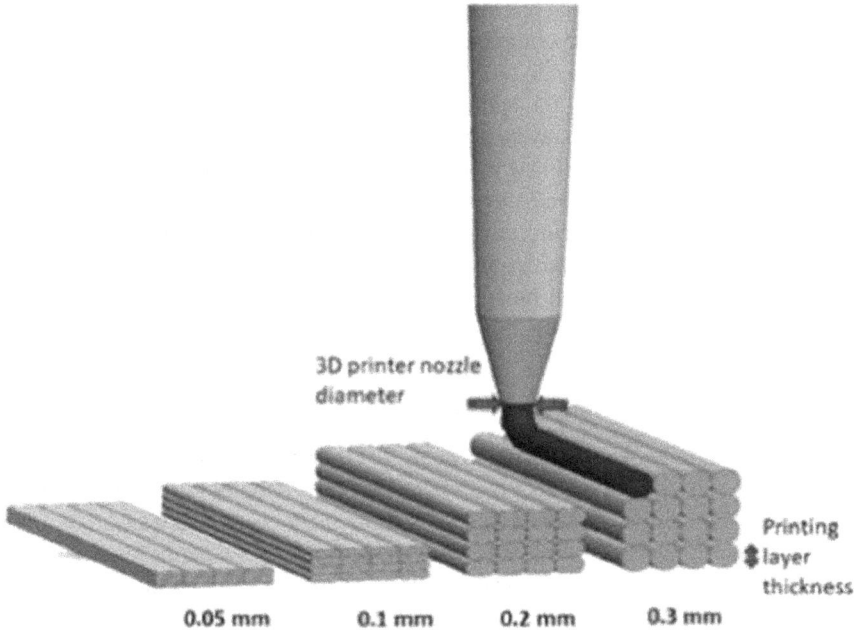

FIGURE 2.9 Schematic representation of printed specimen geometry (Ayrilmis, 2018).

varied from 5% to 50%. Higher cork content enhanced the specific modulus value, particularly the impact strength and specific tensile strength properties, whereas tensile strength and density were reduced. TBC is often used in manufacturing flexible composite filament as a plasticizer.

For FDM filament preparation, Murphy and Collins (2018) employed microcrystalline cellulose (MCC). A coupling agent called titanate is also utilized as a surface modification to improve the dispersion and adherence of MCC. The crystallinity reduces when the percentage of PLA is higher than 3 wt. %, according to an experiment. As the coupling agent reduces MCC's hydrophilicity, water absorption is reduced. As a result, composite components have a higher modulus of storage and a higher degradation temperature than PLA parts. Dong et al. (2017) reinforced the PLA matrix with grafted cellulose nanofibers. Reinforcement percentages of 3% are excellent for storage modulus, strength, and elastic modulus, according to Murphy and Collins (2018).

Lignin has been employed as reinforcement in the preparation of composite FDM filaments as it was shown that lignin and PLA might improve thermal stability and flammability (Domínguez-Robles et al., 2019; Gkartzou et al., 2017; Mimini et al., 2019). However, other studies have shown that the stability of the thermal attribute diminishes as the material's ability to withstand pulling force as well as the elongation at break increases (Gkartzou et al., 2017; Wasti and Adhikari, 2020). Therefore, the characteristics of lignin-blended PLA compounds require more investigation.

On the other hand, Kaygusuz and Özerinç (2019) analyzed the mechanical characteristics of composite items made from FDM filaments by combining 12% of the

PLA with the PHA. It is noted that the reduction in the tensile strength is about 25% in contrast to pure PLA components, while the ductility increased by roughly 160%. The printer's temperature should be between 210 and 240°C, according to the manufacturer's instructions. With the addition of plasticizers, such as the citrates acetyl and tributyl, PHA and PLA may be strengthened (Menčík et al., 2018). Additionally, Ausejo et al. (2018) used the FDM method to create pieces made of PLA/PHA compound in build horizontally and vertically. The researchers concluded that the composite pieces' orientation during construction has a crucial influence on their tensile qualities, morphology, structure, and surface characteristics. Bone scaffolds might benefit from the nontoxic and biocompatible PLA/PHA combination as a filament material.

2.4   FUTURE TRENDS IN FDM BIO-COMPOSITE

Pre-impregnated filaments (Le Duigou et al., 2019) printing of continuous fiber composites using customized printers is a viable current technology. The jute/PLA was found to have a multi-scale debonding, indicating inadequate load transmission and insufficient wetting. Regarding mechanical qualities, bio-composites containing continuous jute or flax threads substantially beat natural discontinuous fiber composites. For example, while PLA/jute composite (6%) had an overall stiffness of 5.11 ± 0.41 GPa and a strength of 57.1 ± 5.33 MPa, PLA/flax composite (30%) had an EL of 23.3 ± 1.8 GPa and 253.7 ± 15 MPa in terms of strength.

The tensile stiffness and strength of 3D-printed flax/PLA bio-composites with the same fiber content are comparable to those of thermally compressed bio-composites. Two scenarios must be considered when comparing composite materials supplemented with continuous synthetic fibers and printing technologies (i.e., pre-impregnated filaments). First, in terms of stiffness and strength, continuous flax fiber composites are inferior to continuous carbon fiber composites . The mechanical performance of continuous flax bio-composites compared to continuous glass/PA composites is promising since continuous flax bio-composites may achieve a comparable range of stiffness. It means that bio-composites made of continuous natural fibers should be considered for semi-structural or structural applications, particularly for stiffness-based design.

Three-dimensional-printed natural fiber composites may benefit from developing specialized and cost-effective processing equipment. Printing speed and fiber content may increase, and processing costs may be reduced using a new bio-composites pellet extrusion nozzle (Zhao et al., 2019). Bio-composites comprising wood and PLA have recently been made possible by large-scale additive printing methods. Due to its rapid deposition rate (up to 50 kg/h) and large building volume (around 27 m^3), this technique has the potential to minimize treatment time and costs. In addition, six-axis printing (polar printing method) instead of three-axis printing (Cartesian printing) has been developed, allowing even more detailed graphics. Additionally, AFP may be performed utilizing a six-axis industrial manipulator (robot arm), albeit their larger expense must be addressed. Integrating a custom system into a gantry-type three-axis printer is currently prohibitively expensive, and it may even need different technical processes (e.g., additional compression).

2.5 CONCLUSION

Compared to synthetic composites research, natural fiber-reinforced composites have been the focus of attention for the last 20 years (e.g., glass or carbon fiber-reinforced composites). Despite this, bio-composites are still in the early stages of research regarding technological advancement and basic knowledge. However, now that bio-composites can be 3D-printed, they can progress at the same pace as synthetic materials.

This study aimed to provide a foundation for future research into the mechanical and actuation capabilities of 3D-printed bio-composites by examining the features of natural fibers. An extensive examination of the literature on printing methods, microstructure, and characteristics of natural fiber bio-composites was carried out to accomplish this aim in detail. Bio-composites and synthetic materials were put side by side. Technical and scientific difficulties, as well as new advancements, have been discussed.

Discontinuous fibers or powder-like reinforced polymers are two common bio-composites in 3D printing. Because of this, the tensile strength of the current product is not comparable to injection or compression molding. It has been shown that natural fibers improve the composite's stiffness but not its tensile strength. In order to lower overall viscosity and improve printability, 3D-printed bio-composites typically have low fiber content (b30 wt. %) and a relatively low aspect ratio (L/d). Because of two factors, first, the porosity of the filament before printing, and, second, the low printing pressure, discontinuous or short fiber-reinforced bio-composites have high porosity content. Porosity makes it more probable that printed samples may soak up a substantial amount of water and hence lose effectiveness when exposed to moist conditions.

ACKNOWLEDGMENT

The authors would like to thank Universiti Teknikal Malaysia Melaka and the Ministry of Higher Education of Malaysia for providing the financial support and grant scheme FRGS/1/2020/TK0/UTEM/02/26 to the principal author in this project.

REFERENCES

D. Ariawan, Z.A. Mohd Ishak, M.S. Salim, R. Mat Taib, M.Z. Ahmad Thirmizir, Wettability and interfacial characterization of alkaline treated kenaf fiber-unsaturated polyester composites fabricated by resin transfer molding, *Polym. Compos.* 38 (3) (2017) 507–515.

J.G. Ausejo, J. Rydz, M. Musioł, W. Sikorska, M. Sobota, J. Wlodarczyk, G. Adamus, H. Janeczek, I. Kwiecie´n, A. Hercog, et al. A comparative study of three-dimensional printing directions: the degradation and toxicological profile of a PLA/PHA blend, *Polym. Degrad. Stab.* 152 (2018) 191–207.

N. Ayrilmis, Effect of layer thickness on surface properties of 3D printed materials produced from wood flour/PLA filament, *Polym. Test.* 71 (2018) 163–166.

A. Azam, S. Khubab, Y. Nawab, J. Madiha, T. Hussain, Hydrophobic treatment of natural fibers and their composites—a review, *J. Ind. Text.* (2016) 1–31.

C. Badouard, F. Traon, C. Denoual, C. Mayer-Laigle, G. Paës, A. Bourmaud, Exploring mechanical properties of fully compostable flax reinforced composite filaments for 3D printing applications, *Ind. Crop. Prod.* 135 (2019) 246–250.

P.K. Bajpai, F. Ahmad, V. Chaudhary, Processing and characterization of bio-composites. *Handb. Ecomater.* (2017) 1–18.

H. Bi, Z. Ren, R. Guo, M. Xu, Y. Song, Fabrication of flexible wood flour/thermoplastic polyurethane elastomer composites using fused deposition molding, *Ind. Crop. Prod.* 122 (2018) 76–84.

A.K. Bledzki, A. Jaszkiewicz, M. Murr, V.E. Sperber, R. Lützendgrf, T. Reußmann, Processing techniques for natural-and wood-fibre composites. In: *Properties and Performance of Natural-Fibre Composites*, Elsevier, Amsterdam, The Netherlands, 2008, pp. 163–192.

E. Chaturvedi, N.S. Rajput, S. Upadhyaya, P. Pandey, Experimental study and mathematical modeling for extrusion using high density polyethylene, *Mater. Today Proc.* 4 (2017) 1670–1676.

V. Chaudhary, P.K. Bajpai, S. Maheshwari, Studies on mechanical and morphological characterization of developed jute/hemp/flax reinforced hybrid composites for structural applications, *J. Nat. Fibers* 15 (2018) 80–97.

H. Chen, *Biotechnology of Lignocellulose*, Springer, Dordrecht, The Netherlands, 2014, ISBN 978-94-007-6897-0.

P. Chen, C. Lu, Q. Yu, Y. Gao, J. Li, X. Li, Influence of fiber wettability on the interfacial adhesion of continuous fiber-reinforced PPESK composite, *J. Appl. Polym. Sci.* 102 (3) (2006) 2544–2551.

G. Cicala, G. Cristaldi, G. Recca, A. Latteri, Composites based on natural fibre fabrics. In: P. Dubrovski (Ed.), *Woven Fabric Engineering*, InTech, London, UK, 2010.

F. Daver, K. Peng, M. Lee, M. Brandt, R. Shanks, Cork–PLA composite filaments for fused deposition modeling, *Compos. Sci. Technol.* 168 (2018) 230–237.

N. Defoirdt, S. Biswas, L. De Vriese, L.Q.N. Tran, J. Van Acker, Q. Ahsan, et al., Assessment of the tensile properties of coir, bamboo and jute fibre, *Compos. Part A Appl. Sci. Manuf.* 41 (2010) 588–595,

D. Depuydt, M. Balthazar, V. Hendrickx, W. Six, E. Ferraris, F. Desplentere, et al., Production and characterization of bamboo and flax fiber reinforced polylactic acid filaments for fused deposition modeling (FDM), *Polym. Compos.* 40(5) (2018), 1951–1963.

A. Dey, N. Yodo, B. Khoda, Optimizing process parameters under uncertainty in fused deposition modeling. In: *Proceeding of the 2019 IIE Annual Conference*, Orlando, FL, 18–21 May 2019, Institute of Industrial and Systems Engineers (IISE): Peachtree Corners, GA, 2019.

A. Dey, I.N. Roan Eagle, N. Yodo, A Review on filament materials for fused filament fabrication. *J. Manuf. Mater. Process.* 5 (2021) 69.

J. Domínguez-Robles, N.K. Martin, M.L. Fong, S.A. Stewart, N.J. Irwin, M.I. Rial-Hermida, R.F. Donnelly, E. Larrañeta, Antioxidant PLA composites containing lignin for 3D printing applications: a potential material for healthcare applications. *Pharmaceutics* 11 (2019) 165.

J. Dong, M. Li, L. Zhou, S. Lee, C. Mei, X. Xu, Q. Wu, The influence of grafted cellulose nanofibers and post extrusion annealing treatment on selected properties of polylactic acid filaments for 3D printing. *J. Polym. Sci. Part B Polym. Phys.* 55 (2017) 847–855.

O. Faruk, A.K. Bledzki, H.P. Fink, M. Sain, Biocomposites reinforced with natural fibers: 2000–2010, *Prog. Polym. Sci.* 37 (11) (2012) 1552–1596.

O. Faruk, A.K. Bledzki, H.-P. Fink, M. Sain, Progress report on natural fiber reinforced composites. *Macromol. Mater. Eng.* 299 (2014) 9–26.

D. Filgueira, S. Holmen, J.Melbø, D.Moldes, A. Echtermeyer, G. Chinga-Carrasco, 3D printable filaments made of biobased polyethylene biocomposites, *Polymers (Basel)* 10 (2018) 314.

P. Fratzl, R. Elbaum, I. Burgert, Cellulose fibrils direct plant organ movements, *Faraday iscuss.* 139 (2008) 275–282.

M. George, M. Chae, D.C. Bressler. Composite materials with bast fibres: structural, technical, and environmental properties, *Prog. Mater. Sci.* 83 (2016) 1–23.

E. Gkartzou, E. Koumoulos, C.A. Charitidis, Production and 3D printing processing of bio based thermoplastic filament, *Manuf. Rev.* 4 (2017) 1.

R. Guo, Z. Ren, H. Bi, Y. Song, M. Xu, Effect of toughening agents on the properties of poplar wood flour/poly (lactic acid) composites fabricated with fused deposition modeling, *Eur. Polym. J.* 107 (2018) 34–45.

M. Ho, H. Wang, J.-H. Lee, C. Ho, K. Lau, J. Leng, D. Hui, Critical factors on manufacturing processes of natural fibre composites, *Compos. Part B Eng.* 43 (2012) 3549–3562.

J. Holbery, D. Houston, Natural-fiber-reinforced polymer composites in automotive applications, *JOM* 58 (11) (2006) 80–86.

S. Izwan, S.M. Sapuan, R.A. Ilyas, A review on tensile properties of sugar palm fibre and its composites, *Prosiding Seminar Enau Kebangsaan 2019*, (2019) 73–77.

J. Jaafar, J.P. Siregar, S. Mohd Salleh, M.H. Mohd Hamdan, T. Cionita, T. Rihayat, Important considerations in manufacturing of natural fiber composites a review, *Int. J. Precis. Eng. Manuf. – Green Technol.* 6 (3) (2019) 647–664.

T. Joffre, R.C. Neagu, S.L. Bardage, E.K. Gamstedt, Modelling of the hygroelastic behavior of normal and compression wood tracheids, *J. Struct. Biol.* 185 (2014) 89–98, https://doi.org/10.1016/j.jsb.2013.10.014.

M.M. Kabir, H. Wang, T. Aravinthan, F. Cardona, K.-T. Lau, Effects of natural fibre surface on composite properties: a review. *Energy Environ. Sustain.* (2011) 94–99. [CrossRef]

M. Kariz, M. Sernek, M. Obućina, M. Kuzman, Effect of wood content in FDM filament on properties of 3D printed parts, *Mater. Today* 14 (2017) 135–140.

P. Kaushik, J. Jaivir, K. Mittal, Analysis of mechanical properties of jute fiber strengthened epoxy/polyester composites, *Eng. Solid Mech.* 5 (2017) 103.

B. Kaygusuz, S. Özerinç, Improving the ductility of polylactic acid parts produced by fused deposition modeling through polyhydroxyalkanoate additions, *J. Appl. Polym. Sci.* 136 (2019) 48154. [CrossRef]

L. Kerni, S. Singh, A. Patnaik, N. Kumar. A review on natural fiber reinforced composites, *Mater. Today: Proc.* 28 (2020) 1616–1621.

T. Khan, H. Sultan, M.T. Bin, A.H. Ariffin, The challenges of natural fiber in manufacturing, material selection, and technology application: a review. *J. Reinforc. Plast. Compos.* 37 (2018) 770e779. https://doi.org/10.1177/0731684418756762.

S. Kyle, Z. Jessop, A. Al-Sabah, I. Whitaker, "Printability" of candidate biomaterials for extrusion based 3D printing: state-of-the-art, *Adv. Healthc. Mater.* (2017) 1700264. https://doi.org/10.1002/adhm.201700264.

A. Le Duc, B. Vergnes, T. Budtova, Polypropylene/natural fibres composites: analysis of fibre dimensions after compounding and observations of fibre rupture by rheo-optics. *Compos. Part Appl. Sci. Manuf.* 42 (2011) 1727–1737. [CrossRef]

A. Le Duigou, A. Barbé, E. Guillou,M. Castro, 3D printing of continuous flax fibre reinforced biocomposite for structural applications, *Mater. Des.* 180 (2019), 107884. https://doi.org/10.1016/j.matdes.2019.107884.

A. Le Duigou, D. Correa, M. Ueda, R. Matsuzaki, M. Castro. A review of 3D and 4D printing of natural fibre biocomposites, Mater. Des. 194 (2020) 108911, 1–26.

A. Lotfi, H. Li, D.V. Dao, G. Prusty, Natural fibre-reinforced composites: a review on material, manufacturing, and machinability, *J. Thermoplast. Compos. Mater.* (2019) 1–47.

P.K. Mallick, *Fiber-Reinforced Composites: Materials, Manufacturing, and Design*, CRC Press, Boca Raton, London, New York, 2007.

P. Menčík, R. Pˇrikryl, I. Stehnová, V. Melˇcová, S. Kontárová, S. Figalla, J. Boˇckaj, Effect of selected commercial plasticizers on mechanical, thermal, and morphological properties of poly (3-hydroxybutyrate)/poly (lactic acid)/plasticizer biodegradable blends for three-dimensional (3D) print, *Materials* 11 (2018) 1893. [CrossRef]

V. Mimini, E. Sykacek, S.N.A.S. Hashim, J. Holzweber, H. Hettegger, K. Fackler, A. Potthast, N. Mundigler, T. Rosenau, Compatibility of kraft lignin, organosolv lignin and lignosulfonate with PLA in 3D printing, *J. Wood Chem. Technol.* 39 (2019) 14–30. [CrossRef]

N. Mohan, P. Senthil, S. Vinodh, N. Jayanth, A review on composite materials and process parameters optimisation for the fused deposition modelling process, *Virtual Phys. Prototyp.* 12 (2017) 47–59. [CrossRef]

C.A. Murphy, M.N. Collins, Microcrystalline cellulose reinforced polylactic acid biocomposite filaments for 3D printing, *Polym. Compos.* 39 (2018) 1311–1320. [CrossRef]

M. Nassar, M.A. El Farahaty, S. Ibrahim, Y.R. Hassan, Design of 3D filament extruder for Fused Deposition Modeling (FDM) additive manufacturing, *Int. Design J.* 9 (2019) 55–62.

U. Nirmal, J. Hashim, M. M. Ahmad, A review on tribological performance of natural fibre polymeric composites, *Tribol. Int.* 83 (2015) 77.

N.M. Nurazzi, M.R.M. Asyraf, A. Khalina, N. Abdullah, H.A. Aisyah, S. Ayu Rafiqah, F.A. Sabaruddin, et al., A review on natural fiber reinforced polymer composite for bullet proof and ballistic applications, *Polymers* 13 (4) (2021) 1–42. https://doi.org/10.3390/polym13040646.

K.L. Pickering, M.G.A. Efendy, T.M. Le, A review of recent developments in natural fibre composites and their mechanical performance, *Compos. Part A* 83 (2016) 98–112.

G. Rajeshkumar, S. Arvindh Seshadri, S. Ramakrishnan, M.R. Sanjay, Suchart Siengchin, K.C. Nagaraja, A comprehensive review on natural fiber/nano-clay reinforced hybrid polymeric composites: materials and technologies, *Polym. Compos.* (2021) 1–15.

asalt fiber reinforced polylactide composites and their feasible evaluation for 3D printing applications. *Compos. Part B: Eng.* 164 (2018) 629–639. https://doi.org/10.1016/j.compositesb.2019.01.085.

S.M. Sapuan, A.R. Mohamed, J.P. Siregar, M.R. Ishak, Pineapple leaf fibers and PALF-reinforced polymer composites, In: *Cellulose Fibers: Bio- and Nano-Polymer Composites*, Springer Berlin Heidelberg, 2011, pp. 325–343.

H.Y. Sastra, J.P. Siregar, S.M. Sapuan, M.M. Hamdan, Tensile properties of Arenga pinnata fiber-reinforced epoxy composites, *Polym. – Plast. Technol. Eng.* 45 (1) (2006) 149–155.

A.R. Sena Neto, M.A.M. Araujo, R.M.P. Barboza, A.S. Fonseca, G.H.D. Tonoli, F.V.D. Souza, L.H.C. Mattoso, J.M. Marconcini, *Comparative study of 12 pineapple leaf fiber varieties for use as mechanical reinforcement in polymer composites*, Elsevier, 2015.

A. Shalwan, B.F. Yousif, *In state of art: mechanical and tribological behaviour of polymeric composites based on natural fibres*, Elsevier, 2013.

H. S. Sharath Shekar, M. Ramachandra, Green composites: a review. *Mater. Today: Proc.* 5 (2018) 2518–2526.

J.P. Siregar, Effects of selected treatments on properties of pineapple leaf fibre reinforced high impact polystyrene composites, *Mat Sci.* (2011).

J.P. Siregar, S.M. Sapuan, M.Z.A. Rahman, H.M.D.K. Zaman, Physical properties of short pineapple leaf fibre (PALF) reinforced high impact polystyrene (HIPS) composites, *Adv. Compos. Lett.* 18 (1) (2009) 25–29.

G. Sodeifian, S. Ghaseminejad, A.A. Yousefi, Preparation of polypropylene/short glass fiber composite as fused deposition modeling (FDM) filament, *Results Phys.* 12 (November 2018) (2019) 205–222. https://doi.org/10.1016/j.rinp.2018.11.065.

D. Stoof, K. Pickering, Sustainable composite fused deposition modeling filament using recycled pre-consumer polypropylene, *Compos. Part B Eng.* 135 (2018) 110–118.

J. Summerscales, N.P.J. Dissanayake, A.S. Virk, W. Hall. A review of bast fibres and their composites. Part 1 – fibres as reinforcements, *Compos. Part A* 41 (10) (2010) 1329–1335.

Md. Syduzzaman, Md. Abdullah Al Faruque, K. Bilisik, M. Naebe, Plant-based natural fibre reinforced composites: a review on fabrication, properties and applications, *Coatings* 10 (2020) 973, 1–34.

Y. Tao, H. Wang, Z. Li, P. Li, S.Q. Shi, Development and application of wood flour-filled polylactic acid composite filament for 3D printing, *Materials* 10 (2017) 339. [CrossRef]

A. Thygesen, A.B. Thomsen, G. Daniel, H. Lilholt, Comparison of composites made from fungal defibrated hemp with composites of traditional hemp yarn, *Ind. Crops Prod.* 25 (2) (2007) 147–159.

T.N. Tran, I.S. Bayer, J.A. Heredia-Guerrero, M. Frugone, M. Lagomarsino, F. Maggio, et al., Cocoa shell waste biofilaments for 3D printing applications, *Macromol. Mater. Eng.* 302(11) (2017), 1–10.

S. Wasti, S. Adhikari, Use of biomaterials for 3D printing by fused deposition modeling technique: a review, *Front. Chem.* 8 (2020), 1–14. [CrossRef]

M. Westman, L. Fifield, K. Simmons,S. Laddha, & T.A. Kafentzis,Natural fiber composites: a review, *J. Phycol.* 35(4) (2010), 806–814.

S. Witayakran, W. Smitthipong, R. Wangpradid, R. Chollakup, P.L. Clouston, *Natural Fibre Composites: Review of Recent Automotive Trends*, Elsevier, London, UK, 2017, 166–174.

X.F. Wu, Y.A. Dzenis. Droplet on a fiber: geometrical shape and contact angle. *Acta Mech.* 185(3–4) (2006) 215–225.

J. Zhan, J. Li, G. Wang, Y. Guan, G. Zhao, J. Lin, H. Naceur, D. Coutellier, Review on the performance, foaming and injection molding simulation of natural fiber composites, *Polym. Compos.* 588 (2020), 1–20.

X. Zhao, H. Tekinalp, X. Meng, D. Ker, B. Benson, Y. Pu, et al., Poplar as biofiber reinforcement in composites for large-scale 3D printing, *ACS Appl. Bio. Mater.* (2019).

M.H. Zin, K. Abdan, N. Mazlan, E.S. Zainudin, K.E. Liew, The effects of alkali treatment on the mechanical and chemical properties of pineapple leaf fibres (PALF) and adhesion to epoxy resin. *IOP Conf. Ser. Mater. Sci. Eng.* 368 (2018) 012035. [CrossRef]

3 Biodegradable Natural Fiber Polymer Composite as Future 3D Printing Feedstock
A Review

Hamat S., Ishak M.R., Sapuan S.M., and Yidris N.

3.1 INTRODUCTION

Three-dimensional printing is also known as fused filament fabrication (FFF) or fused deposition modeling (FDM); it has played a valuable role in this additive manufacturing (AM) technology era by offering a simple and low-cost alternative to more conventional manufacturing techniques, such as molding (Kazmer and Colon 2020), forming (Klimyuk, Serezhkin, and Plokhikh 2020), and computer numerical control machining (Kalsoom, Nesterenko, and Paull 2018; Low et al. 2017), while it continues to affect manufacturing and prototyping areas.

In contrast to conventional methods in which the material is removed in order to achieve the desired models, FDM technology that was invented in the 1980s by Scott Crump, co-founder and chairman of Stratasys Ltd. (Kocovic 2017), is obtained by adding the materials in a layered sequence using the slicing manufacturing technology software for 3D structures directly from a computer-aided design (CAD) file, heating and extruding the thermoplastic filament, and depositing layers of semi-liquid beads along an STL-defined extrusion path (Tofail et al. 2018). Thus, it gives AM technology the flexibility in printing models with various geometry and complexity with lesser production cost and time. A few industries were started with various geometry and complexity with lesser production cost and time. A few industries started to utilize this technology in producing 3D objects, products, parts, or components that are as competent as the existing products. In fact, 3D printing provides easier and cheaper production but with good quality as another option in fabricating parts (Jaisingh Sheoran and Kumar 2020).

There are a number of 3D printing techniques available. The major difference is how layers are built to create components. For example, selective laser sintering (SLS), FDM (Figure 3.1), and stereolithography (SLA) are the most commonly used technologies for 3D printing (Quan et al. 2020). According to Chacón et al. (2017), FDM is a common rapid prototyping (RP) technology that is widely used in

DOI: 10.1201/9781003362128-3

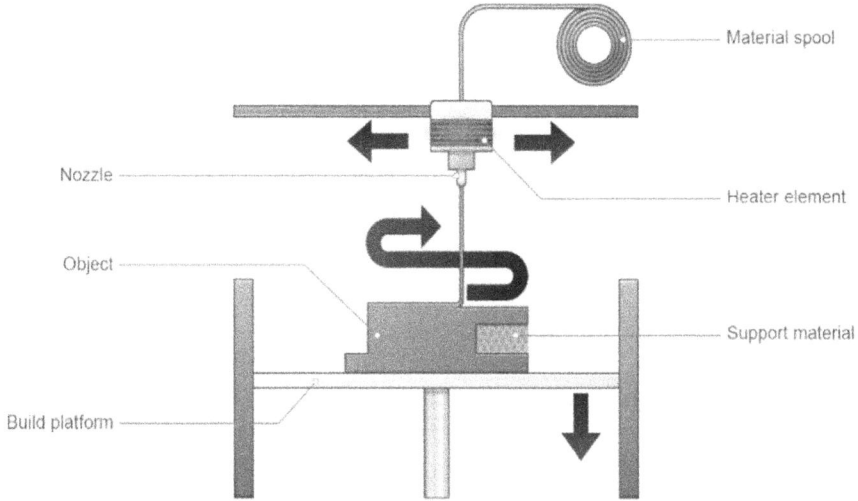

FIGURE 3.1 FDM fundamental process (Dassault System 2018).

industries to construct complicated geometrical functional components in a short time and which ultimately depends on their parameters and structures (Chacón et al. 2017).

Three-dimensional printing technology has evolved in recent years and it is becoming a leading manufacturing technology. It opens up new opportunities and provides many possibilities for industries seeking to improve manufacturing efficiency, particularly concerning the new development of feedstock (Shahrubudin, Lee, and Ramlan 2019). According to the Wohlers Report (2019), the overall AM market is predicted to rise from $US15.8 billion in 2019 to $US35.6 billion in 2024, including manufacturing machines and related products, software and services (AMFG 2019; Wohlers Associates 2019). As Industry 4.0 is evolving, 3D printing has the potential to have a longstanding impact in the sustainable manufacturing business arena (Godina et al. 2020).

Indeed, this smart manufacturing facilitates a highly flexible production process that is capable of rapidly altering individualized mass production in high-quality and product customization. Clearly, AM capabilities are one of the crucial elements of the Industrial 4.0 revolution due to the ease with which new products can be manufactured on the premises. In this way, customers, factories, and designer roles will be significantly redefined in the future of manufacturing (Mehrpouya et al. 2019).

During the COVID-19 pandemic, AM also played a vital role in the supply chain of medical industries that suffered from a massive shortage of critical protective products and medical equipment (Attaran 2020). For example, it allowed the manufacture of these components on demand to reduce the supply chain gap in mass-produced items and then delivered them to hospitals by rapidly producing the items of personal protective equipment (PPE) such as face mask parts, coronavirus nasal test swabs, ventilator components, and face shields (Arora et al. 2020; Sinha, Bourgeois, and Sorger 2020).

Despite being an evolving technology with some constraints in part size, construction time, and material feedstock, various polymer, and metal, AM technologies are applied in complex and functional components (Gadagi and Lekurwale 2020). Of all the materials used for different 3D printing applications, polymer composites filament is one of the fastest-growing. Materials such as natural or synthetic polymers, metals, ceramics, resins, composites, or still living cells are used for filament production nowadays (Yaragatti and Patnaik 2020).

The common materials used for AM technology are polylactic acid (PLA) and acrylonitrile butadiene styrene (ABS) (Saxena and Kamran 2016). However, as the risk of global warming increases, it has led to a rise in environmental awareness, which comes along with new environmental regulations (Khosravani and Reinicke 2020). Environmentally friendly materials are now more widely considered following health awareness across the globe. Therefore, for example, PLA filament is widely used nowadays as it is one of the most popular biodegradable materials derived from renewable resources, including plant-based resources such as sugar cane (Zeidler et al. 2018).

According to Balla et al. (2019) and Valino et al. (2019), there are several benefits to using AM to manufacture polymer filament composites on a macro- and nanoscale, which can increase adhesion between fillers and the matrix. For example, the ability of AM to create complex geometrical parts with good mechanical properties is unachievable in traditional manufacturing ; other benefits include high accuracy of fabrication, flexible processing parameters that make parts with better performance, tailorable properties, cost, and time saving (Balla et al. 2019; Valino et al. 2019).

Nonetheless, 3D printing has its drawbacks as well. One of the disadvantages is that the technology itself is expensive. Pirjan and Petrosanu (2013) have mentioned that 3D printing devices and the input materials come at a very high cost when it comes to the mass production of simple objects. Although its flexibility enables developers to print complex objects, it is only beneficial and profitable in printing small amounts of complex objects (Pirjan and Petrosanu 2013). Other than that, 3D printing also creates too many plastic by-products that end up in landfills. Even though reusing materials reduces the waste, but the machines still yield excess plastic wastes due to support materials and degradation of mechanical properties issues (Uitterhaegen et al. 2018).

This review chapter describes the biodegradable 3D printing feedstocks that use polymers and natural fibers, which focuses more on biodegradable polymers and natural fibers, the role of biodegradable polymers involved in the growth of 3D printing technologies, and recent developments in the use of biodegradable polymer composites for FFF.

3.2 BIODEGRADABLE POLYMERS, PROPERTIES, AND APPLICATIONS

Biodegradation is a two-step process that occurs due to exposure to enzymes and chemical deterioration induced by living organisms like fungi, bacteria, and algae (Subach 1997). In the initial step, polymers break down into smaller molecular components through either non-biological processes such as oxidation, photodegradation, or hydrolysis (abiotic responses), or biological processes involving degradation by microorganisms (biotic responses). Following this primary breakdown, the polymer components can undergo further transformation through bio-assimilation by microorganisms and ultimately mineralization (Vroman and Tighzert 2009). Specifically, the biodegradation of biodegradable polymers happens in vivo; it can be either enzymatic or non-enzymatic and it produces biocompatible or safe by-products (Sin and Tueen 2019c). Biodegradability depends not just on the polymer but moreover on its chemical structure and the natural debasing conditions (Prajapati et al. 2019).

Since the beginning of life on the planet, biodegradable polymers already existed; they are mainly used for packaging, medicine, agriculture, and more. Recently, there have been developments in the field of biodegradable polymers. There are two classes of biodegradable polymers: synthetic and natural polymers. The polymers can be sourced either from biological resources (renewable resources) or from petroleum resources (non-renewable resources). Generally, natural polymers have more benefits rather than synthetic polymers (Vroman and Tighzert 2009).

The formation of natural polymers during the growth cycles of all organisms produces biodegradable polymers, also called natural biodegradable polymers or biopolymers. Two primary renewable sources of biopolymers can be derived from protein-based material or polysaccharides such as starch or cellulose as these two are the most characteristic family of natural polymers. Recently, lipids have been discovered as another resource for biopolymers. Regular chemical modifications of natural polymers serve two primary purposes: enhancing their mechanical properties and controlling their degradation rate (Vroman and Tighzert 2009).

The demand for biodegradable biomaterials arises from the need for materials suitable for specific applications that neither induce a sustained inflammatory response nor yield harmful degradation byproducts (Fathi and Barar 2017). Biodegradation is pivotal in polymers as it allows the production of substances that are non-hazardous to humans and safe for environmental disposal. Consequently, biodegradable materials are cost-effective, sourced from nature, and amenable to straightforward processing and customization (Prajapati et al. 2019).Researchers who lack expertise in sugar chemistry often encounter challenges when dealing with the variability of sugar molecules. Polysaccharides, which are polymers composed of multiple sugar units, are also referred to as saccharide units. The chemical name for commonly used table sugar is sucrose; glucose and fructose are monosaccharides while sucrose is a disaccharide composed of glucose and fructose (Zhang et al. 2014). Polysaccharides are composed of monosaccharide units linked together by O-glycosidic bonds in their backbone. This structural composition imparts several valuable characteristics to polysaccharides, including stability, hydrophilicity, non-toxicity, biocompatibility, biodegradability, and ease of modification. These qualities make them particularly suitable for applications like drug delivery and targeting (Huh et al. 2017; Prajapati et al. 2019).

Cellulose and starch are frequently used for materials application. However, there is more focus on complex carbohydrate polymers made by bacteria and fungi, mostly polysaccharides such as hyaluronic acid, curdlan, xanthan, and pullulan. The latter polymers are made up from more than one type of carbohydrate unit and some have branched structures that cause enzymes to catalyze hydrolysis reactions; the biodegradation of each type of polysaccharide might differ and may not be modifiable.

Hydrolytic degradation is the process of breakdown of a phosphoester bond in a polymer, which helps in forming the protein-based biomaterials backbone and side chains. This process is a naturally controlled degradation process (Bauer et al. 2017; Prajapati et al. 2019). Proteins are thermoplastic heteropolymers that represent polar and non-polar α-amino acids. The ability of amino acids to form intermolecular linkages leads to different interactions, which result in chemical functionalities and functional properties. For example, proteins such as fibrous proteins (i.e., silk, wool, and collagen) are neither soluble nor fusible (Puppi and Chiellini 2020). Enzymes, like protease, and an amine hydrolysis reaction are needed for the biodegradation of proteins. The rate of biodegradation is controlled by the grafting of protein (Koyamatsu et al. 2014; Vroman and Tighzert 2009).

These materials have functional properties which are highly dependent on the hydrophilic behavior of proteins, structural heterogeneity, and thermal sensitivity (Roy 2010), for example, biopolymers from protein-based substances such as gelatin, albumin, and collagen (Prajapati et al. 2019). Fatty acids such as plant oils and animal fat are in common use (e.g., flax (linseed); tung oils are drying oils used in paints, enamels, and varnishes) or are used in soaps, cosmetics, detergents, and lubricant applications. Light and resistant composite materials can be achieved through a mixture of thermoset and natural fibers where the thermoset can be obtained from plant oils (Mehta, Bhardwaj, and Gupta 2017). For high-volume applications, new low-cost composites are produced from the combination of bio-based resins and natural fibers (poultry and plant) or lignin (Mohammed et al. 2015). This combination is widely used in agricultural equipment, automotive sheet-molding compounds (SMCs), civil and rail infrastructures, housing, the construction industry, and marine applications (Li et al. 2020; Mehta, Bhardwaj, and Gupta 2017).

Plant oil is the largest source of global lipid production, accounting for about 80% of the total. Within this category, soybean and palm oils play crucial roles as significant contributors. Interestingly, European oils such as rapeseed, sunflower, and linseed are particularly notable for their high content of unsaturated fatty acids, comprising approximately 90% of their composition. Triglycerides are the best candidate because of their high level of unsaturation and because they are made up of active sites such as double bonds, ester groups, allylic carbons, and carbons alpha to the ester group. The active sites from the synthesis of petrochemical-based polymers occur through the same synthetic techniques as those for producing polymerizable groups on the triglyceride. Polymer formation can happen in castor oils that contain ricinoleic acid presenting a hydroxyl group. Based on these, plant triglycerides are able to produce polyolefins, polyurethane, polyesters, or polyamide resins (Mehta, Bhardwaj, and Gupta 2017).

There are two categories of synthetic polymers which have hydrolyzable backbone functions, such as ester, amide, and urethane, or polymers with carbon backbones added with antioxidants. Synthetic polymers with hydrolyzable backbones are vulnerable to biodegradation under certain conditions. There are different categories of polymers such as polyanhydrides, polyamides, polyesters, polyurethanes and polyureas, poly(amide-enamine)s (Nair and Laurencin 2007), and phosphorous-based ones (Alizadeh-Osgouei, Li, and Wen 2019). Non-biodegradable synthetic polymers, primarily derived from petroleum resources, often exhibit resistance to degradation. This resistance stems from their limited accessibility to environmental microbes, primarily due to the presence of significant amounts of other moieties, including hydrocarbons. As a result, these polymers typically undergo only partial biodegradation (Subach 1997).

Additives are added to ease the biodegradation process. Another method is the addition of antioxidants into the polymer chains in order to degrade polyolefins, under UV; the antioxidant-induced degradation is called photo-oxidation. However, this biodegradability is still disputable and as such non-biodegradable synthetic polymers are favored (Vroman and Tighzert 2009). The polymers contain a large number of other moieties, including hydrocarbons such as acrylic polymers, cellulose derivatives, silicones, ethyl vinyl cetate, polaxamers, poloxamine, and polyvinyl pyrrolidone (Bharadwaz and Jayasuriya 2020).

The polymer backbone of polyesters is made of an aliphatic ester bond and is mostly hydrophobic and with an ester bond produced from bulk erosion (Zhang, Tan, and Li 2018). In drug module applications such as vaccines, proteins, peptides, macromolecules, and biomolecules, polyesters can be used in numerous devices (Prajapati et al. 2019). The development of carrier systems is well-established over the time of polymeric materials such as PLA, poly(glycolic acid) (PGA), and their copolymer PLA-co-glycolic acid (PLGA) (Jana et al. 2021) (Figure 3.2).

PGA was initially used as a polymer for biomedical purposes. Because PGA has good mechanical properties, is non-toxic and compatible with physiological conditions, and is the easiest linear and aliphatic polyester, it increases the effectiveness of drug formulations for the purpose of delivering drugs and provides controlled drug release patterns (Prajapati et al. 2019). Polyglycolide has a melting point in the range of 225–230°C, which results in a glass transition temperature of between 35 and 40°C, and PGA is insoluble in water due to high crystal levels, around 45–55%; also, PGA fibers have high strength and modulus (7 GPa) and high stiffness (Ayyoob, Lee, and Kim 2020; Roy 2010).

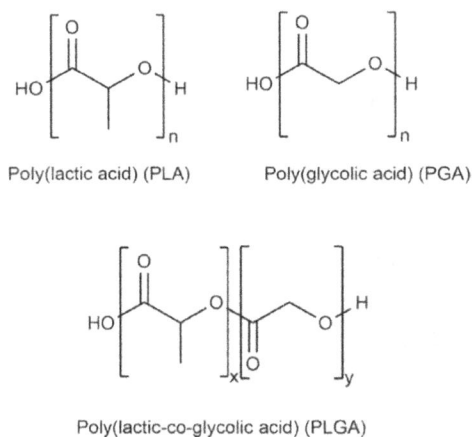

Poly(lactic acid) (PLA)   Poly(glycolic acid) (PGA)

Poly(lactic-co-glycolic acid) (PLGA)

FIGURE 3.2 Chemical structure of PLA, PGA, and PLGA (Jana et al. 2021).

Polylactides also known as Polylactic acid (i.e., PLA) is one of the families of bio-degradable polyesters (Sin and Tueen 2019c). It is the most popular and widely used family compared to commercial polymers because it is easily derived from renewable resources such as starch, which is both disposable and biodegradable. Also, polylactides have lower toxicity and high mechanical performance (Roy 2010; Sin and Tueen 2019a). Moreover, PLA is a better organic solvent than PGA (Pachence, Bohrer, and Kohn 2007).

In 1974, PLA was introduced and sold as Vicryl in the USA. It was first used as a suture material and was combined with PGA (Mehta et al. 2005). PLA was produced from the esterification of lactic acid by fermentation. The micro-organism can be *Lactobacilli*, *Pediococci*, or certain fungi such as *Rhizopus oryzae* (Mehta, Bhardwaj, and Gupta 2017). Lactic acid (2-hydroxy propionic acid) is the single monomer of PLA obtained through fermentation or chemical synthesis. In contrast, the bacterial (homofermentative and heterofermentative) fermentation of carbohydrates produces two configurations: L(+) and D(-) stereoisomers (Roy 2010; Zhong et al. 2020).

The formation of amorphous polymers produced from the polymerization of a racemic mixture is (D,L)-lactide and mesolactide. Thus, L-lactide is the natural isomer, which is similar to polyglycolide; poly(L-lactide) (PLLA) is also a crystalline polymer (~37% crystallinity) and the molecular weight and polymer processing parameters control the degree of crystallinity. L-lactide has a glass transition temperature of 60–65°C and a melting temperature of approximately 175°C (Roy 2010). Poly(L-lactide) has good tensile strength; its low extension and high modulus (4.8 GPa) make it a slow-degrading polymer compared to polyglycolide (Sin and Tueen 2019b).

However, PLA deforms at higher stress levels due to low resistance to oxygen permeation and brittleness with less than 10% elongation at break. Y. Ma et al. (Yang et al. 2019) had introduced a solution for this challenge and applied it for food packaging in combination with PLA and exfoliated clay in conjunction with thermoplastic starch. These have been proven to produce materials with improved mechanical strength, satisfactory oxygen barrier, and better water resistance as compared with pure PLA (Zhong et al. 2020).

Biodegradable materials such as PLA and PGA are commonly used in medicine and surgical devices and for drug delivery, which provides remarkable bioavailability; non-immunogenic and non-toxic block polymer have been used for sustained/controlled release pattern, and their copolymers are PLGA. The glass transition temperature of PLGA copolymer ranges between 45°C and 55°C. They are soluble in tetrahydrofuran, acetone, chloroform, ethyl acetate, hexafluoroisopropanol dichloromethane, benzyl alcohol, and ethyl acetate. PLGA is obtainable either as acid or as ester. PLGA is able to quickly degrade as compared to homopolymers such as PLA and PLG (Samadi et al. 2019).

3.3 NATURAL FIBERS, PROPERTIES, AND APPLICATIONS

Natural fiber is considered to be an ancient natural material and was used in daily activities in different fields. Natural fibers or plant fibers are a bundle of dead plant cells elongated together to form a strain by pectin and other non-cellulosic components. The plant fiber consists of two cell walls. The primary wall is the cell wall of the growing plant cell, which consists of 90% polysaccharide and 10% gly-coprotein (Goriparthi, Suman, and Mohan Rao 2012). The secondary cell wall is formed by successive layer deposition, and consists of three sublayers (Figure 3.3).

The second sublayer (S2) is the thickest among the three sublayers, while the second layer consists of almost 80% of the secondary cell wall (McNeil et al. 1984). Cellulose is the major component in the plant fiber. The microfibril of crystalline cellulose consists of aggregated chains of beta-4-linked D-glucan while the hemicel-lulose consists of short and branched polysaccharides chain attached to the cellulose after the pectin is removed. Lignin is a stiff cell wall that acts as a protective barrier for the cell wall (Bhattacharyya, Subasinghe, and Kim 2015).

Different plant fibers have their own chemical properties. In order to use the plant fiber effectively for composite applications, it is important to understand the individual properties and growing conditions of widely used plant-based fibers. The most commonly used natural fibers in industries are flax, jute, cotton, and kenaf. The natural fiber is used to produce the bio-composite, which is a mixture of natural fiber with synthetic polymers for applications such as energy absorption in insulators, noise-absorbing panels, or for producing synthetic reinforcement fibers (Peças et al. 2018).

Natural fiber also produces nanocomposites used in the biological field. Chitin is one of the natural fibers used in biomaterials for bone filling material and anti-tumor agents (Ji and Gao 2010). Chitin is also used to extract pollutants from water and is used as a biosensor in the food industry (Kılınç, Durmuşkahya, and Seydibeyoğlu 2017). Cotton is one of the most commonly used fibers in society. When cotton is ripe, the fruit of the cotton bursts, exposing a tuft of cotton fiber ranging in length from 25 to 0 mm in diameter. Cotton consists of three wall layers, according to Varghese and Mittal (2017); the primary wall, which is the outermost layer, act as a protective layer and is made from crystallized cellulose fibrils. The secondary cell wall contains closely packed parallel fibrils made up of cellulose which are winded spirally. The third layer is the lumen in which the absorption of water occurs. Cotton contains 80–90% of cellulose and 4–6% of hemicellulose (Varghese and Mittal 2017). Cotton is widely used in the textile industry and fash-ion field (Sfiligoj et al. 2013).

Jute fiber is one of the important fibers other than cotton. The harvesting of jute fiber is through jute retting during which the dry jute plant fiber is extracted; it com-prises 4% of the fresh jute plant and is about 1.5–3 m in length (Singh et al. 2018). Jute fiber is a shiny and soft fiber that can form into a strong string or threads for making fabrics; hence, jute fiber is normally used in the fashion field, and for floor covering and making luggage (Wu, Misra, and Mohanty 2020).

Flax fiber is a cellulose fiber whose composite is more crystalline; it is stiff and has a hard characteristic and hence wrinkles easily (Baley et al. 2020). Flax fiber consists of 70–75% of cellulose and 15% of hemicellulose. Flax fiber is normally used in seed/oil and fiber production as flax seed has a high oil content of about 40–45% (Singh et al. 2018). The flax plant grows to the height to about 80–150 cm, but only the center section of the plant is used to produce flax fiber. Flax fiber is

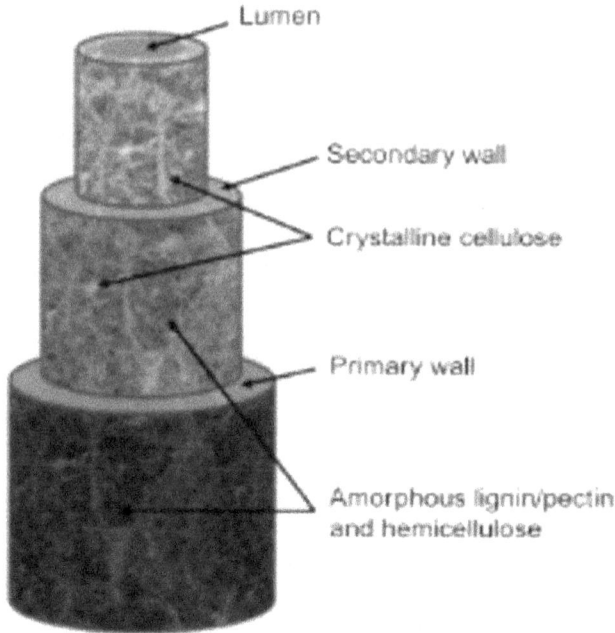

FIGURE 3.3   Natural fiber structure (McNeil et al. 1984).

flexible and yet very strong but with low stretchability. Flax fiber has weak resistance against acid solutions but has better resistance in alkaline solutions. As flax fiber is softer than cotton and jute fiber due to lesser cellulose content, it is normally used to manufacture towels, carpet, ropes, and fabric (Müssig and Haag 2015).

Kenaf plant is a common natural fiber crop that can be found annually in warm countries such as in Asian countries and is native to Africa. Kenaf is considered an eco-friendly crop due to its ability to reduce carbon dioxide, which is the highest among any plant; therefore, it can prevent global warming (Wu, Misra, and Mohanty 2020). Kenaf fiber comprises two types of fibers: short and polygonal fibers located at the lignum region of the kenaf plant and long fibers at the cortical layer of the plant. Kenaf fiber is normally used for manufacturing paper products, livestock feed, building materials, and absorbents (Akhtar et al. 2016). It is also used as a reinforced fiber for the reinforcement of polymers.

3.4   POLYMER COMPOSITE OF FFF

Fused filament fabrication (FFF) is also called fused deposition modeling (FDM), which is a material extrusion method to print 3D models based on a 3D CAD model. Composites refer to the mixture of two or more constituent materials that have substantially different properties in order to produce a single substance to provide better efficiency relative to the individual components (Akhtar et al. 2016). Polymers such as PLA and ABS are commonly used in the FFF process. In order to increase the tensile properties of the pure polymer, short fiber is normally added to the polymer, which forms a reinforced polymer that is stronger than the pure polymer. PLA is considered a natural-based biodegradable polymer, while ABS is considered a non-biodegradable polymer (Yaragatti and Patnaik 2020).

However, a study by Sang et al. (2019) during the development of PLA mixed short carbon fiber filament polymer composites has reported that propagated porosities are large at a high-volume fraction of fiber in a polymer (Figure 3.4). This may lead to insufficient interfacial bonding, which results in low-efficient load transfer, reduced tensile strength, and increased ductility (Sang et al. 2019). Kumar et al. (2020) had concluded in their comprehensive review of FFF operating capability using thermoplastic material that PLA gained better biocompatibility in the tissue-engineering field when chitosan and hydroxyapatite particles were used as reinforcement (Kumar et al. 2020).

The earlier study by Wang et al. (2017) on the overview of polymer composite materials properties in 3D printing for biomedical, electronics, and automotive fields highlights that although 3D printing is now widely used, the material to use as feedstock was limited to only low glass transition temperature thermoplastic polymers. A few photopolymers and powdered materials may not fulfill the various requirements for industrial applications and thus the performance of polymer composites can be improved by adding reinforcement, but the printed composites in FFF may have lower mechanical strength when compared to the polymer composite manufactured by using the traditional manufacturing method (Wang et al. 2017).

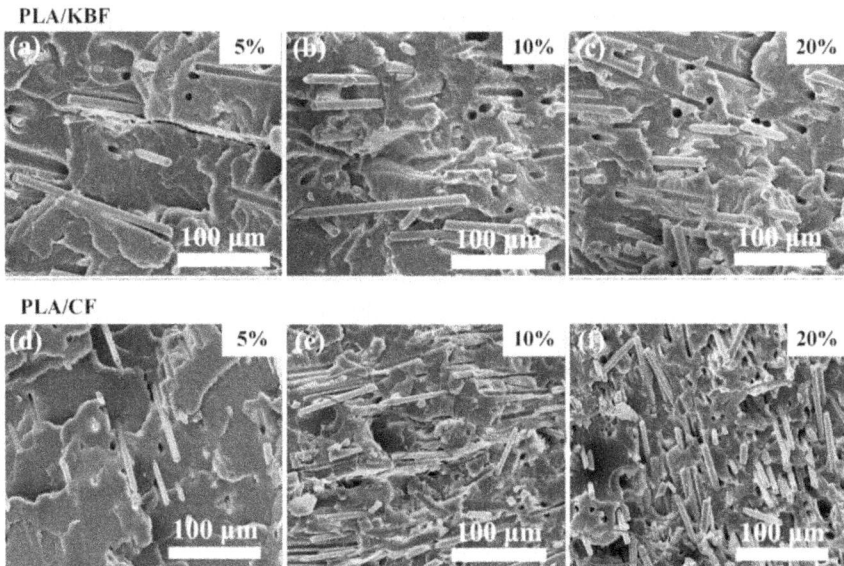

FIGURE 3.4 Fracture morphologies of porosities in high weight fraction of fibers (Sang et al. 2019).

3.5   BIODEGRADABLE POLYMER AND NATURAL FIBER FOR FFF

Natural fibers like jute fiber and kenaf fiber are customarily used as the reinforcement for polymer composites. Jute fiber is one of the strongest lignocellulosic vegetable bast fibers. The individual effects of alkali-, permanganate-, peroxide-, and silane-functionalized jute fibers in PLA for both treatments were successful in enriching the interface bonding on the PLA matrix of jute fibers, thereby enhancing the tensile and flexural properties of the resulting composite. But sadly, after altering the jute fiber, the impact strength of the composite showed a declining tendency (Goriparthi, Suman, and Mohan Rao 2012).

Kenaf fiber is typically used in the reinforcement of PU, epoxy resin, PP, and polyester. Balla et al. (2019) studied the characteristics and properties of different types of natural fibers for reinforcement in biodegradable polymers, and concluded that the replacement of synthetic fibers with natural fibers in polymer composites may not be successful due to the poor mechanical properties of natural fibers. In their 2020 study, Dickson, Abourayana, and Dowling investigated the enhancement of mechanical properties through the use of 3D printing technology, specifically Fused Filament Fabrication (FFF). They focused on printing polymer composites reinforced with both short and continuous natural fibers. Their findings indicated that incorporating either short or continuous fibers could enhance the mechanical properties of the polymers. However, it's worth noting that the presence of porosity in the printed components limited their practical applications.

3.6   CHALLENGES AND FUTURE OPPORTUNITIES

As discussed previously, biodegradable polymer natural fibers have been investigated for their properties and for future replacement of petrochemical materials and as the main reinforcement material for composites. The nontoxicity and recyclability properties of biodegradable polymer composites are the main advantages for a researcher to develop new eco-friendly material for future industrial use (Liu et al. 2019).

Although the biodegradable polymer composite is a sustainable choice for becoming the feedstock for 3D printing in the future, there are still some challenges to developing the feedstock for the FFF process. First, the effect of the fiber properties such as fiber orientation and fiber matrix adhesion on the finishing of the product must be considered and there might be nozzle clogging if an inappropriate combination of natural fiber concentration is used with the polymer. The properties of biodegradable polymers, such as weight, might vary from those of conventional polymers, which can be unfavorable in many practical scenarios (Shanmugam et al. 2021).

Next, the biodegradable polymer natural fibers were mostly plant-based and produced from an organic source, which increased the chance of contamination by pesticides. FFF involves some process such as mixing of natural fibers with polymers; compounding and extrusions of the filament may end up destroying the characteristic of the natural fiber because of the high temperature and pressure involved in the preparation process of FFF feedstock (Boparai et al. 2016).

However, the use of natural-derived biopolymers, such as lignocellulosic biomass, starch, algae, and chitosan-based materials in native or functional products as feedstocks for various AM technologies in the future, has a great deal of potential for producing 3D printing filaments , which, in turn, promotes further academic and industrial research and development (R&D) toward biodegradable products (Park et al. 2017).

3.7 CONCLUSION

Although there are many challenges and limitations in applying biodegradable material as future 3D printing feedstock, they have been gaining more attention the past few years with increasing awareness of environmental protection. In the future, using eco-friendly material for production and the usage of biodegradable material and natural fibers reinforcement in AM will become more common in industries and households. While natural and synthetic polymer composites were succeeded applied in FFF, including fibers and particle, the application of FFF for total biodegradable feedstock systems needs to be optimized for each new type of 3D printing filament to be developed. Even with filler addition, there is no assurance that the thermo-mechanical properties of the printed material are bound to improve. In a nutshell, the extension of FFF to filled biodegradable polymers and natural fibers holds the promise of improved properties of 3D printing fabricated parts and novel applications. Although the required optimization of process parameters depends upon the materials system and 3D printing techniques, it leads to a challenge for the biodegradable FFF of filled natural fibers and defines the need for future research for green 3D printing.

ACKNOWLEDGMENT

This research was supported and funded by Fundamental Research Grants Scheme (FRGS) under a grant number of FRGS/1/2019/TK03/UNIMAP/03/5 from the Ministry of Higher Education Malaysia. The authors would like to thank Universiti Malaysia Perlis (UniMAP) for the scholarship granted to the first author and Universiti Putra Malaysia (UPM) for the facilities provided.

REFERENCES

Akhtar, Majid Niaz, Abu Bakar Sulong, M. K. Fadzly Radzi, N. F. Ismail, M. R. Raza, Norhamidi Muhamad, and Muhammad Azhar Khan. 2016. "Influence of Alkaline Treatment and Fiber Loading on the Physical and Mechanical Properties of Kenaf/Polypropylene Composites for Variety of Applications." *Progress in Natural Science: Materials International* 26 (6): 657–64. https://doi.org/10.1016/j.pnsc.2016.12.004.

Alizadeh-Osgouei, Mona, Yuncang Li, and Cuie Wen. 2019. "A Comprehensive Review of Biodegradable Synthetic Polymer-Ceramic Composites and Their Manufacture for Biomedical Applications." *Bioactive Materials* 4 (1): 22–36. https://doi.org/10.1016/j.bioactmat.2018.11.003.

AMFG. 2019. "State of the 3D Printing Industry Survey 2019 Am Service Providers Market Trends, Expert Insights and Industry Perspectives."

Arora, Pawan K., Ranjan Arora, Abid Haleem, and Harish Kumar. 2020. "Application of Additive Manufacturing in Challenges Posed by COVID-19." *Materials Today: Proceedings*, no. xxxx: 8–10. https://doi.org/10.1016/j.matpr.2020.08.323.

Attaran, Mohsen. 2020. "3D Printing Role in Filling the Critical Gap in the Medical Supply Chain during COVID-19 Pandemic." *American Journal of Industrial and Business Management* 10 (5): 988–1001. https://doi.org/10.4236/ajibm.2020.105066.

Ayyoob, Muhammad, Seungmook Lee, and Young Jun Kim. 2020. "Well-Defined High Molecular Weight Polyglycolide-b-Poly(L-)Lactide-b-Polyglycolide Triblock Copolymers: Synthesis, Characterization and Microstructural Analysis." *Journal of Polymer Research* 27 (5). https://doi.org/10.1007/s10965-019-2001-4.

Baley, Christophe, Moussa Gomina, Joel Breard, Alain Bourmaud, and Peter Davies. 2020. "Variability of Mechanical Properties of Flax Fibres for Composite Reinforcement. A Review." *Industrial Crops and Products* 145 (November 2019): 111984. https://doi.org/10.1016/j.indcrop.2019.111984.

Balla, Vamsi Krishna, Kunal H. Kate, Jagannadh Satyavolu, Paramjot Singh, and Jogi Ganesh Dattatreya Tadimeti. 2019. "Additive Manufacturing of Natural Fiber Reinforced Polymer Composites: Processing and Prospects." *Composites Part B: Engineering* 174 (May): 106956. https://doi.org/10.1016/j.compositesb.2019.106956.

Bauer, Kristin N., Hisaschi T. Tee, Maria M. Velencoso, and Frederik R. Wurm. 2017. "Main-Chain Poly(Phosphoester)s: History, Syntheses, Degradation, Bio-and Flame-Retardant Applications." *Progress in Polymer Science* 73: 61–122. https://doi.org/10.1016/j.progpolymsci.2017.05.004.

Bharadwaz, Angshuman, and Ambalangodage C. Jayasuriya. 2020. "Recent Trends in the Application of Widely Used Natural and Synthetic Polymer Nanocomposites in Bone Tissue Regeneration." *Materials Science and Engineering C* 110 (May 2019): 110698. https://doi.org/10.1016/j.msec.2020.110698.

Bhattacharyya, Debes, Aruna Subasinghe, and Nam Kyeun Kim. 2015. *Natural Fibers: Their Composites and Flammability Characterizations. Multifunctionality of Polymer Composites: Challenges and New Solutions.* Elsevier Inc. https://doi.org/10.1016/B978-0-323-26434-1.00004-0.

Boparai, K. S., R. Singh, F. Fabbrocino, and F. Fraternali. 2016. "Thermal Characterization of Recycled Polymer for Additive Manufacturing Applications." *Composites Part B: Engineering* 106: 42–47. https://doi.org/10.1016/j.compositesb.2016.09.009.

Chacón, J. M., M. A. Caminero, E. García-Plaza, and P. J. Núñez. 2017. "Additive Manufacturing of PLA Structures Using Fused Deposition Modelling: Effect of Process Parameters on Mechanical Properties and Their Optimal Selection." *Materials and Design* 124: 143–57. https://doi.org/10.1016/j.matdes.2017.03.065.

Dassault System. 2018. "3D Printing – Additive (Material Extrusion – FDM)." 2018. https://make.3dexperience.3ds.com/processes/material-extrusion.

Dickson, Andrew N., Hisham M. Abourayana, and Denis P. Dowling. 2020. "3D Printing of Fibre-Reinforced Thermoplastic Composites Using Fused Filament Fabrication—A Review." *Polymers* 12 (10). https://doi.org/10.3390/POLYM12102188.

Fathi, Marziyeh, and Jaleh Barar. 2017. "Perspective Highlights on Biodegradable Polymeric Nanosystems for Targeted Therapy of Solid Tumors." *BioImpacts* 7 (1): 49–57. https://doi.org/10.15171/bi.2017.07.

Gadagi, Basavraj, and Ramesh Lekurwale. 2020. "Materials Today : Proceedings A Review on Advances in 3D Metal Printing." *Materials Today: Proceedings*, no. xxxx. https://doi.org/10.1016/j.matpr.2020.10.436.

Godina, Radu, Inês Ribeiro, Florinda Matos, Bruna T. Ferreira, Helena Carvalho, and Paulo Peças. 2020. "Impact Assessment of Additive Manufacturing on Sustainable Business Models in Industry 4.0 Context." *Sustainability (Switzerland)* 12 (17): 1–21. https://doi.org/10.3390/su12177066.

Goriparthi, Bhanu K., K. N. S. Suman, and Nalluri Mohan Rao. 2012. "Effect of Fiber Surface Treatments on Mechanical and Abrasive Wear Performance of Polylactide/Jute Composites." *Composites Part A: Applied Science and Manufacturing* 43 (10): 1800–808. https://doi.org/10.1016/j.compositesa.2012.05.007.

Huh, Myung Sook, Eun Jung Lee, Heebeom Koo, Ji Young Yhee, Keun Sang Oh, Sohee Son, Sojin Lee, Sun Hwa Kim, Ick Chan Kwon, and Kwangmeyung Kim. 2017. "Polysaccharide-Based Nanoparticles for Gene Delivery." *Topics in Current Chemistry* 375 (2): 65–66. https://doi.org/10.1007/s41061-017-0114-y.

Jaisingh Sheoran, Ankita, and Harish Kumar. 2020. "Fused Deposition Modeling Process Parameters Optimization and Effect on Mechanical Properties and Part Quality: Review and Reflection on Present Research." *Materials Today: Proceedings* 21: 1659–72. https://doi.org/10.1016/j.matpr.2019.11.296.

Jana, Pradip, Mousumi Shyam, Sneha Singh, Venkatesan Jayaprakash, and Abhimanyu Dev. 2021. "Biodegradable Polymers in Drug Delivery and Oral Vaccination." *European Polymer Journal* 142 (September 2020): 110155. https://doi.org/10.1016/j.eurpolymj.2020.110155.

Ji, Baohua, and Huajian Gao. 2010. "Mechanical Principles of Biological Nanocomposites." *Annual Review of Materials Research* 40: 77–100. https://doi.org/10.1146/annurev-matsci-070909-104424.

Kalsoom, Umme, Pavel N. Nesterenko, and Brett Paull. 2018. "Current and Future Impact of 3D Printing on the Separation Sciences." *TrAC – Trends in Analytical Chemistry* 105: 492–502. https://doi.org/10.1016/j.trac.2018.06.006.

Kazmer, David O., and Austin Colon. 2020. "Injection Printing: Additive Molding via Shell Material Extrusion and Filling." *Additive Manufacturing* 36 (July). https://doi.org/10.1016/j.addma.2020.101469.

Khosravani, Mohammad Reza, and Tamara Reinicke. 2020. "On the Environmental Impacts of 3D Printing Technology." *Applied Materials Today* 20: 100689. https://doi.org/10.1016/j.apmt.2020.100689.

Kılınç, A. Ç., C. Durmuşkahya, and M. Ö. Seydibeyoğlu. 2017. "Natural Fibers. Fiber Technology for Fiber-Reinforced Composites.Pdf."

Klimyuk, Daniil, Mikhail Serezhkin, and Andrey Plokhikh. 2020. "Application of 3D Printing in Sheet Metal Forming." *Materials Today: Proceedings*, no. xxxx: 1–5. https://doi.org/10.1016/j.matpr.2020.08.155.

Kocovic, Petar. 2017. "3D Printing and Its Impact on the Production of Fully Functional Components." http://services.igi-global.com/resolvedoi/resolve.aspx?doi=10.4018/978-1-5225-2289-8.

Koyamatsu, Yuichi, Taisuke Hirano, Yoshinori Kakizawa, Fumiyoshi Okano, Tohru Takarada, and Mizuo Maeda. 2014. "PH-Responsive Release of Proteins from Biocompatible and Biodegradable Reverse Polymer Micelles." *Journal of Controlled Release* 173 (1): 89–95. https://doi.org/10.1016/j.jconrel.2013.10.035.

Kumar, Sudhir, Rupinder Singh, T. P. Singh, and Ajay Batish. 2020. "Fused Filament Fabrication: A Comprehensive Review." *Journal of Thermoplastic Composite Materials*, 1–21. https://doi.org/10.1177/0892705720970629.

Li, Mi, Yunqiao Pu, Valerie M. Thomas, Chang Geun Yoo, Soydan Ozcan, Yulin Deng, Kim Nelson, and Arthur J. Ragauskas. 2020. "Recent Advancements of Plant-Based Natural Fiber-Reinforced Composites and Their Applications." *Composites Part B: Engineering* 200 (August). https://doi.org/10.1016/j.compositesb.2020.108254.

Liu, Jun, Lushan Sun, Wenyang Xu, Qianqian Wang, Sujie Yu, and Jianzhong Sun. 2019. "Current Advances and Future Perspectives of 3D Printing Natural-Derived Biopolymers." *Carbohydrate Polymers* 207 (June 2018): 297–316. https://doi.org/10.1016/j.carbpol.2018.11.077.

Low, Ze Xian, Yen Thien Chua, Brian Michael Ray, Davide Mattia, Ian Saxley Metcalfe, and Darrell Alec Patterson. 2017. "Perspective on 3D Printing of Separation Membranes and Comparison to Related Unconventional Fabrication Techniques." *Journal of Membrane Science* 523 (May 2016): 596–613. https://doi.org/10.1016/j.memsci.2016.10.006.

McNeil, M., A. G. Darvill, S. C. Fry, and P. Albersheim. 1984. "Structure and Function of the Primary Cell Walls of Plants." *Annual Review of Biochemistry* 53: 625–63. https://doi.org/10.1146/annurev.bi.53.070184.003205.

Mehrpouya, Mehrshad, Amir Dehghanghadikolaei, Behzad Fotovvati, Alireza Vosooghnia, Sattar S. Emamian, and Annamaria Gisario. 2019. "The Potential of Additive Manufacturing in the Smart Factory Industrial 4.0: A Review." *Applied Sciences (Switzerland)* 9 (18). https://doi.org/10.3390/app9183865.

Mehta, Akshita, Kamal Kumar Bhardwaj, and Reena Gupta. 2017. *Biodegradable Polymers for Industrial Applications. Green Polymeric Materials: Advances and Sustainable Development.* https://doi.org/10.1201/9781439823699.

Mehta, Rajeev, Vineet Kumar, Haripada Bhunia, and S. N. Upadhyay. 2005. "Synthesis of Poly(Lactic Acid): A Review." *Journal of Macromolecular Science – Polymer Reviews* 45 (4): 325–49. https://doi.org/10.1080/15321790500304148.

Mohammed, Layth, M. N. M. Ansari, Grace Pua, Mohammad Jawaid, and M. Saiful Islam. 2015. "A Review on Natural Fiber Reinforced Polymer Composite and Its Applications." *International Journal of Polymer Science* 2015. https://doi.org/10.1155/2015/243947.

Müssig, J., and K. Haag. 2015. "The Use of Flax Fibres as Reinforcements in Composites." *Biofiber Reinforcements in Composite Materials.* https://doi.org/10.1533/9781782421276.1.35.

Nair, Lakshmi S., and Cato T. Laurencin. 2007. "Biodegradable Polymers as Biomaterials." *Progress in Polymer Science (Oxford)* 32 (8–9): 762–98. https://doi.org/10.1016/j.progpolymsci.2007.05.017.

Pachence, James M., Michael P. Bohrer, and Joachim Kohn. 2007. "Biodegradable Polymers." *Principles of Tissue Engineering*, 323–39. https://doi.org/10.1016/B978-012370615-7/50027-5.

Park, Sung Bin, Eugene Lih, Kwang Sook Park, Yoon Ki Joung, and Dong Keun Han. 2017. "Biopolymer-Based Functional Composites for Medical Applications." *Progress in Polymer Science* 68: 77–105. https://doi.org/10.1016/j.progpolymsci.2016.12.003.

Peças, Paulo, Hugo Carvalho, Hafiz Salman, and Marco Leite. 2018. "Natural Fibre Composites and Their Applications: A Review." *Journal of Composites Science* 2 (4): 66. https://doi.org/10.3390/jcs2040066.

Pirjan, Alexandru, and Dana-Mihaela Petrosanu. 2013. "The Impact of 3D Printing Technology on the Society and Economy." *Journal of Information Systems & Operations Management* 7 (2): 360–70. https://ideas.repec.org/a/rau/journl/v7y2013i2p360-370.html.

Prajapati, Shiv Kumar, Ankit Jain, Aakanchha Jain, and Sourabh Jain. 2019. "Biodegradable Polymers and Constructs: A Novel Approach in Drug Delivery." *European Polymer Journal* 120 (March): 109191. https://doi.org/10.1016/j.eurpolymj.2019.08.018.

Puppi, Dario, and Federica Chiellini. 2020. "Biodegradable Polymers for Biomedical Additive Manufacturing." *Applied Materials Today* 20. https://doi.org/10.1016/j.apmt.2020.100700.

Quan, Haoyuan, Ting Zhang, Hang Xu, Shen Luo, Jun Nie, and Xiaoqun Zhu. 2020. "Photo-Curing 3D Printing Technique and Its Challenges." *Bioactive Materials* 5 (1): 110–15. https://doi.org/10.1016/j.bioactmat.2019.12.003.

Roy, Ipsita. 2010. "Biodegradable Polymers." *Journal of Chemical Technology and Biotechnology* 85 (6): 731. https://doi.org/10.1002/jctb.2420.

Samadi, Kosar, Michelle Francisco, Swati Hegde, Carlos A. Diaz, Thomas A. Trabold, Elizabeth M. Dell, and Christopher L. Lewis. 2019. "Mechanical, Rheological and Anaerobic Biodegradation Behavior of a Poly(Lactic Acid) Blend Containing a Poly(Lactic Acid)-Co-Poly(Glycolic Acid) Copolymer." *Polymer Degradation and Stability* 170: 109018. https://doi.org/10.1016/j.polymdegradstab.2019.109018.

Sang, Lin, Shuangfeng Han, Zhipeng Li, Xiaoli Yang, and Wenbin Hou. 2019. "Development of Short Basalt Fiber Reinforced Polylactide Composites and Their Feasible Evaluation for 3D Printing Applications." *Composites Part B: Engineering* 164 (October 2018): 629–39. https://doi.org/10.1016/j.compositesb.2019.01.085.

Saxena, Abhishek, and Medhavi Kamran. 2016. "A Comprehensive Study on 3D Printing Technology." *MIT International Journal of Mechanical Engineering* 6 (2): 63–69. https://www.researchgate.net/publication/310961474.

Sfiligoj, M., S. Hribernik, K. Stana, and T. Kree. 2013. "Plant Fibres for Textile and Technical Applications." *Advances in Agrophysical Research*. https://doi.org/10.5772/52372.

Shahrubudin, N., T. C. Lee, and R. Ramlan. 2019. "An Overview on 3D Printing Technology: Technological, Materials, and Applications." *Procedia Manufacturing* 35: 1286–96. https://doi.org/10.1016/j.promfg.2019.06.089.

Shanmugam, Vigneshwaran, Deepak Joel Johnson Rajendran, Karthik Babu, Sundarakannan Rajendran, Arumugaprabu Veerasimman, Uthayakumar Marimuthu, Sunpreet Singh, et al. 2021. "The Mechanical Testing and Performance Analysis of Polymer-Fibre Composites Prepared through the Additive Manufacturing." *Polymer Testing* 93 (October 2020): 106925. https://doi.org/10.1016/j.polymertesting.2020.106925.

Sin, Lee Tin, and Bee Soo Tueen. 2019a. *Degradation and Stability of Poly(Lactic Acid)*. *Polylactic Acid*. https://doi.org/10.1016/b978-0-12-814472-5.00007-8.

———. 2019b. *Mechanical Properties of Poly(Lactic Acid)*. *Polylactic Acid*. https://doi.org/10.1016/b978-0-12-814472-5.00005-4.

———. 2019c. *Overview of Biodegradable Polymers and Poly(Lactic Acid)*. *Polylactic Acid*. https://doi.org/10.1016/b978-0-12-814472-5.00001-7.

Singh, Harpreet, Jai Inder Preet Singh, Sehijpal Singh, Vikas Dhawan, and Sunil Kumar Tiwari. 2018. "A Brief Review of Jute Fibre and Its Composites." *Materials Today: Proceedings* 5 (14): 28427–37. https://doi.org/10.1016/j.matpr.2018.10.129.

Sinha, Michael S., Florence T. Bourgeois, and Peter K. Sorger. 2020. "Personal Protective Equipment for COVID-19: Distributed Fabrication and Additive Manufacturing." *American Journal of Public Health* 110 (8): 1162–64. https://doi.org/10.2105/AJPH.2020.305753.

Subach, Daniel J. 1997. "Biodegradable Polymers." *Chemist* 74 (3): 7–9. https://doi.org/10.1533/9780857097149.31.

Tofail, Syed A. M., Elias P. Koumoulos, Amit Bandyopadhyay, Susmita Bose, Lisa O'Donoghue, and Costas Charitidis. 2018. "Additive Manufacturing: Scientific and Technological Challenges, Market Uptake and Opportunities." *Materials Today* 21 (1): 22–37. https://doi.org/10.1016/j.mattod.2017.07.001.

Uitterhaegen, E., J. Parinet, L. Labonne, T. Mérian, S. Ballas, T. Véronèse, O. Merah, et al. 2018. "Performance, Durability and Recycling of Thermoplastic Biocomposites Reinforced with Coriander Straw." *Composites Part A: Applied Science and Manufacturing* 113 (August): 254–63. https://doi.org/10.1016/j.compositesa.2018.07.038.

Valino, Arnaldo D., John Ryan C. Dizon, Alejandro H. Espera, Qiyi Chen, Jamie Messman, and Rigoberto C. Advincula. 2019. "Advances in 3D Printing of Thermoplastic Polymer Composites and Nanocomposites." *Progress in Polymer Science* 98: 101162. https://doi.org/10.1016/j.progpolymsci.2019.101162.

Varghese, Anish M., and Vikas Mittal. 2017. *Surface Modification of Natural Fibers. Biodegradable and Biocompatible Polymer Composites: Processing, Properties and Applications*. Elsevier Ltd. https://doi.org/10.1016/B978-0-08-100970-3.00005-5.

Vroman, Isabelle, and Lan Tighzert. 2009. "Biodegradable Polymers." *Materials* 2 (2): 307–44. https://doi.org/10.3390/ma2020307.

Wang, Xin, Man Jiang, Zuowan Zhou, Jihua Gou, and David Hui. 2017. "3D Printing of Polymer Matrix Composites: A Review and Prospective." *Composites Part B: Engineering* 110: 442–58. https://doi.org/10.1016/j.compositesb.2016.11.034.

Wohlers Associates. 2019. "Wohlers Report 2019: 3D Printing and Additive Manufacturing State of the Industry." Fort Collins, Colorado.

Wu, Feng, Manjusri Misra, and Amar K. Mohanty. 2020. "Sustainable Green Composites from Biodegradable Plastics Blend and Natural Fibre with Balanced Performance: Synergy of Nano-Structured Blend and Reactive Extrusion." *Composites Science and Technology* 200 (April): 108369. https://doi.org/10.1016/j.compscitech.2020.108369.

Yang, Chunxiang, Haibing Tang, Yifen Wang, Yuan Liu, Jing Wang, Wenzheng Shi, and Li. 2019. "Development of PLA-PBSA Based Biodegradable Active Film and Its Application to Salmon Slices." *Food Packaging and Shelf Life* 22 (September). https://doi.org/10.1016/j.fpsl.2019.100393.

Yaragatti, Neha, and Amar Patnaik. 2020. "A Review on Additive Manufacturing of Polymers Composites." *Materials Today: Proceedings*, no. xxxx. https://doi.org/10.1016/j.matpr.2020.10.490.

Zeidler, Henning, Diana Klemm, Falko Böttger-Hiller, Sebastian Fritsch, Marie Joo Le Guen, and Sarat Singamneni. 2018. "3D Printing of Biodegradable Parts Using Renewable Biobased Materials." *Procedia Manufacturing* 21: 117–24. https://doi.org/10.1016/j.promfg.2018.02.101.

Zhang, Xing, Beng Hoon Tan, and Zibiao Li. 2018. "Biodegradable Polyester Shape Memory Polymers: Recent Advances in Design, Material Properties and Applications." *Materials Science and Engineering C* 92 (October 2017): 1061–74. https://doi.org/10.1016/j.msec.2017.11.008.

Zhang, Zheng, Ophir Ortiz, Ritu Goyal, and Joachim Kohn. 2014. *Biodegradable Polymers. Handbook of Polymer Applications in Medicine and Medical Devices.* Elsevier Inc. https://doi.org/10.1016/B978-0-323-22805-3.00013-X.

Zhong, Yajie, Patrick Godwin, Yongcan Jin, and Huining Xiao. 2020. "Biodegradable Polymers and Green-Based Antimicrobial Packaging Materials: A Mini-Review." *Advanced Industrial and Engineering Polymer Research* 3 (1): 27–35. https://doi.org/10.1016/j.aiepr.2019.11.002.

4 Development of 3D Printing Filament Material Using Recycled Polypropylene (rPP) Reinforced with Coconut Fiber

Yusliza Yusuf, Nuzaimah Mustafa, Mastura M.T.,
Muhamad Ariffadzilah Mohd Latip,
and Dwi Hadi S.

4.1 INTRODUCTION

Diversified industries, including 3D printing, are paying increasing consideration to the development of sustainable and eco-friendly materials. Utilizing recycled materials combined with natural reinforcements is a promising approach. In this context, the development of filament material for 3D printing from recycled polypropylene (rPP) reinforced with coconut fiber arises as a compelling option.

PP is a thermoplastic with desirable properties such as chemical resistance, impact strength, and light weight. However, the extensive consumption of PP and the resulting accumulation of plastic refuse pose environmental problems. Recycling PP not only addresses the waste problem, but also contributes to the circular economy by reducing reliance on virgin plastics.

In order to improve the mechanical properties and durability of PP, scientists have turned to natural reinforcements such as coconut fiber. High strength, low density, and biodegradability are only a few of the advantages of coconut fiber, which is extracted from coconut shells (Sengupta & Basu, 2016). By including coconut fiber as a reinforcement in the 3D printing filament, it is possible to improve the general performance of the printed items while lowering their environmental impact.

Numerous applications are possible when rPP and coconut fiber are combined. Industries like automotive, aerospace, and consumer goods have a strong demand for materials that are mechanically reliable and sustainably derived. This need is met by the development of a 3D printing filament material produced from rPP and coconut fiber, which offers a performance-unaffected greener alternative to conventional polymers.

DOI: 10.1201/9781003362128-4

In this chapter, we elaborate about the establishment of a filament material for 3D printing process made from rPP reinforced with coconut fiber. We scrutinize the manufacturing process, mechanical properties, and the possible applications of this innovative composite material. In addition, the advantages and disadvantages of incorporating natural reinforcement into 3D printing filaments are discussed. By understanding the potential of this novel material, we expect to contribute to the development of 3D printing technologies that are more sustainable and effective.

4.2 rPP AND ITS APPLICATION

rPP is a variety of plastic made from re-using PP. PP is a thermoplastic polymer used extensively in a variety of industries due to its versatile properties, which include high strength, durability, chemical resistance, and low lost. When PP products have reached the end of their useful existence, recycling then into rPP conserves resources, reduces waste, and reduces environmental impact.

Part of the recycling process is collecting used PP items, such as packaging containers, car parts, household items, and textiles. Then, these things are sorted, cleaned, and worked on to get rid of anything that isn't right. After the PP has been cleaned, it is melted and made into pellets or flakes. These can be used as raw materials to make new goods. rPP can be used in many markets and businesses. The following are common applications and purposes of rPP (Mailto et al., 2022; Meran et al., 2008):

Packaging materials: rPP is frequently utilized in packaging applications, such as food containers, beverage bottles, trays, and films. It has excellent barrier properties, impact resistance, and durability, making it appropriate for protecting and preserving a variety of products.

Automotive industries: rPP is used to manufacture interior components such as dashboards, door panels, moldings, and battery enclosures. It provides lightweight, cost-effective solutions that meet the required strength and rigidity standards.

Furniture business: rPP is used by the furniture industry to manufacture chairs, tables, cabinets, and other household objects. rPP possesses exceptional properties, such as durability, weather resistance, and a variety of color options, allowing for numerous design opportunities.

Construction industries: rPP is utilized in the production of pipelines, connections, insulation materials, roofing sheets, and geo synthetics. Due to its exceptional chemical resistance, light weight, and durability, it is utilized in these applications.

Toys and consumer products: rPP is used to make toys, household items, and consumer goods such as storage boxes, organizers, and gardening tools. It is an attractive option for manufacturers due to its adaptability and low cost.

Utilizing rPP offers numerous benefits. It helps divert plastic waste from landfills and incineration while reducing the consumption of virgin materials, energy, and greenhouse gas emissions associated with the production of new plastics. Additionally, rPP promotes a circular economy by enhancing PP materials' durability and conserving natural resources.

Despite the fact that recycling PP is advantageous, it is essential to note that the procedure has limitations. Contamination, improper sifting, and the presence of other plastics can impact the quality of recycled materials (Meran et al., 2008). To maximize the potential of rPP and other recycled plastics, it is necessary to promote proper waste management, improve recycling infrastructure, and increase public awareness of the significance of recycling.

4.3 BIOCOMPOSITES MATERIALS

Biocomposites are composite materials made of natural fiber and petroleum-derived non-biodegradable polymers like PP, polyethylene (PE), and epoxies or biopolymers like polylactic acid (PLA). Biocomposites can also be derived from biopolymers and synthetic fibers such as glass and carbon, which are partially eco-friendly biocomposite materials. The environmentally benign or green composites are derived from natural fibers and crop or bio-derived plastics, such as biopolymers (Mohanty et al., 2005). Figure 4.1 depicts the composition of a typical biocomposite, which consists of at least one phase of biological origin.

Sources of natural fibers are agricultural byproducts, which are prevalent in agricultural sectors. The bio-fibers used for reinforcement consist of both non-wood and wood fibers. Figure 4.2 demonstrates the classification of bio-reinforcing fibers. Fiber and matrix properties have a significant impact on the composite's mechanical properties. The modulus of tensile strength is more sensitive to matrix properties than it is to the fiber properties. According to Jagadeesh et al. (2022) a biocomposite's tensile strength can be enhanced by a strong interface, a low stress concentration, and the correct orientation of the fibers.

In this chapter, the biocomposite being discussed is coconut fiber reinforced thermoplastic biocomposite. It contains some eco-friendly biocomposites. Therefore, the properties of the fibers, the aspect ratio, and the fiber-matrix interface in the composite are crucial parameters that need to be considered in order to achieve consistent performance in the thermoplastic composite's product properties.

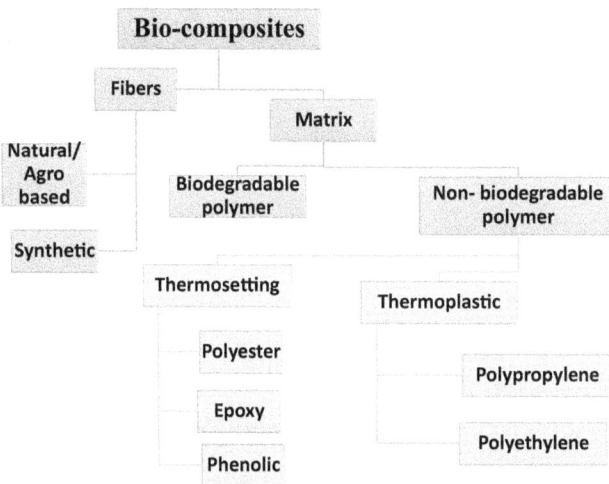

FIGURE 4.1 Biocomposite composition (Mohanty et al., 2005).

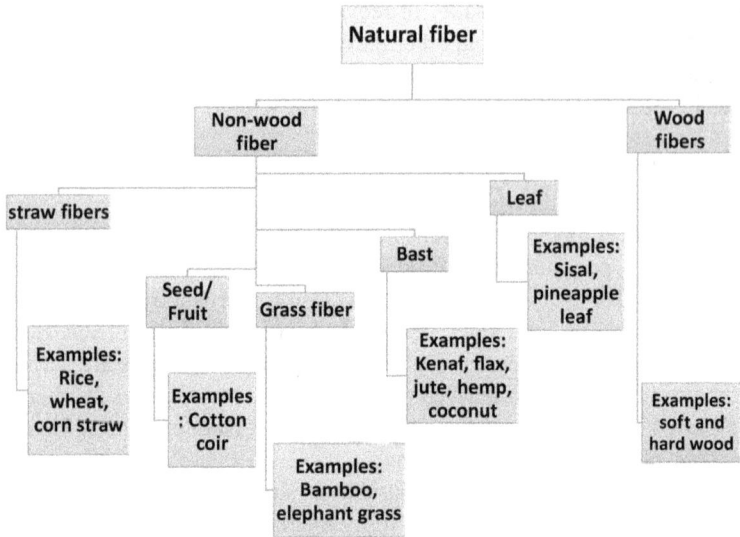

FIGURE 4.2   Classification of bio-reinforcing fibers in composite manufacturing (Sahari & Sapuan, 2011).

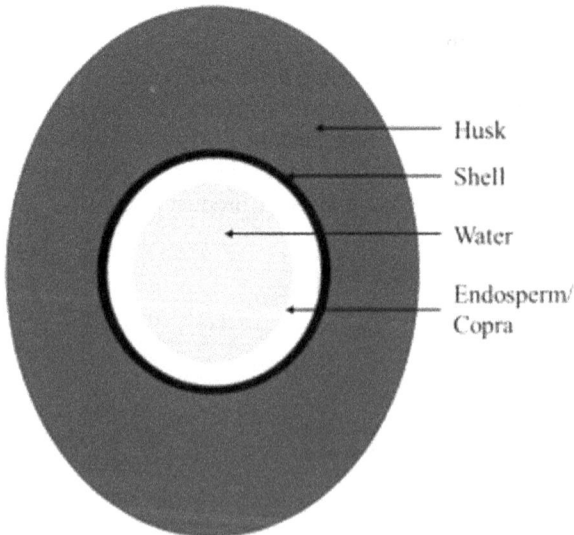

FIGURE 4.3   Coconut fruit cross-section (Patil & Benjakul, 2018).

4.4   COMPOSITES WITH COCONUTS FIBER REINFORCEMENT

Coconut fiber, also known as coir fiber, is a natural fiber derived from coconut shells (Figure 4.3). A coconut tree may produce 50–100 coconut fruits annually. The extracted fiber from theshells is known as coir fiber. The endocarp and exocarp

FIGURE 4.4   Coconut fiber also referred to as coir fiber.

layers of coconut fruits are utilized to extricate the fiber (Figure 4.4). Upon extraction from coconut husks and subsequent cleansing, the extracted coir fibers are typically golden or brown-reddish in hue. The diameter of coir fiber strands is typically ranges between a value of 0.01–0.04 inches.Each coconut husk contains 20–30% fibers of varying lengths. Due to the various sections of the coconut palm tree, such as the petiole bark, leaf sheath, and leaf midrib, it can also be considered an integral renewable resource for fiber production. Indonesia, Sri Lanka, Brazil, the Philippines, Vietnam, Thailand, Malaysia, Bangladesh, and India produce the majority of palm coconuts.

Due to its distinct properties and sustainable nature, coconut fiber has gained popularity as a reinforcement material in composites. Incorporating coconut fiber into composites enhances their mechanical properties, reduces their environmental impact, and opens up a variety of application possibilities (Martinelli et al., 2023). The characteristics and advantages of coconut fiber as a composite reinforcement material are outlined below (Sengupta & Basu, 2016):

High tensile strength: Due to its high tensile strength, coconut fiber is an outstanding reinforcement material for composites. The fiber reinforces the composite material by dispersing and absorbing tension, which ultimately contributes to the material's increased overall strength and durability.

Lightweight properties: In addition to its low weight, coconut fiber can be effectively incorporated into the production of composite materials. When incorporated into composite materials, coconut fibers can help reduce the total weight of the material. This results in enhanced energy efficiency, ease of handling, and the potential for weight reduction in industries such as aerospace and automotive.

Renewable and sustainable: Utilizing coconut fiber as a reinforcement material contributes to a more sustainable environment and decreases reliance on non-renewable resources.

Biodegradable: Coconut fiber is biodegradable, which means that it can decompose without harming the ecosystems around it. This characteristic is particularly desirable in situations where the disposal of composites at the end of their useful lifetimes is a consideration.

Good thermal and acoustic insulation: Coconut fiber has inherent thermal acoustic insulation characteristics. When incorporated into composites, it can contribute to improved insulation performance, making it appropriate for applications requiring temperature or sound control, such as construction materials and automobile interiors.

Good chemical resistance: Coconut fiber has a high level of chemical resistance, including to alkaline substances. This quality makes it appropriate for use in chemically corrosive environments and industries.

Cost-effectiveness: Coconut fiber is a material that can be used for reinforcing and it has a lower cost per unit than synthetic fibers. Its availability to composite producers as a byproduct of the coconut industry results in a cost-effective option for those manufacturers, contributing to a reduction in total manufacturing costs as a result.

Numerous and expanding applications exist for composites reinforced with coconut fibers. In the automotive industry, coconut fiber composites can be used to produce interior components like door panels, dashboards, and seat backs. The lightweight and robust character of coconut fiber contributes to an increase in fuel efficiency and a decrease in carbon emissions. In building and construction, coconut fiber composites are used in the construction industry for products such as panels, boards, roofing sheeting, and insulation materials (Ahmad et al., 2022). They offer enhanced tensile strength, thermal insulation, and water resistance. Moreover, coconut fiber composites are used to make tennis instrument frames, golf club shafts, and bicycle frames. They are suited for these applications due to their combination of durability, light weight, and adaptability.

In other applications, coconut fiber composites can be used to make furniture, flooring, and other domestic items due to their natural appearance and resilience, which makes them an appealing option for eco-friendly and sustainable design. The use of coconut fiber as a reinforcing material in composites provides numerous benefits, such as increased strength, decreased weight, enhanced sustainability, and cost-effectiveness. However, it is essential to note that the performance of coconut fiber composites can vary based on variables such as fiber treatment, fiber length, and matrix material (Faridul Hassan et al., 2021). Therefore, proper processing techniques and optimization of the fiber-matrix interface are essential for attaining the desired mechanical properties and performance in composites reinforced with coconut fiber. Table 4.1 shows the comparison done for the mechanical and physical properties of coconut fiber to others that are closely related, such as jute and sisal (Sengupta & Basu, 2016).

TABLE 4.1
Comparison properties parameter for different types of fiber (Sengupta & Basu, 2016)

Property parameter	Coconut	Jute	Sisal
Physical parameters			
Diameter (µm)	320	60	190
Length (mm)	183	60	770
Length-diameter ratio	750	1,000	4,052.6
Linear density (tex)	59.2	3.8	29.67
True density (gcm^{-3})	1.40	1.48	1.45
Apparent density (gcm^{-3})	1.17	1.23	1.2
Bulk density	0.43	0.48	0.45
Mechanical behavior			
Breaking tenacity (cN/tex)	11.25	33.2	28.2
Breaking extension (%)	21.5	1.8	2.76
Initial modulus (cN/tex)	200	1,900	1,100
Specific work of rupture (mJ/tex-m)	13.4	2.8	4.41
Flexural rigidity (mN-mm^2)	1,100	22.1	284.1
Coefficient of friction	0.35	0.45	0.56
Moisture relationship			
Moisture regain at 65% RH (%)	11.7	13.5	10.92
Longitudinal swelling (%)	0.6	0.07	0.08
Transverse swelling (%)	15	25	15
Water imbibition (%)	58	92	64
Vertical wicking length after 24 h (mm)	2	7	17
Electrical properties			
Mass specific resistance (Ω-kg m^{-2})	4.0	1.83	2.96

4.5 ADVANTAGES AND DISADVANTAGES OF EXISTING FILAMENT MATERIALS FOR COMMERCIAL 3D PRINTING

Nowadays, commercial 3D printing employs a variety of filament materials, each with its own benefits and drawbacks. Acrylonitrile butadiene styrene (ABS), PLA, and PP are the most commonly used filament materials.

ABS filament is an impact-resistant, long-lasting material suitable for functional prototypes and end-use parts. It has a high melting point between 210°C and 250°C, making it more resistant to heat than other materials (Dey et al., 2021). ABS also provides superior layer adhesion and can be readily post-processed via sanding and acetone smoothing. ABS emits fumes during printing that, if not properly ventilated, can be disagreeable or even hazardous (Tambrallimath et al., 2019). Besides, for improved adhesion, a heated bed is required, and deformation can still occur. Additionally, ABS has a higher printing temperature, rendering it incompatible with printers with limited temperature capabilities.

PLA is extensively utilized due to its printability, low cost, and minimal warping. It is eco-friendly because it is derived from renewable resources (such as cornstarch or sugarcane) (Tümer & Erbil, 2021). PLA is available in a variety of colors and is biodegradable. It is more durable than ABS, does not distort when printed, and is normally simple to work with, though it can cause extruder jams on occasion. Alternatively, PLA filament has weak mechanical properties. PLA is not well-known for its durability or thermal stability. As a result of the low printing temperature of PLA, the resulting prints are rapidly warped (Bhagia et al., 2021). The brief lifespan of PLA is a natural result of the biodegradability of the material. PLA prints are not intended for long-term use because, over time, they degrade to their basic components and lose strength.

PP has an advantage over ABS in that it can be printed at relatively high temperatures. This indicates that prints created with PP filament are suitable for high-temperature applications, such as food contact or exposure to the elements. PP is a chemically stable and food-safe polymer owing to its predominantly straight-chain carbon-based chemical structure and saturated bonds (Meran et al., 2008). As a robust material that can withstand repeated stress and breaking, PP filament is also durable and resistant to impacts. Due to its mechanical strength and chemical resistance, PP is utilized in numerous industries for pipes, sheets, and enormous chemical storage containers. PP has a relatively low melting point compared to other thermoplastics used in engineering. This renders it unsuitable for applications requiring components that can withstand high temperatures. In addition, its low surface energy and high shrinkage can result in distortion, curling, or poor print quality if not addressed properly. Successful printing frequently requires the use of specialized printers and print settings, such as heated build plates and an enclosure.

4.6   PRODUCING FILAMENT FOR 3D PRINTING FROM rPP REINFORCED WITH COCONUT FIBER

Coconut fiber is being utilized more frequently to create thermoplastic matrix composites because of their distinctive qualities, such as their light weight, low cost, biodegradability, and good mechanical properties. The ideal procedure for fibers and thermoplastic matrices is melt mixing, or compounding followed by extrusion or injection molding. To create composite pellets, the fibers are typically combined with the polymer matrix using a twin-screw extruder. To get the correct final shape, these pellets are treated using extrusion or injection molding.

The fiber and matrices must first go through a number of procedures, including cleaning and washing. This crucial phase gets rid of any extra materials or particles that can damage the composite process. The fibers and matrices are then ground and crushed to speed up the sieved process and produce the required size/particle. To ensure that the substance is suitable for compounding, treatment and testing are required. Thermogravimetric analysis (TGA) and scanning electron microscopy (SEM) tests are performed on the composite pellet made from PP and coconut fiber in order to analyze and study the characterized physical and thermal properties of the PP reinforced with coconut fiber pellet. Figure 4.5 depicts the overall flow process of producing 3D printing filament using rPP reinforced with coconut fiber and Figure 4.6 shows the process flow involved for preparing the coconut fiber for the manufacturing of 3D printing filaments.

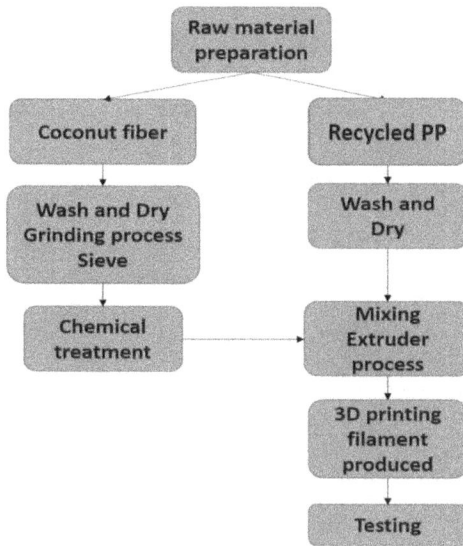

FIGURE 4.5 Overall flow process of producing 3D printing filament using rPP reinforced coconut fiber.

Coconut fiber following its the cleaning and drying process.

Low-speed granulator is used to pulverize the coconut fibers. The tangled fibers are effectively reduced to an average length and diameter.

Even though the fiber length is mostly uniform after the grinding process, there are contaminants in the fibers, such as small leaves, as well as variations in the fiber diameter. Therefore, a sieve analysis is conducted to remove contaminants.

The coconut fiber undergoes a sieving process, resulting in a fined powder.

Next, the coconut fiber is treated with sodium hydroxide (NaOH). This procedure is also known as mercerization or alkaline treatment. This process enhances the properties of coconut fiber for a variety of applications. Before the treatment, the coconut fiber is typically cleansed and dried to eliminate impurities and moisture. The coconut fiber is then immersed in a specific concentration and temperature solution of NaOH (Figure 4.7).

Coconut fiber is immersed in NaOH solution. For alkali treatment, the ratio is 6% NaOH to 94% distilled water.

It is then rinsed with distilled water to remove any remaining NaOH. The drying process is undertaken in an oven at a temperature of 104°C for 24 hours.

The concentration and duration of the treatment can vary based on the intended results and application-specific requirements. The solution of NaOH reacts with the lignin component of the coconut fiber. This reaction causes lignin to inflate, degrade, and partially dissolve (Martinelli et al., 2023). It results in the separation of individual fiber bundles and the elimination of non-cellulosic substances, such as pectin and hemicellulose. After the desired amount of time has passed, the coconut fiber is thoroughly rinsed with water to remove any remaining NaOH. Among the effects of implementing NaOH treatments are increased fiber strength and durability, improved

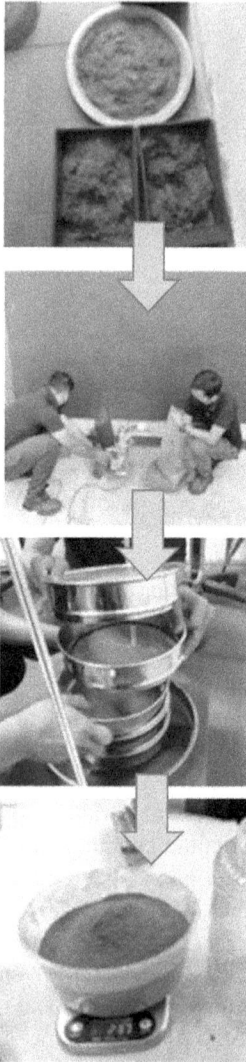

Coconut fiber after cleaning and drying process.

Low-speed granulator was used to pulverise the coconut fibers. The tangled fibers were effectively reduced to an average length and diameter.

Even though the fiber length was mostly uniform after the grinding process, there were contaminants in the fibers, such as small leaves, as well as variations in the fiber diameter. Therefore, a sieve analysis was conducted to remove contaminants

Coconut fiber fined powder after the sieving process

FIGURE 4.6   The procedure for preparing the coconut fiber.

fiber surface properties resulting from an increase in surface irregularity, and the creation of more active bonding sites, which may improve fiber compatibility and adhesion properties.

The filament is produced using the extrusion process. Coconut fiber and rPP are weighed according to their fiber loading and both materials are inserted onto the machine via hopper. The twin-extruder machine is equipped with a chamber into which the compounding ingredients are poured. Two rotors in the chamber generate high shear forces that disperse the fillers and other raw materials within the polymer. Figure 4.8 shows the produced filament using the extrusion process.

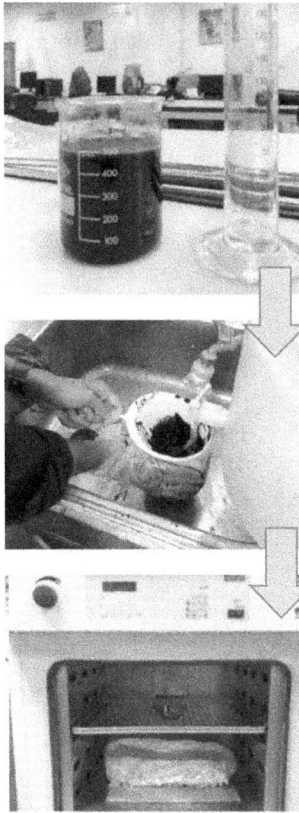

Coconut fiber immersed in NaOH solution.

For alkali treatment, the ratio is 6% NaOH to

94% distilled water.

Rinsed with distilled water to remove any
remaining NaOH

Drying process in oven at a temperature of

104 °C for 24 hours

FIGURE 4.7   Coconut fiber NaOH treatment process.

The analysis of the tensile test indicates that fiber loading composition has a significant influence on the tensile strength properties of filaments. Increasing the proportion of coconut fiber in NaOH-treated coconut fiber results in an increase in the filaments' tensile strength (Figure 4.9). The presence of untreated coconut fiber decreases the filaments' tensile strength, resulting in a different outcome. Figure 4.10 shows the tensile strength results conducted for 3D printing filament.

In addition, water absorption analysis reveals that the water absorption rate is proportional to the quantity of fiber in the filament, such that the greater the amount of fiber added, the more the water absorbed. Figure 4.11 illustrates the water absorption of the filament at various compositions of fiber loading as well as the fiber treatment process.

The TGA test on untreated coconut fiber revealed that degradation begins at 33.11°C and persists until 126.50°C, resulting in a loss of 32.753% of the fiber content in the presence of water. The second phase of degradation begins at a temperature of 244.05°C, resulting in a 62.184% loss of fiber and involving hemicellulose and lignocellulose, respectively. Final decomposition occurs at 286.73°C, leading to the

(a)

(b)

FIGURE 4.8   (a) Extrusion process producing filament. (b) The rPP reinforced with coconut fiber filament produced.

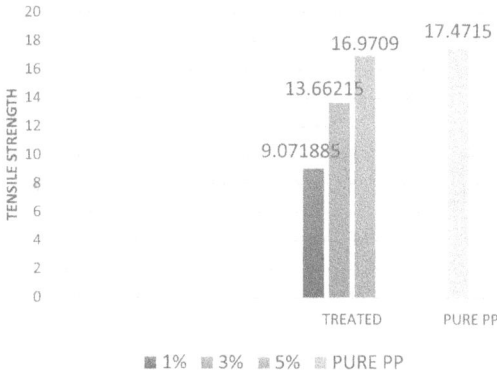

FIGURE 4.9 Tensile strength result for 3D printing filament.

FIGURE 4.10 The tensile strength results conducted on 3D printing filament produced.

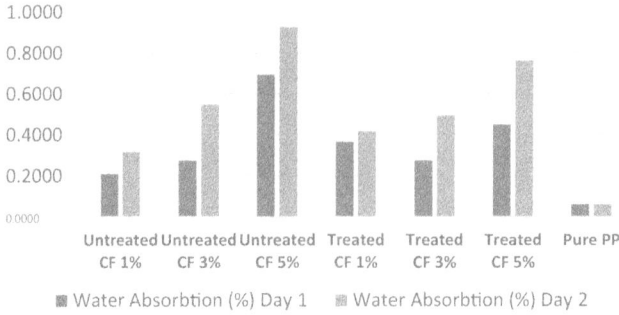

FIGURE 4.11   Water absorption percentage for rPP reinforced coconut fiber filament.

formation of particles. Meanwhile, for TGA analysis of treated coconut fiber reveals that degradation occurs between 98.19 and 233.33°C, with a loss of 14.812% of the fiber content when the fiber is in a saturated state. The second phase of decomposition begins at 355.27°C and results in 52.418% fiber loss, with hemicellulose and lignocellulose as the primary constituents, and final decomposition occurs at 420°C, resulting in the formation of ashes. This shows that chemically treated coconut fiber is more thermally stable than the untreated fiber.

By coating the fiber to resist thermal decomposition, the treated fiber effectively increased thermal stability with a wider temperature range and higher temperature during degradation (Nurazzi et al., 2021).

Figure 4.12 depicts a scanning electron micrograph of rPP filaments reinforced with coconut fiber at different compositions and treatment processes. For untreated fiber, there were numerous voids between the fiber and the matrix. However, treated fiber demonstrated a smaller opening between fiber and matrix. In composite manufacturing, voids influence a variety of mechanical properties, such as interlaminar shear, tensile, compressive, and flexural strength, as well as fracture toughness and fatigue life.

The interlaminate shear strength of a composite decreases by about 7% for each 1% of void, regardless of the resin, fiber type, or fiber surface treatment (Mahmud et al., 2021). Various post studies investigate the effect of voids and also moistures on the mechanical properties of fiber and resin composites. Due to the natural fibers' intrinsic voids, which have an impact on the composites' transverse failure, the void content in natural fiber reinforced composites is a matter of concern. The investigations also came to the conclusion that the void content has a linear or second-order polynomial relationship with stiffness or a reduction in strength. Micro-voids may emerge after the composites have been cured if air or other volatile chemicals are trapped inside them during the injection of the fibers into the resin. Poor mechanical characteristics caused by these micro-voids cause the composites to break suddenly. A post-extrusion hot-drawing method can be used to modify the voids' forms from spherical to elongated and lower their content. The amount of voids may rise linearly with different types of manufacturing techniques used to generate samples, according to another noteworthy observation (Dong & Takagi, 2014).

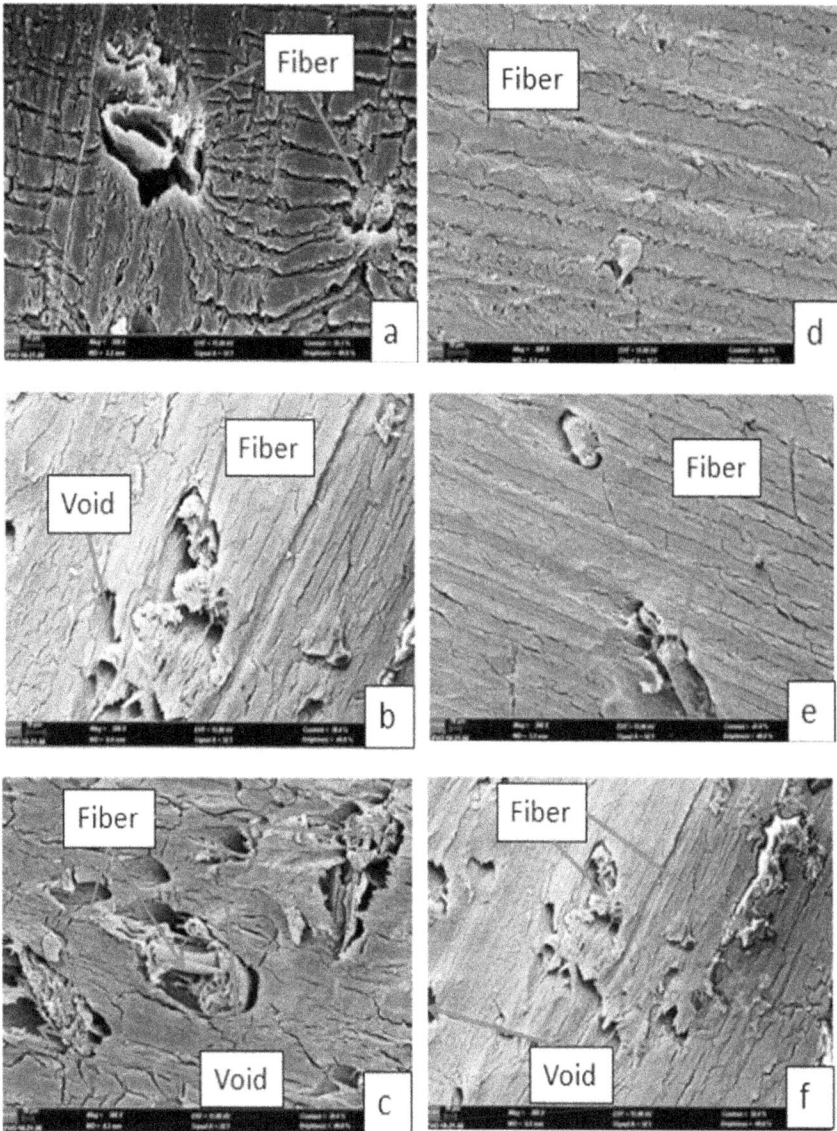

FIGURE 4.12 SEM micrograph of surface structure of wood dust reinforced with recycle PP (a) 1%; (b) 3%; (c) 5% for untreated filaments; and (d) 1%; (e) 3%; (f) 5% for treated filaments with NaOH.

On the other hand, moisture is present in natural fiber reinforced composites due to the hydrophilic and moisture-absorbing properties of lignocellulose fibers. Between the macromolecules in the cell wall of the plant fiber, there are a lot of hydrogen bonds. When moisture from the air comes in touch with the fiber, the hydrogen bond dissolves, and hydroxyl groups then establish new hydrogen bonds with water molecules (Zuhairah et al., 2021). The cross-section of the fiber becomes the main entrance to the penetrating water. When hydrophilic fiber is reinforced with hydrophobic resin fiber, the result is matrix swelling. Poor fiber-matrix bonding, dimensional instability, matrix cracking, and poor mechanical characteristics of composites are the results of this. Therefore, removing or at the very least reducing moisture from fibers is a crucial stage in the creation of composites.

4.7   CONCLUSION

In conclusion, the creation of a filament material for 3D printing that makes use of rPP and is reinforced with coconut fiber has a great deal of potential for the achievement of a manufacturing process that is both more sustainable and more efficient. This novel strategy not only lessens the amount of waste produced from plastic, but it also results in increased durability and is environment-friendly as a result of the use of recovered plastics and natural fibers. This material has the potential to revolutionize the way we create items while simultaneously minimizing the impact those products have on the environment as technology continues to improve and additional investments are made.

The utilization of rPP helps to solve the growing issues regarding plastic pollution and waste management by reducing the demand for new plastic manufacture. This is one of the ways in which these concerns can be addressed. Because of the filament material's improved mechanical qualities as a result of the incorporation of coconut fiber as a reinforcing agent, the material can be used for a wide variety of applications.

However, it is crucial to acknowledge that additional research and development are required in order to optimize the mechanical characteristics, printability, and cost-effectiveness of the filament material. In order to satisfy the ever-increasing demand for environmentally friendly materials on the market, it is essential to ensure that the manufacturing process can be easily scaled up.

ACKNOWLEDGMENT

This work was supported by the Ministry of Higher Education, Malaysia, Universiti Teknikal Malaysia Melaka internal grant (PJP/2020/FTKMP/PP/S01737).

REFERENCES

Ahmad, Jawad, Ali Majdi, Amin Al-Fakih, Ahmed Farouk Deifalla, Fadi Althoey, Mohamed Hechmi El Ouni, and Mohammed A. El-Shorbagy. 2022. "Mechanical and Durability Performance of Coconut Fiber Reinforced Concrete: A State-of-the-Art Review." *Materials* 15, no. 10: 3601. https://doi.org/10.3390/ma15103601
Bhagia, Samarthya, Kamlesh Bornani, Ruchi Agrawal, Alok Satlewal, Jaroslav Ďurkovič, Rastislav Lagaňa, Meher Bhagia, Chang Geun Yoo, Xianhui Zhao, Vlastimil Kunc, Yunqiao Pu, Soydan Ozcan, and Arthur J. Ragauskas. 2021. "Critical Review of FDM 3D Printing of PLA Biocomposites Filled with Biomass Resources, Characterization, Biodegradability, Upcycling and Opportunities for Biorefineries." *Applied Materials Today* 24: 101078. https://doi.org/10.1016/j.apmt.2021.101078.

Dey, Arup, Isnala Nanjin Roan Eagle, and Nita Yodo. 2021. "A Review on Filament Materials for Fused Filament Fabrication." *Journal of Manufacturing and Materials Processing* 5, no. 3: 69. https://doi.org/10.3390/jmmp5030069

Dong, Chensong, and Hitoshi Takagi. 2014. "Flexural Properties of Cellulose Nanofiber Reinforced Green Composites." *Composites Part B: Engineering* 58: 418–421. https://doi.org/10.1016/j.compositesb.2013.10.032.

Faridul Hasan, K.M., Péter György Horváth, Miklós Bak, and Tibor Alpár. 2021. "A State-of-the-Art Review on Coir Fiber-Reinforced Biocomposites." *Royal Society of Chemistry* 11: 10548–10571. https://dou.org/ 10.1039/d1ra00231g

Jagadeesh, Praveenkumara, Madhu Puttegowda, Pawinee Boonyasopon, Sanjay Mavinkere Rangappa, Anish Khan, and Suchart Siengchin. 2022. "Recent Developments and Challenges in Natural Fiber Composites: A Review." *Polymer Composites* 43, no. 5: 2545–2561.

Mahmud Zuhudi, Nurul Zuhairah, Afiq Faizul Zulkifli, Muzafar Zulkifli, Ahmad Naim Ahmad Yahaya, Nurhayati Mohd Nur, and Khairul Dahri Mohd Aris. 2021. "Void and Moisture Content of Fiber Reinforced Composites." *Journal of Advanced Research in Fluid Mechanic and Thermal Science* 87, no. 3: 78–93. https://doi.org/10.37934/arfmts.87.3.7893.

Maitlo, G., I. Ali, H.A. Maitlo, S. Ali, I.N. Unar, M.B. Ahmad, D.K. Bhutto, R.K. Karmani, S.U.R. Naich, R.U. Sajjad, et al. 2022. "Plastic Waste Recycling, Applications, and Future Prospects for a Sustainable Environment." *Sustainability* 14: 11637. https://doi.org/10.3390/ su141811637

Martinelli, Flávia Regina Bianchi, Francisco Roger Carneiro Ribeiro, Markssuel Teixeira Marvila, Sergio Neves Monteiro, Fabio da Costa Garcia Filho, and Afonso Rangel Garcez de Azevedo. 2023. "A Review of the Use of Coconut Fiber in Cement Composites." *Polym*ers 15, no. 5: 1309. https://doi.org/10.3390/polym15051309

Meran, C., O. Ozturk, and M. Yuksel. 2008. "Examination of the Possibility of Recycling and Utilizing Recycled Polyethylene and Polypropylene." *Materials and Design* 29, no. 3: 701–705. https://doi.org/10.1016/j.matdes.2007.02.007

Mohanty, Amar K., Manjusri Misra, and Lawrence T. Drzal. 2005. *Natural Fibers, Biopolymers, and Biocomposites*. CRC Press. Taylor & Francis Group.

Nurazzi, N.M., M.R.M. Asyraf, M. Rayung, M.N.F. Norrrahim, S.S. Shazleen, M.S.A. Rani, A.R. Shafi, H.A. Aisyah, M.H.M. Radzi, F.A. Sabaruddin, R.A. Ilyas, E.S. Zainudin, and K. Abdan. 2021. "Thermogravimetric Analysis Properties of Cellulosic Natural Fiber Polymer Composites: A Review on Influence of Chemical Treatments." *Polymers* 13, no. 16: 2710. https://doi.org/10.3390/polym13162710

Patil, Umesh, and Soottawat Benjakul. 2018. "Coconut Milk and Coconut Oil: Their Associated with Protein Functionality." *Journal of Food Science* 83, no. 1. doi: 10.1111/1750-3841.14223

Sahari, J., and S.M. Sapuan. 2011. "Natural Fiber Reinforced Biodegradable Polymer Composites." *Reviews on Advanced Materials Science* 30: 166–174.

Sengupta, S., and G. Basu. 2016. "Properties of Coconut Fiber." *Encyclopedia of Renewable and Sustainable Materials* 2: 263–281. https://doi.org/10.1016/b978-0-12-803581-8.04122-9

Tambrallimath, V., Keshavamurthy, R., Saravanbavan, D., Kumar, G. S. P., & Kumar, M. H. 2019. "Synthesis and Characterization of Graphene Filled PC-ABS Filament for FDM Applications." *AIP Conference Proceedings*, 2057 (June). https://doi.org/10.1063/1.5085610

Tümer, E.H., & H.Y. Erbil. 2021. "Extrusion-Based 3D Printing Applications of PLA Composites: A Review." *Coatings* 11, no. 4: 1–42. https://doi.org/10.3390/coatings11040390

5 Advances in Polylactic Acid Composites with Biofiller as 3D Printing Filaments in Biomedical Applications

Azmah Hanim Mohamed Ariff,
Mohammad Akram Syazwie Mohd Areff, Che Nor
Aiza Jaafar, Zulkiflle Leman, and Recep Calin

5.1 BIODEGRADABLE POLYMER

Synthetic plastic was regarded as the material of the future in the previous century, but it is already a significant environmental problem. The delayed breakdown process and the usage of non-renewable raw materials in synthesis are the main downsides of synthetic polymers. Experts estimate that polymer manufacture consumes up to 7% of global oil and gas supply (Čolnik et al., 2020). Polymers that are biodegradable or manufactured from renewable resources provide an alternate option. As a result, the manufacture and usage of bio-based and biodegradable polymer materials is increasing considerably, which can help to mitigate environmental concerns associated with waste polymer materials. The use of maize, soy, sugarcane, potatoes, rice, or wheat, as well as seeds, is the subject of study on renewable resources. The biodegradable polymer has several benefits for the environment because it degrades naturally and is simple to recycle (Narayanan et al., 2015). Given that they are made primarily of plant components, biodegradable polymers provide a number of benefits. As a result, it is no longer essential to use the chemical fillers found in products made of synthetic plastic. Utilizing biodegradable products could lessen reliance on fossil fuels and help to keep the environment cleaner. Biodegradation occurs as a result of enzyme action and/or chemical breakdown connected with living organisms. Polymer biodegradation mechanisms and estimate approaches have been examined. Polymer biodegradability is governed not only by its origin, but by its chemical composition and the environment degrade conditions (Jakubowicz, 2003). Applications for biodegradable polymers include food packaging, agricultural films, and medical equipment. There are two main sources of biodegradable polymers: those produced from renewable resources and those derived from petroleum resources.

DOI: 10.1201/9781003362128-5

5.1.1 BIODEGRADABLE POLYMERS DERIVED FROM PETROLEUM RESOURCES

Acidic synthetic polymers with hydrolyzable functions like ester, amide, and urethane, as well as those with carbon backbones and additives like antioxidants, are examples of biodegradable polymers generated from petroleum resources. For instance, synthetic polymers having carbon backbones, hydrolyzable synthetic polymers, and polymers containing additives.

First, let us discuss polymers that contain additives. In order to speed up biodegradation, additives are added. One example is the degrading of polyolefins, which involves adding antioxidants to the polymer chains. The reason for this is that polyolefins are resistant to hydrolysis, oxidation, and biodegradation thanks to photoinitiators and stabilizers. They can be made oxo-degradable by adding pro-oxidant chemicals. These additives are created using metal alloys, such as Mn^{2+}/Mn^{3+}. The polyolefin will then degrade as a result of a free radical chain reaction (Jakubowicz, 2003).

Next is synthetic polymers with hydrolyzable backbones. Under certain conditions, polymers containing hydrolyzable backbones are biodegradable. Polyesters, polyamides, polyurethanes and polyureas and polyanhydrides are examples of these (Nair & Laurencin, 2007). Due to their significant diversity and synthetic adaptability, aliphatic polyesters are the biodegradable polymer class that has received the most research. Based on how the components of polyesters are joined together, two categories can be distinguished. The first class consists of polyhydroxyalkanoates. These hydroxyacid polymers, HO-R-COOH, are produced. Poly(ethylene glycol) and poly(glycolic acid) are two examples. Poly(alkene dicarboxylates) make up the second class (Vroman & Tighzert, 2009). The simplest basic aliphatic polyester, polyglycolide (PGA), is made by polymerizing a cyclic lactone called glycolide through ring opening. PGA has superior mechanical attributes. However, due to its low solubility and rapid rate of breakdown, which results in acidic byproducts, its biological applications are restricted. Additionally, poly(butylene succinate) (PBS) and its copolymers are included under the category of aliphatic polyesters. Three thermoplastic polymers developed by Showa High Polymer in 1990 include PBS, EnPol, and Bionolle. They are produced through the polycondensation of glycols and aliphatic dicarboxylic acids like succinic and adipic acid. PBS has better processability than PLA and PGA and is also more versatile. For some uses (such as foams and stretched blown bottles), another polymer with a long chain branch has been developed. Surface modification through plasma treatment was used because PBS lacked the biocompatibility and bioactivity required for medical applications (Wang et al., 2009).

Other type of synthetic polymers with hydrolyzable backbones is aromatic copolyesters. A blend of aliphatic and aromatic monomers makes up aliphatic-aromatic copolyesters. They frequently have terephthalic acid as their base. These polyesters' mechanical characteristics are inferior to those of non-biodegradable polymers. Poly(butylene adipate-co-terephthalate) is the copolyester that has been examined the most often (PBAT) (Liao et al., 2001). Polyamide is a type of synthetic polymer. The same amide is bonded in polyamides as it is in polypeptides. Using 1,2-ethanediol, glycine, and diacids with two to eight different methylene groups, poly(ester-amide)s have been created. It has been discovered that amide and ester group copolymers break down quickly. The phenylalanineglycine ratio might be changed to regulate the pace of breakdown. A special type of polymer with a broad variety of physical and chemical attributes is polyurethane. To suit the requirements of contemporary technologies including coatings, adhesives, fiber, foams, and thermoplastic elastomers, it has undergone substantial customization. Three components are used to make polyurethanes: a polyol, a chain extender, and a diisocyanate. By selecting the right soft section, one may tune the deterioration rate (Kim et al., 2003).

Lastly, biodegradable derived from petroleum resources include synthetic polymers with carbon backbones. Hydrolysis is often not an issue for vinyl polymers. They must undergo an oxidation process in order to degrade biologically. Majority of biodegradable vinyl polymers include a functional group that is easily oxidizable, and a catalyst is added to encourage oxidation or photooxidation. Since polyvinyl alcohol dissolves in water, it is frequently utilized. Both enzymes and microorganisms may quickly biodegrade it (Nair & Laurencin, 2007).

5.1.2 BIODEGRADABLE POLYMERS DERIVED FROM RENEWABLE RESOURCES

Biodegradable polymers created from renewable resources have drawn a lot of attention recently. This expanding interest has been driven by worries about fossil fuel extraction and growing environmental consciousness. There aren't many examples of biodegradable polymers that come from renewable resources like bacterial and natural polymers. In nature, all phases of species development result in the production of natural biodegradable polymers. The polysaccharide family, which includes starch and cellulose, is the most characteristic group of these natural polymers. Such polymers are frequently chemically altered in order to enhance their mechanical characteristics. For example, protein contained in thermoplastic heteropolymers. Protein biodegradation is an amine hydrolysis process carried out by enzymes such as protease. Protein grafting is a method for regulating the pace of biodegradation. Protein can be classified into protein from animal sources and from vegetable sources. Collagen is the major protein present in connective tissues of animals. It is composed of a number of polypeptides, most of which contain glycine and proline. Enzymatic digestion of collagen is possible, and it has special biological characteristics (Nair & Laurencin, 2007).

Polysaccharides are biodegradable polymers made from natural sources. Polysaccharides are obtained from two sources: marine and vegetable. Polysaccharides from marine sources are chitin or chitosan. Crab, shrimp, crawfish, and insect shells commonly contain chitin. It may be thought of as amino cellulose. A different supply of chitin may be available through the growth of fungus, according to recent developments in fermentation technology. Due to its limited solubility, chitin is frequently replaced in a variety of applications (Je & Kim, 2006). Starch is a polysaccharide that comes from a vegetable. Starch is a well-known hydrocolloid biopolymer. The hydrophilic granules that make up this substance are produced by agricultural plants. The most popular sources of starch include rice, wheat, maize, potatoes, and maize. Depending on the source of the starch, different molar weights and relative amounts of amylose and amylopectine exist (Vroman & Tighzert, 2009). Another option to synthetic polymers is thermoplastic starch, often known as plasticized starch (Netravali & Chabba, 2003).

Bacterial polymers is another biodegradable polymer derived from renewable resources. These are obtained from polymerization of fermented monomers under different environmental conditions. There are two types of bacterial polymers: semi-synthetic polymers and microbial polymers. For semi-synthetic polymers, Cargill Dow Polymers was the first to produce a new line of polylactic acid (PLA). Lactic acid generated by starch fermentation is used to make PLA. This biotechnological process produces lactic acid that is virtually entirely L-lactic acid. Microbes may degrade PLA even if it is not soluble in water (Mecking, 2004). PLA is typically produced by combining D- or L-lactic acid or by lactide ring-opening polymerization. Because of the -CH3 side groups, PLA is a hydrophobic polymer. Because of the steric shielding effect of the methyl side groups, it is more resistant to hydrolysis than PGA (when compared to synthetic polymer). To improve crystallization and chain mobility, PLA can be polymerized. For plasticization, oligomeric acid, citrate ester, or low molecular polyethylene glycol are utilized. High molecular weight PLAs are produced using ring-opening polymerization. This method also enables control of the final properties of PLA by varying the ratios of the two enantiomers. The crystallinity of PLA affects how quickly it deteriorates. Several copolymers of lactide and glycolide have been investigated as bioresorbable implant materials since PLA degrades significantly more slowly than PGA (Wang et al., 2009).

There are two forms of microbial polymers: polyhydroxyalkanoates (PHA) and polyhydroxybutyrate (PHB). PHA is an intracellular reserve substance that bacteria can accumulate in. Depending on the organism, PHA can account for anywhere between 30% and 80% of the total dry weight of a cell when there is an abundant supply of carbon and little or no nitrogen present. PHB is a biotechnologically produced polyester that can be degraded under various environmental conditions by a number of microorganisms, including bacteria, fungi, and algae. The hydrolytic breakdown produces 3-hydroxy butyric acid, a substance that is typically found in blood, at a rather slow rate (Vroman & Tighzert, 2009).

5.2   POLYLACTIC ACID (PLA)

Developed from lactic acid, PLA is a linear aliphatic thermoplastic polyester. It is produced by fermenting 100% renewable and biodegradable plant materials, including the starches in corn or rice. Dextrose can be created by chemically converting maize or other carbohydrate sources. There are four different ways to make PLA: ring-opening polymerization, azeotropic hydration condensation, lactide formation polymerization, and direct condensation polymerization. However, the most typical method for producing PLA is ring-opening polymerization (Auras et al., 2004; Xiao et al., 2006).

Because they break down through the hydrolysis of the ester groups in their backbones and perhaps through microbial attack, PLA polymers are biodegradable and recyclable. Factors like crystallinity, molecular weight, shape, water diffusion rate, and stereoisomeric concentration have an impact on the rate at which PLA degrades. Because of its slow rate of degradation, PLA is useful for some biological applications but troublesome for consumer goods like packaging films. For instance, PLA hydrolyzes to hydroxyl acid when it is used in the human body, and this hydroxyl acid is subsequently combined with tricarboxylic acid. It is then removed from the body (Lasprilla et al., 2012).

Additionally, according to Murariu and Dubois (2016), the melting temperature of commercially available PLA polymer is 160–180°C, while the glass transition temperature is 60–65°C. It is extremely delicate, with significantly less than 10% elongation at break, but having a high tensile strength of 17–74 MPa (Rasal et al., 2010). PLA's mechanical strength is inferior than that of PC and acrylonitrile buta-diene styrene (ABS) polymers. Because of its more linear molecular chain structure, PLA offers greater mechanical strength and prevents chain tangling. Antioxidants, heat stabilizers, light stabilizers, impact modifiers, and a range of other chemicals are used to enhance the properties of commercial PLA.

Several production methods, including injection molding, film extrusion, blow molding, thermoforming, film forming, and fiber spinning, can be adapted to PLA's strong thermal processability. However, depending on the process, a number of vari-ables (D-isomer content, molecular weight distribution) must be controlled. L-, D-, or meso-lactide stereoisomers can be included into the polymer backbone to pro-vide different PLA materials for different applications (Lim et al., 2008). The most used method is extrusion because it provides uniform mixing of the pellets at high temperatures. PLA can also be dissolved in chloroform or other solvents including dichloromethane, methylene chloride, or acetonitrile to produce films with a high level of transparency and gloss (Hughes et al., 2012).

Recently PLA gained much attention in the 3D printing industry to produce PLA filaments that can be easily biodegradable. Three-dimensional printing is evolving as a favored method to produce components in automobiles, aerospace, packaging, and biomedicals due to its advantages such as low initial investment, design freedom for complex shapes and profiles, and shorter time for product development (Palaniyappan & Sivakumar, 2023). Polymers commonly used with 3D printing are thermoplastics materials such as PLA, nylon, polypropylene, and ABS. The melting temperature of PLA, which is lower than that of other polymers used in the fused deposition modeling (FDM) printing industry and needs less energy usage during printing, is 150–160°C. This gives PLA an advantage over other materials. In contrast to other feedstock polymers like PS, PP, and PE, PLA has a strong Young's modulus, and ten-sile and flexural properties. It is also suitable for bioproducts because it is affordable, widely available, sustainable, and biodegradable (Hamad et al., 2015; Palaniyappan & Sivakumar, 2023).

5.3   APPLICATION OF PLA

The applications for PLA are more diverse today than ever before. For instance, it can be used in the packaging, automotive, and healthcare industries. Bottles, labels, cups, flexible and water-resistant films, pastry wrapping, and disposable plates can all be made from it. In tissue engineering, it serves as a binding agent for the production of flexible and waterproof films, bottles, labels, and cups, as well as bakery packaging and disposable dishes. In tissue engineering for biomedicine, it also functions as a binding agent (Madhavan et al., 2010). The compatibility of material properties for converters, technical viability of alternative processing methods, range of applications, and commercial viability are the primary factors that will determine whether or not it is commercially viable. PLA is the most advanced and flexible polymer derived from a natural source that can be processed using the same equipment as polymers derived from petroleum. By varying the molecular weight and copolymerizing it with various polymers, it can be put to a variety of purposes (Domenek & Ducruet, 2016). Around the world, PLA is used in a variety of industries, including textile, packaging, automotive, building, and biomedicine.

5.3.1   PLA in Textile Industry

The first application for PLA is in the textile industry. For this reason, several PLA composites were created in order to increase productivity, decrease energy and water consumption, and minimize waste in order to reduce environmental impact and increase production. It's crucial to create intelligent and useful textiles to avoid wasting water, electricity, chemicals, etc. (Tümer & Erbil, 2021). PLA fibers are becoming more popular and accepted in a variety of commercial textile industries due to their ease of melt processing, distinctive property spectrum, derivation from renewable sources, and potential for composting and recycling. PLA also has a wide range of possible applications in the form of fibers and nonwoven textiles, including upholstery, disposable clothing, awnings, feminine hygiene products, and diapers (Shen et al., 2009). Because of a variety of wet processing applications (pretreatment, dyeing, and subsequent finishing treatments), the PLA fibers are put under chemical and physical pressures. According to Domenek and Ducruet (2016), the development of PLA stereo complexes may be able to lessen the amount that PLA fibers and fabrics shrink after being ironed.

For example, when it comes to the direct approach of polymer material into textiles using 3D printing, it is extremely difficult to maintain mechanical resistance, durability, and comfort levels that are on par with or better than those of simple textile substrates. Non-conductive PLA and conductive PLA-carbon composite filaments were added to white woven PET materials through the 3D printing process (Eutionnat-Diffo et al., 2019). Due to the PET fabric's reduced flexibility and diffusion, the PLA deposition technique affects the tensile properties of the 3D-printed textile, which leads to poor adherence of the polymer textile. The tensile characteristics of the extruded polymeric materials were unaffected by the addition of conductive additives to PLA. FDM is demonstrated in Figure 5.1.

PLA textiles, on the other hand, have the softness and feel of natural fibers like cotton, silk, and wool but the quality, cost, and simplicity of care of synthetics. Athletic apparel applications benefit from the silky feel, stretch, toughness, and water vapor permeability of PLA textiles since they allow for ventilation.

5.3.2   PLA in the Packaging Industry

While PLA's rapid hydrolytic disintegration is helpful for medical applications, it is a significant drawback for usage in food packaging. As a thermoplastic packaging material, PLA satisfies a variety of criteria and has been suggested as a commodity

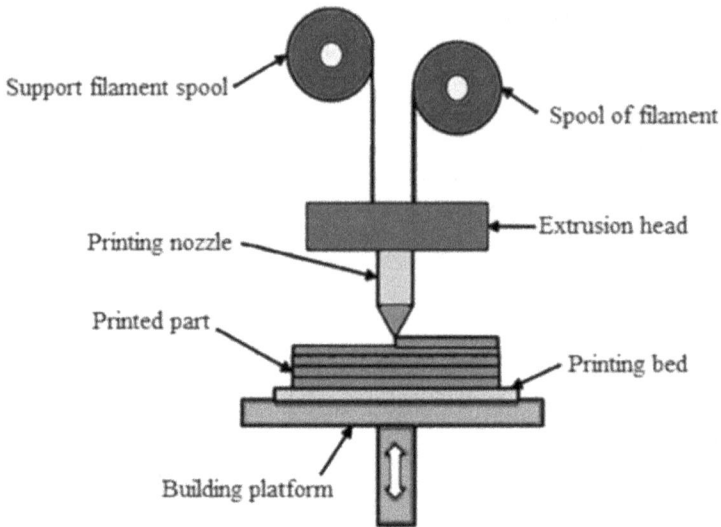

FIGURE 5.1   Fused deposition modeling process (Mogan et al., 2023).

resin for general packaging applications (Finglas et al., 2003). Thermoformed and/ or extruded PLA products have lately been developed for popular applications such as cups, overwrap, blister packaging, food and beverage containers (Groot et al., 2010). Haugaard et al. (2003) discovered that PLA cups were just as effective as high-density polyethylene cups at preventing quality changes in an orange juice simulant and a dressing during storage. The properties of PLA may even be better for yogurt packaging due to the lower oxygen permeabilities of PLA cups compared to PS cups. A Danish dairy manufacturer has employed PLA, which is advertised as being bio-degradable for yogurt cups, made from high-impact polystyrene (Süfer, 2017). These investigations demonstrate that PLA, especially for small containers that are not currently recycled, has the potential to replace polyolefins and PS, two polymers that are frequently used in food applications.

Consumers are demanding natural foods with few to no preservatives and little to no microbial contamination while using sustainable packaging, which has led to an increase in the usage of PLA in antimicrobial packaging (Rhim et al., 2009). Antipack TM is a film formed from a PLA-/starch-based polymer that has been impregnated with an antibacterial component. It was developed by Handary in Belgium and is a commercialized antimicrobial PLA packing product. By dispersing chitosan-containing natamycin over the surfaces of solid foods like cheese, fruits, vegetables, meat, and chicken, this product is said to prevent the growth of yeast and mold during storage (Süfer, 2017).

5.3.3   PLA in the Automotive Industry

The manufacture of automobiles and component parts strengthened plastic polymers with natural resources. Hemp, sisal, flax, wood, and kenaf are used in a biodegradable or non-biodegradable matrix. A fully bio-based solution can be produced by combining biofibers with a polymer made from renewable resources, such as PLA (Peças et al., 2018). A twin-screw extruder was used to make PLA-flax fiber composites with flax fiber contents of 30% and 40% by weight in a study by Oksman et al. Although the incorporation of plasticizers had no effect on impact strength, the mechanical properties of these composites are adequate for use in automobile panels (Oksman et al., 2003). In a different 3D printing investigation, the strength and elastic modulus of the printed specimens increased as the thickness of the various printing layers decreased. The effects of additional materials, printing layer thickness, and compression characteristics show that the maximum elastic modulus was determined to be 1.69 GPa, with a high carbon fiber modulus ratio (Mei et al., 2019). Carbon fibers, graphene, and silicon carbide nanowires were also used in 3D printing.

In addition, to increase their enticing appeal, vehicle prototypes could mimic textural effects (sparkle or graininess), metallic, or gonio-appearance. In a related study, diffractive pigments and gray metallic PLA were combined, and statistically designed experiments were carried out to identify the best printing parameters that would affect the final gonio-appearance of the 3D-printed samples. According to Micó-Vicent et al. (2019), the layer height was found to be crucial in enhancing the flop or dazzling effects.

5.3.4 PLA in Building

Although biopolymers are exceedingly uncommon in the building business, PLA development may be possible in the future. Expandable foams are frequently created using polymers derived from fossil fuels (such polystyrene or polypropylene), and they are mostly employed as insulation materials. A Dutch company started producing expandable PLA, a biodegradable and bio-based alternative to EPS-foam, in 2010. Additionally, PLA has been used as a binder in Buitex's Isonat Nat'isol, a hemp-based building insulator. In industries where flame retardancy is required, such as transportation, electric and electronic equipment, and others, PLA may eventually replace traditional polymers. When the same kind of additives are employed, the mode of action is similar to that of other polymers. PLA may have improved mechanical, thermal, and flame retardant properties when combined with nanofillers made of graphite and silica (Fukushima et al., 2010).

5.4 APPLICATION OF PLA IN BIOMEDICINE

In addition to being ecologically safe and biocompatible, PLA is also thermoplastically producible. Cells hydrolyze and metabolize the immunogenic soluble oligomers to destroy PLA in human and animal bodies (Tyler et al., 2016). According to Moravej and Mantovani (2011), PLA clouds are employed in plastic surgery as spinal cages, tacks, and pins for ligament attachment, anchors, and to fixate craniomaxillofacial bones. In the design and development of therapeutic applications, PLA-PLGA microspheres containing bioactive compounds, for instance, are well known for their significance (Qi et al., 2019). The Food and Drug Administration (FDA) has approved PLGA for use in human sutures, bone implants, screws, and the encapsulation of vaccine antigens for sustained medication delivery (Sadeghi et al., 2021). Plasticizers are necessary for mechanical property profiles, particularly improved flexibility and impact resistance, to improve the application of PLA. The PLA plasticizer must be biodegradable, safe for contact with food, and safe for disposal. PEG is currently used most frequently to increase the hydrophobicity of PLA and is a hydrophilic polymer (Ebrahimi & Ramezani Dana, 2022). PLA has many uses in the biomedical field, including as a drug carrier and in tissue engineering and orthopedics.

The initial biomedical use is as a medication carrier. The biodegradability of PLA is one property that makes it a good candidate for drug delivery systems. In extracellular conditions, PLA disintegrates readily, and the rate of breakdown can be regulated to produce the desired effect. For continuous delivery of medications, the kinetics of this breakdown may be expanded. This gives the medication ample time to work, which is important because this treatment method could be hampered by metabolic processes (Tyler et al., 2016). For instance, certain leukemia cells secrete chemicals that lessen the potency of anti-cancer medications. Daunorubicin can overcome such inhibitors when combined with PLA (Lv et al., 2022). Furthermore, the blood-brain barrier (BBB) has been successfully crossed by PLA. According to reports, PLA composites have purportedly enhanced the diffusion of medication particles across the BBB (Tyler et al., 2016).

The use in tissue engineering comes next. When using different tissue engineering and tissue regeneration therapy approaches, PLA is crucial. As an illustration, consider its ability to encourage hard tissue regeneration during bone grafting procedures. The integration of tissue-engineered bone with natural bone is the main goal of the current approach. It has been demonstrated that the chemical produced stimulates angiogenesis and osteogenesis in the tissues around it (Wang et al., 2019). In particular, glass, such as calcium phosphate glass, can help PLA scaffolds attain this goal. The foam forming method may be used to create these scaffolds. This process requires injecting the glass into a PLA solution in order to make it easier to connect these two pieces together. Because of this, calcium phosphate glass can develop uniformly. The PLA/calcium phosphate glass further provides the necessary mechanical strength and the proper amount of cell-material interaction. These composites shown better biogenesis and nutrient sequestration as compared to unmodified PLA. With the inclusion of bioglass, the porosity of the PLA composite rose from 93% to 97%. The PLA/calcium phosphate composite was able to achieve a value of 20.2 MPa in comparison to the PLA control (Navarro et al., 2004).

Last but not the least, orthopedics is where PLA is used in biomedicine. Devices made of biological orthopedic materials may benefit from PLA. Resorbable fracture healing plates have used PLA suture devices since their introduction in 1973. Due to its radiolucent characteristics, polymer fixation offers greater quality imaging while also lowering the possibility of artifact development (Pawar et al., 2014). Although the breakdown byproducts of PLA copolymers are not hazardous, they might make the resorption site more acidic. To encourage cell proliferation, osteogenic or anabolic bioactive may be added to orthopedic implantables made of PLA. A possible treatment for osteopenia is provided by biodegradable fixing devices (Gupta et al., 2007). Blends of PLA have demonstrated effectiveness in improving tendon and ligament attachment and healing. ACL repair has been made easier by the use of PLA and copolymer-based pins, suture hooks, and screws. A highly specialized and customized porous scaffold or cell carrier made of PLA has been created to restore extracellular ligamentous matrix elements (Narayanan et al., 2016).

Based on all the information above, we can conclude that in the applications of PLA in biomedicine and others, it is nearly impossible to use PLA in its pure form in any of these usages. The PLA needs to be modified to achieve certain character-istics that fit its application. One way to achieve this is by adding in filler into the PLA material. For 3D printing method, filler can be added into PLA and then made into filament for subsequent manufacturing process. Three-dimensional printing is currently being explored for fabrication of scaffolds in biomedical applications since it enables fabrication of complex geometry with controlled porosity. This is vital for tissue engineering scaffold applications in biomedicine to enable appropriate cell penetration, waste removal, and fluid flow (Wang et al., 2017). In terms of fillers, there are different forms of fillers for reinforcements, such as particles, fiber, and flakes. Between these three forms, particle type reinforcements are more suitable for 3D printing method with filaments. This is as a result of fewer side wall defects and nozzle blockage during printing (Palaniyappan & Sivakumar, 2023). The chapter's next section talks about particle filler and how it affects the PLA composite material that is created.

5.5   PLA WITH FILLER

PLA is one of the most desirable choices to replace petroleum-based polymers due to its widespread usage in many industries. However, it still has a number of significant flaws, including a low melt strength and poor toughness. Various modification strate-gies have been suggested to address these shortcomings (Murariu & Dubois, 2016). Nowadays, the creation of PLA-based nanocomposites has made use of a variety of inorganic fillers, including carbon nanotubes (CNTs), graphene, natural rubber, sili-cates, and cellulose (Yang et al., 2016). Researchers have focused on silica particles as one of the greatest possibilities for reinforcing materials among the numerous types of inorganic nanomaterials because of its low cost, high strength modulus, and functional variety. Excellent filler dispersion has already been achieved using a variety of techniques. These comprise conventional synthetic operations like co-extrusion, melt-extrusion, and solution casting. In addition, specific coupling agents such oleic acid, rubber, L-lactic acid oligomers, and 2-methacryloyloxyethyl isocya-nate are added to create interfacial modifications (Hajibeygi & Shafiei-Navid, 2019).

A filler is typically combined with other materials to lower cost or to change the compound's physical, mechanical, optical, and other qualities. For instance, the use of fillers can lower costs and improve mechanical properties, such as increas-ing modulus/stiffness at ambient temperature or raised temperatures, especially in PLA grades that permit a high degree of crystallinity, if the fillers do not alter the molecular, thermal, or mechanical properties of PLA (Murariu & Dubois, 2016). Previously synthetic fibers were added to PLA to improve its properties and perfor-mance; however, synthetic fibers rely on non-renewable resources and are not biode-gradable at the end of their service life. Hence, more economical natural fibers and particles as fillers in PLA would provide lower overall price, improve properties of the composite, and result in biodegradable and sustainable products. Sources from marine, wood, agricultural residues, and other natural waste products would be good candidates as fillers in PLA. Most importantly, for biomedical applications, toxicity and biocompatibility also need to be studied (Yang et al., 2023).

Talc is one of the fillers most frequently used for blending with PLA. Talc may be added to PLA to boost stiffness and heat resistance, enhance dimensional stability and barrier qualities, speed up PLA crystallization, decrease molding time and manufacturing costs, and more. Extensive research has been undertaken regarding the use of PLA-talc composites. First, the composition of talc can impact the crystallization of the PLA material. In this instance, the addition of 2% talc to the PLA cut the isothermal crystallization's duration in half and reduced it by over 65-fold (Yang et al., 2010). According to Petchwanna et al. (2014), talc was shown to be a useful nucleating agent for enhancing PLA crystallization rate. Higher levels of crystallinity were induced slightly by the finer talc. This is the impact of fillers with different particle sizes (1, 5 and 30 m) on PLA when talc concentrations up to 10 wt% were present. The manufacturing of talc uses PLA, a crucial raw material that can be extremely effective in the nucleation and for significantly raising the degree of crystallinity. It is commonly believed that the PLA and talc macromolecular chains have a strong affinity for one another (Ouchiar et al., 2015). In addition, talc can make more sophisticated, densely packed composites and boost thermal stability. In a 2014 study by Shakoor and Thomas, the mechanical and thermomechanical characteristics of PLA filled with up to 30% ultra-fine talc were described. This mixture enhances the strength of the PLA-talc compound from 4.1 GPa for pure PLA to 9.8 GPa for composites. Dynamic mechanical analysis (DMA) may be used to confirm the findings on increasing storage moduli.

Next, PLA can combine with carbonaceous filler. To create micro- or (nano) composites with carbonaceous fillers of various geometries, PLA was utilized as the polymeric matrix. The composites may exhibit enhanced nucleating, mechanical, thermal, and FR characteristics, as well as customized electrical and thermal conductivity. They may show promise as components for sensors that are sensitive to strain, temperature, or organic solvents (Sullivan et al., 2014). Graphene nanosheets, which make up graphite, have distinct structural characteristics and physical characteristics. Its efficacy is obviously higher when the graphite layers are divided and functionalized, which is possible using various methods for dividing and functionalizing graphite layers, to achieve higher efficacy. By using a variety of methods, derivatives of graphite and graphene have been combined with PLA matrix to create nanocomposites (Fukushima et al., 2010). Different addition methods may be used to melt-blend PLA with commercial extended graphite (EG) micro fillers (mean diameter (d50) of 36 μm, size of primary particles of 35 nm). The PLA-EG composite has better features compared to the clean PLA matrix such as stiffness, crystallization kinetics, increased thermal stability, antistatic to conductive electrical properties correlated with filler loading. As an interesting observation about the shape of composites, TEM analysis showed that the PLA matrix contained graphene layers (Murariu et al., 2010). As shown by DMTA results in this study, melt-blending PLA with EG increased storage modulus, which together with increased crystallinity enables the use of PLA-EG composites in applications requiring higher usage temperatures.

Also being researched is PLA combined with hydroxyapatite (HA) filler. Recently, PLA-HA has drawn the most attention for use in biomedicine. It has the advantage of combining the osteoconductivity and bone-bonding capabilities of HA with the absorbency and processing simplicity of PLA and its copolymers. The impact of filler content on the shape and physical characteristics of PLA-HA composites has been researched by Tazibt et al. (2023). Samples of various compositions, ranging from HA 0% to 15% with a 5% interval for each composition, were used in this experiment. Based on the findings, PLA/HA (5 wt%) has somewhat higher tensile strength and elastic modulus than plain PLA, and even higher than other composites, as shown in Figure 5.2. In comparison to neat PLA, which has values of 2,963 and 62 MPa for its elastic modulus and tensile strength, PLA/HA (5 wt%) has an elastic modulus and tensile strength of 3,057 and 64 MPa, respectively. However, the tensile properties of PLA composites start to decline at 5 wt%, becoming more noticeable at 15 wt%. By functioning as stress concentrators, the filler aggregates impede the stress transmission between the PLA matrix and HA. Filling aggregates disrupt the matrix continuity and reduce the polymer chains' ability to endure stress. The PLA matrix reduced contact area and physical flaws in the composite samples that are caused by the substantial aggregation of HA particles there (Wan et al., 2015).

Studies on PLA that included starch as fillers discovered that while the two materials have different mechanical and barrier properties, their combination could produce films with improved functional capabilities. However, PLA and starch are thermodynamically incompatible since PLA is hydrophobic and starch is very hydrophilic. Starch or PLA can be modified to modulate their hydrophobicity in order to

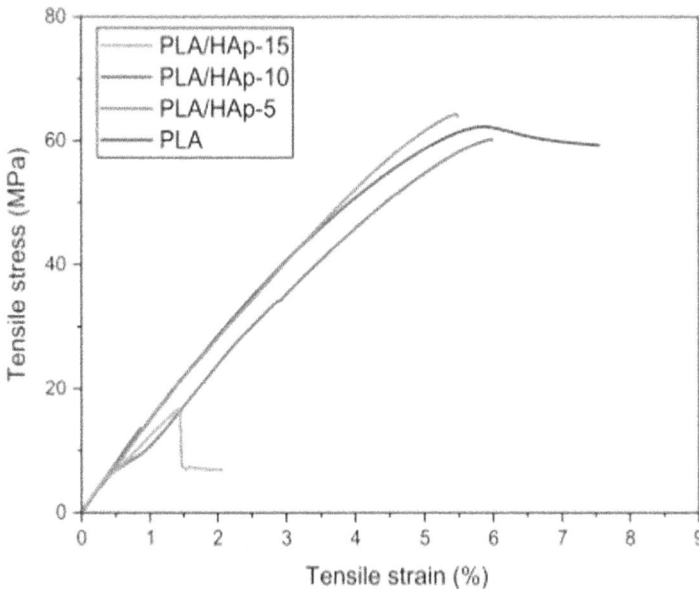

FIGURE 5.2 Stress-strain curve of the neat PLA and PLA/HA particles at various filler contents (Tazibt et al., 2023).

increase their compatibility (for instance, via plasma treatment) (Turalija et al., 2016). However, there has been some interest in the research community over the use of PLA-starch mixtures.

In a 2009 study, Orozco et al. (2009) created PLA-g-MA copolymers using the grafting initiator dicumyl peroxide (DCP) and PLA-starch copolymers using reactive blending with starch percentages ranging from 0% to 60%. The SEM analysis found that MA had a plasticizing effect and enhanced PLA and starch compatibility, resulting in a stable and homogeneous interface without stress fractures, holes, or cavities at the interface of the two polymers. Tung oil anhydride (TOA) was used as a bio-based reactive plasticizer by Xiong et al. (2013) in PLA-starch blends via the prepared reaction of MA. They noted improved compatibility between the two polymers as shown by SEM micrographs and Fourier transform infrared (FTIR) spectra, as well as increased elongation at break from 6% to 31%. In order to increase the hydrophobicity and PLA compatibility of waxy and high amylose starches Wokadala et al. (2014) investigated PLA-starch blends without compatibilizers. The modified mixture showed improved mechanical performance despite a reduction in polymer heat stability, and SEM micrographs showed a more uniform structure as a result of the starch modification. This research also showed that the tensile properties of starch-PLA mix films are influenced by the amylose/amylopectin content of the starch. Due to amylose's propensity to self-aggregate, composites made from butyl-etherified high amylose starch performed worse than those made from butyl-etherified waxy starch in terms of elongation at break and tensile strength at higher starch levels.

Research by Abdullah et al. (2019) shows the mechanical analysis of PLA and starch composite. In accordance with the findings, starch-PLA 0% has the lowest tensile strength; adding PLA to starch-PLA 3% and starch-PLA 10% increases the tensile strength values. However, in the percentage of rising PLA, the elongation at break of starch-PLA 3% and starch-PLA 10% is decreased. It was demonstrated how incorporating PLA may increase tensile strength while lowering elongation at break. PLA can enhance bioplastic composites because of its brittle mechanical properties and high tensile strength (55.4 MPa); it also improves the tensile strength of bioplastic composites through tensile strength. This is consistent with earlier research showing that starch-based bioplastics have more bending capabilities than PLA because these materials feature amylose/amylopectin linkages which provide flexural qualities and the addition of glycerol to these materials further increases their flexibility (Phetwarotai et al., 2012). Figure 5.3 shows the mechanical properties for this study.

However, doing so necessitates the use of reactive processes, which in turn may result in reactant residues in the films that need to be assessed for their ability to be consumed. Important results suggested that blending PLA and starch might be a good alternative method to improve the film properties of net PLA or starch films. The strategy of the bi- or multilayer films did not call for as much work to improve the interfacial characteristics of the polymers because there is just contact at the layer interface. However, from a mechanical standpoint, it is crucial to ensure that the materials that come into contact with other layers have good interfacial adhesion (Muller et al., 2017).

Shell waste from the marine industry can also be used to make biofillers. According to Yan and Chen (2015), this sector generated six to eight million tons of crab shells

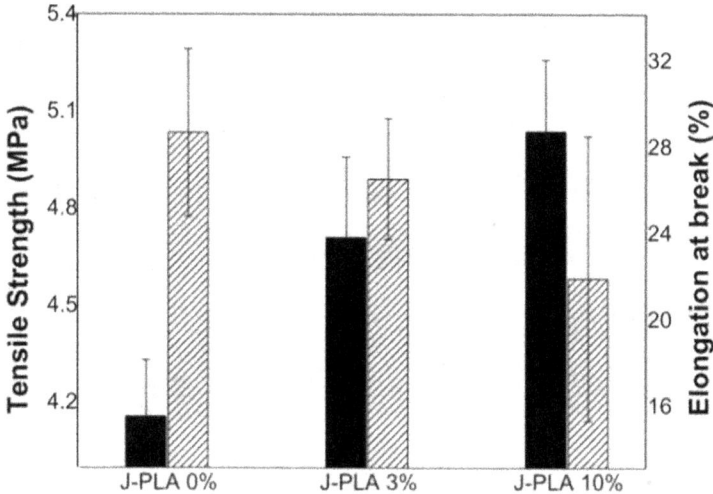

FIGURE 5.3   Mechanical properties of bioplastic composites (Abdullah et al., 2019).

and other shell trash each year. For instance, it has been discovered that crab shells are a good source of calcium carbonate, chitin, and other vital elements, making them suitable for use as biofillers in polymeric composites. In earlier research to evaluate crab shells as a possible material for bone tissue engineering, Wilson et al. (2012) discovered that it has the ability to start bone growth in a form that may be osteoinductive, making it ideal to be used in advanced functional bone implants. In a study by Palaniyappan and Sivakumar (2023), crab shell particles were added into PLA to study its effect on physical, mechanical properties, and the filament stability in terms of surface roughness and diameter deviation. This is because, for high-quality 3D-printed product, it depends on the quality of the filaments used. The crab shells were washed and reduced in size through ball milling technique. They were then extruded to produce filaments using bench type single-screw extruder. To reduce internal stress and surface deviation, the extruded filaments were submerged in water before being wound in the winding spool using a semi-automatic filament winder. According to the results, samples containing 8 wt% of crab shell particles had the highest tensile strength. The addition of crab shells increased the surface roughness and diameter deviation of the extruded filament, as illustrated in Figure 5.4 (Palaniyappan & Sivakumar, 2023).

Crab shells as filler in PLA were the subject of another study by Yang et al. in 2023, which is more thorough with regard to the evaluation of biocompatibility and antibacterial characteristics. In this study, the crab shell particles were mixed with PLA to create filaments after being chemically treated with hydrochloric acid, sodium hydroxide, and ethanol. It was discovered that chemical processing helps to keep chitin as the primary component of the crab shell powders while removing calcium carbonate and protein. The polymeric composite's crystallinity is increased by the addition of crab shell powders, and tests show that the resulting composite has acceptable antibacterial activity against *E. coli* and good biocompatibility.

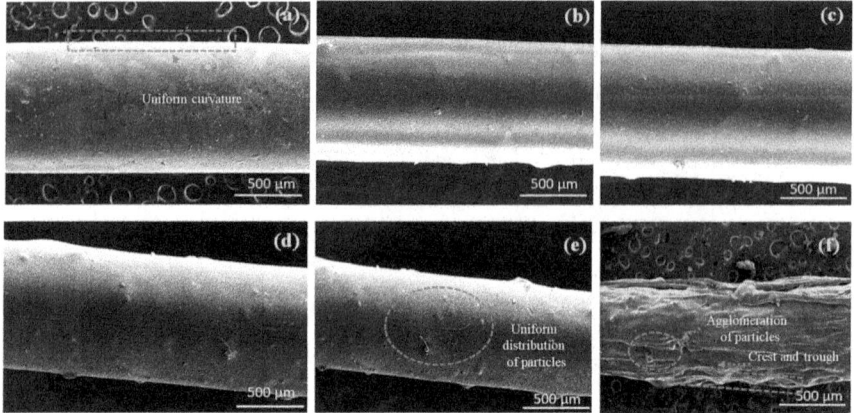

FIGURE 5.4 SEM image of the extruded polymeric filaments: (a) neat PLA, (b) 2%, (c) 4%, (d) 6%, (e) 8%, and (f) 10% (Palaniyappan & Sivakumar, 2023).

TABLE 5.1
Filament extrusion parameters for developed PLA with various biofillers in recent published papers

Matrix material	Filler material	Extruder temperature and screw speed	Tensile strength (MPa)	Melting temperature (°C)	References
PLA	Crab shell particles	180°C, screw rotational speed 40 mm/s	61.3 MPa with 8 wt% crab shell	—	Palaniyappan and Sivakumar (2023)
PLA	Chitin and starch	120–200°C and feeding rate 30 set	55±30 MPa	—	Olaiya et al. (2020)
Starch	PLA	80–130°C and screw speed 35–50 rpm	4.2±0.9 MPa	94.3±13.3	Abdullah et al. (2019)
PLA	Potato thermoplastic	140–176°C and screw speed 35 rpm	9.37±15.37 MPa	166.1±5.8	Haryńska et al. (2021)
PLA	Waste paper	175–180°C	—	165.3±1.4	Tao et al. (2021)

Table 5.1 shows examples of a few chosen papers on PLA with biofillers and its parameters for filament extrusion. It was noticed that the melting temperature of the end product generally decreases with increasing composition filler material. For example, melting point of 144.1°C was attained when polypropylene glycol was applied in a research done by Kanabenja et al. (2022). Similarly, PLA with melting point as low as 94.3°C was also acquired by Abdullah et al. (2019) when a PLA composition of 30 wt.% was adapted. Therefore, this further affirms the direct relationship between composition and melting point, which allows us to have a better idea in further fabrication and testing phases. Apart from that, it is interesting to know that the tensile strength of the specimens tends to react inversely to melting point where a higher melting point typically leads to lower mechanical tensile strength.

5.6 CONCLUSIONS

PLA has been established as one of the well-known renewable resource polymers. It belongs to the thermoplastic category, making it a suitable candidate for 3D printing, which is a method of manufacturing for biomedical components. In biomedical application it can be summarized that pure PLA is not sufficient to reach the characteristics required. Fillers were added to improve some of its properties and performance. There are synthetic and biofillers. To ensure sustainability and biodegradability, biofillers were explored. Different types of biofillers were discussed in this chapter and its influence on the PLA composites were mentioned. It can be concluded that further studies are needed to be carried out to ensure its biocompatibility for long-term application in the human body, controlled biodegradation in situations that require that function, sufficient mechanical strength, and other characteristics to function.

ACKNOWLEDGMENTS

This study is supported by the Research Management Center, Universiti Putra Malaysia (UPM/GP-IPB/2020/9688700), and the Department of Mechanical and Manufacturing Engineering, Faculty of Engineering, Universiti Putra Malaysia.

REFERENCES

Abdullah, A. H. D., Fikriyyah, A. K., Putri, O. D., & Puspa Asri, P. P. (2019). Fabrication and characterization of poly lactic acid (PLA)-starch based bioplastic composites. *IOP Conference Series: Materials Science and Engineering*, 553(1). https://doi.org/10.1088/1757-899X/553/1/012052

Auras, R., Harte, B., & Selke, S. (2004). An overview of polylactides as packaging materials. *Macromolecular Bioscience*, 4(9), 835–864. https://doi.org/10.1002/mabi.200400043

Čolnik, M., Hrnčič, M. K., Škerget, M., & Knez, Ž. (2020). Biodegradable polymers, current trends of research and their applications, a review. *Chemical Industry and Chemical Engineering Quarterly*, 26(4), 401–418. https://doi.org/10.2298/CICEQ191210018C

Domenek, S., & Ducruet, V. (2016). Characteristics and applications of PLA. *Biodegradable and Biobased Polymers for Environmental and Biomedical Applications*, 171–224. https://doi.org/10.1002/9781119117360.ch6

Ebrahimi, F., & Ramezani Dana, H. (2022). Poly lactic acid (PLA) polymers: from properties to biomedical applications. *International Journal of Polymeric Materials and Polymeric Biomaterials*, 71(15), 1117–1130. https://doi.org/10.1080/00914037.2021.1944140

Eutionnat-Diffo, P. A., Chen, Y., Guan, J., Cayla, A., Campagne, C., Zeng, X., & Nierstrasz, V. (2019). Stress, strain and deformation of poly-lactic acid filament deposited onto polyethylene terephthalate woven fabric through 3D printing process. *Scientific Reports*, 9(1), 1–18. https://doi.org/10.1038/s41598-019-50832-7

Finglas, P., Yada, R., Schubert, H., Ax, K., Behrend, O., Brien, J. O., Vitapole, D., Buckle, K., Jelen, P., & Knorr, D. (2003). Review of possible applications of nanosecond electron beams for sterilization in industrial poultry farming. 14(1), 11431.

Fukushima, K., Murariu, M., Camino, G., & Dubois, P. (2010). Effect of expanded graphite/layered-silicate clay on thermal, mechanical and fire retardant properties of poly(lactic acid). *Polymer Degradation and Stability*, 95(6), 1063–1076. https://doi.org/10.1016/j.polymdegradstab.2010.02.029

Groot, W., Van Krieken, J., Sliekersl, O., & De Vos, S. (2010). Production and purification of lactic acid and lactide. In *Poly(Lactic Acid): Synthesis, Structures, Properties, Processing, and Applications*, pp. 3–18. https://doi.org/10.1002/9780470649848.ch1

Gupta, B., Revagade, N., & Hilborn, J. (2007). Poly(lactic acid) fiber: An overview. *Progress in Polymer Science (Oxford)*, 32(4), 455–482. https://doi.org/10.1016/j.progpolymsci.2007.01.005

Hajibeygi, M., & Shafiei-Navid, S. (2019). Design and preparation of poly(lactic acid) hydroxyapatite nanocomposites reinforced with phosphorus-based organic additive: Thermal, combustion, and mechanical properties studies. *Polymers for Advanced Technologies*, 30(9), 2233–2249. https://doi.org/10.1002/pat.4652

Hamad, K., Kaseem, M., Yang, H. W., Deri, F., & Ko, Y. G. (2015). Properties and medical applications of polylactic acid: A review. *Express Polymer Letters*, 9(5), 435–455. https://doi.org/10.3144/expresspolymlett.2015.42

Haryńska, A., Janik, H., Sienkiewicz, M., Mikolaszek, B., & Kucińska-Lipka, J. (2021). PLA-potato thermoplastic starch filament as a sustainable alternative to the conventional PLA filament: Processing, characterization, and FFF 3D printing. *ACS Sustainable Chemistry and Engineering*, 9(20), 6923–6938. https://doi.org/10.1021/acssuschemeng.0c09413

Haugaard, V. K., Danielsen, B., & Bertelsen, G. (2003). Impact of polylactate and poly(hydroxybutyrate) on food quality. *European Food Research and Technology*, 216(3), 233–240. https://doi.org/10.1007/s00217-002-0651-6

Hughes, J., Thomas, R., Byun, Y., & Whiteside, S. (2012). Improved flexibility of thermally stable poly-lactic acid (PLA). *Carbohydrate Polymers*, 88(1), 165–172. https://doi.org/10.1016/j.carbpol.2011.11.078

Jakubowicz, I. (2003). Evaluation of degradability of biodegradable polyethylene (PE). *Polymer Degradation and Stability*, 80(1), 39–43. https://doi.org/10.1016/S0141-3910(02)00380-4

Je, J. Y., & Kim, S. K. (2006). Antioxidant activity of novel chitin derivative. *Bioorganic and Medicinal Chemistry Letters*, 16(7), 1884–1887. https://doi.org/10.1016/j.bmcl.2005.12.077

Kanabenja, W., Passarapark, K., Subchokpool, T., Nawaaukkaratharnant, N., Román, A. J., Osswald, T. A., Aumnate, C., & Potiyaraj, P. (2022). 3D printing filaments from plasticized Polyhydroxybutyrate/Polylactic acid blends reinforced with hydroxyapatite. *Additive Manufacturing*, 59(September). https://doi.org/10.1016/j.addma.2022.103130

Kim, B. K., Seo, J. W., & Jeong, H. M. (2003). Morphology and properties of waterborne polyurethane/clay nanocomposites. *European Polymer Journal*, 39(1), 85–91. https://doi.org/10.1016/S0014-3057(02)00173-8

Lasprilla, A. J. R., Martinez, G. A. R., Lunelli, B. H., Jardini, A. L., & Filho, R. M. (2012). Poly-lactic acid synthesis for application in biomedical devices – A review. *Biotechnology Advances*, 30(1), 321–328. https://doi.org/10.1016/j.biotechadv.2011.06.019

Liao, Y., Wang, Q., Xia, H., Xu, X., Baxter, S. M., Slone, R. V., Wu, S., Swift, G., & Westmoreland, D. G. (2001). New polymer syntheses. CIX. Biodegradable, alternating copolyesters of terephthalic acid, aliphatic dicarboxylic acids, and alkane diols. *Journal of Polymer Science, Part A: Polymer Chemistry*, 39(19), 3371–3382. https://doi.org/10.1002/pola.1320

Lim, L. T., Auras, R., & Rubino, M. (2008). Processing technologies for poly(lactic acid). *Progress in Polymer Science (Oxford)*, 33(8), 820–852. https://doi.org/10.1016/j.progpolymsci.2008.05.004

Lv, X., Wang, S., Shan, P., Zhao, Y., & Zuo, L. (2022). A machine learning based method for automatic differential scanning calorimetry signal analysis. *Measurement: Journal of the International Measurement Confederation*, 187(August 2021), 110218. https://doi.org/10.1016/j.measurement.2021.110218

Madhavan, N. K., Nair, N. R., & John, R. P. (2010). An overview of the recent developments in polylactide (PLA) research. *Bioresource Technology*, 101(22), 8493–8501. https://doi.org/10.1016/j.biortech.2010.05.092

Mecking, S. (2004). Nature or petrochemistry? – Biologically degradable materials. *Angewandte Chemie – International Edition*, 43(9), 1078–1085. https://doi.org/10.1002/anie.200301655

Mei, H., Yin, X., Zhang, J., & Zhao, W. (2019). Compressive properties of 3D printed polylactic acid matrix composites reinforced by short fibers and SiC nanowires. *Advanced Engineering Materials*, 21(5). https://doi.org/10.1002/adem.201800539

Micó-Vicent, B., Perales, E., Huraibat, K., Martínez-Verdú, F. M., & Viqueira, V. (2019). Maximization of FDM-3D-objects gonio-appearance effects using PLA and ABS filaments and combining several printing parameters: "A case study." *Materials*, 12(9). https://doi.org/10.3390/ma12091423

Mogan, J., Harun, W. S. W., Kadirgama, K., Ramasamy, D., Foudzi, F. M., Sulong, A. B., Tarlochan, F., & Ahmad, F. (2023). Fused deposition modelling of polymer composite: A progress. *Polymers*, 15, 28. https://doi.org/10.3390/polym15010028

Moravej, M., & Mantovani, D. (2011). Biodegradable metals for cardiovascular stent application: Interests and new opportunities. *International Journal of Molecular Sciences*, 12(7), 4250–4270. https://doi.org/10.3390/ijms12074250

Muller, J., González-Martínez, C., & Chiralt, A. (2017). Combination of poly(lactic) acid and starch for biodegradable food packaging. *Materials*, 10(8), 1–22. https://doi.org/10.3390/ma10080952

Murariu, M., Dechief, A. L., Bonnaud, L., Paint, Y., Gallos, A., Fontaine, G., Bourbigot, S., & Dubois, P. (2010). The production and properties of polylactide composites filled with expanded graphite. *Polymer Degradation and Stability*, 95(5), 889–900. https://doi.org/10.1016/j.polymdegradstab.2009.12.019

Murariu, M., & Dubois, P. (2016). PLA composites: From production to properties. *Advanced Drug Delivery Reviews*, 107, 17–46. https://doi.org/10.1016/j.addr.2016.04.003

Nair, L. S., & Laurencin, C. T. (2007). Biodegradable polymers as biomaterials. *Progress in Polymer Science*, 32(8–9), 762–798. https://doi.org/10.1016/j.progpolymsci.2007.05.017

Narayanan, G., Vernekar, V. N., Kuyinu, E. L., & Laurencin, C. T. (2016). Poly (lactic acid)-based biomaterials for orthopaedic regenerative engineering. *Advanced Drug Delivery Reviews*, 107, 247–276. https://doi.org/10.1016/j.addr.2016.04.015

Narayanan, T. S., Park, I., & Lee, M. (2015). *Surface Modification of Magnesium and its Alloys for Biomedical Applications: Modification and Coating Techniques* (Woodhead Publishing Series in Biomaterials) (1st ed.). Woodhead Publishing.

Navarro, M., Ginebra, M. P., Planell, J. A., Zeppetelli, S., & Ambrosio, L. (2004). Development and cell response of a new biodegradable composite scaffold for guided bone regeneration. *Journal of Materials Science: Materials in Medicine*, 15(4), 419–422. https://doi.org/10.1023/B:JMSM.0000021113.88702.9d

Netravali, A. N., & Chabba, S. (2003). Composites get greener. *Materials Today*, 6(4), 22–29. https://doi.org/10.1016/S1369-7021(03)00427-9

Oksman, K., Skrifvars, M., & Selin, J. F. (2003). Natural fibres as reinforcement in polylactic acid (PLA) composites. *Composites Science and Technology*, 63(9), 1317–1324. https://doi.org/10.1016/S0266-3538(03)00103-9

Olaiya, N. G., Nuryawan, A., Oke, P. K., Khalil, H. P. S. A., Rizal, S., Mogaji, P. B., Sadiku, E. R., Suprakas, S. R., Farayibi, P. K., Ojijo, V., & Paridah, M. T. (2020). The role of two-step blending in the properties of starch/chitin/polylactic acid biodegradable composites for biomedical applications. *Polymers*, 12(3). https://doi.org/10.3390/polym12030592

Orozco, V. H., Brostow, W., Chonkaew, W., & López, B. L. (2009). Preparation and characterization of poly(lactic acid)-G-maleic anhydride + starch blends. *Macromolecular Symposia*, 277(1), 69–80. https://doi.org/10.1002/masy.200950309

Ouchiar, S., Stoclet, G., Cabaret, C., Georges, E., Smith, A., Martias, C., Addad, A., & Gloaguen, V. (2015). Comparison of the influence of talc and kaolinite as inorganic fillers on morphology, structure and thermomechanical properties of polylactide based composites. *Applied Clay Science*, 116–117, 231–240. https://doi.org/10.1016/j.clay.2015.03.020

Palaniyappan, S., & Sivakumar, N. K. (2023). Development of crab shell particle reinforced polylactic acid filaments for 3D printing application. *Materials Letters*, 341, 134257. https://doi.org/10.1016/j.matlet.2023.134257.

Pawar, R. P., Tekale, S. U., Shisodia, S. U., Totre, J. T., & Domb, A. J. (2014). Biomedical applications of poly(lactic acid). *Recent Patents on Regenerative Medicine*, 4(1), 40–51. https://doi.org/10.2174/2210296504666140402235024

Peças, P., Carvalho, H., Salman, H., & Leite, M. (2018). Natural fibre composites and their applications: A review. *Journal of Composites Science*, 2(4), 1–20. https://doi.org/10.3390/jcs2040066

Petchwattana, N., Covavisaruch, S., & Petthai, S. (2014). Influence of talc particle size and content on crystallization behavior, mechanical properties and morphology of poly(lactic acid). *Polymer Bulletin*, 71(8), 1947–1959. https://doi.org/10.1007/s00289-014-1165-7

Phetwarotai, W., Potiyaraj, P., & Aht-Ong, D. (2012). Characteristics of biodegradable polylactide/gelatinized starch films: Effects of starch, plasticizer, and compatibilizer. *Journal of Applied Polymer Science*, 0(0), 1–11. https://doi.org/10.1002/app

Qi, F., Wu, J., Li, H., & Ma, G. (2019). Recent research and development of PLGA/PLA microspheres/nanoparticles: A review in scientific and industrial aspects. *Frontiers of Chemical Science and Engineering*, 13(1), 14–27. https://doi.org/10.1007/s11705-018-1729-4

Rasal, R. M., Janorkar, A. V., & Hirt, D. E. (2010). Poly(lactic acid) modifications. *Progress in Polymer Science (Oxford)*, 35(3), 338–356. https://doi.org/10.1016/j.progpolymsci.2009.12.003

Rhim, J. W., Hong, S. I., & Ha, C. S. (2009). Tensile, water vapor barrier and antimicrobial properties of PLA/nanoclay composite films. *LWT*, 42(2), 612–617. https://doi.org/10.1016/j.lwt.2008.02.015

Sadeghi, I., Byrne, J., Shakur, R., & Langer, R. (2021). Engineered drug delivery devices to address Global Health challenges. *Journal of Controlled Release*, 331(September 2020), 503–514. https://doi.org/10.1016/j.jconrel.2021.01.035

Shakoor, A., & Thomas N. L. (2014). Talc as a nucleating agent and reinforcing filler in poly(lactic acid) composites. *Polymer Engineering and Science*, 54 (1), 65–70. https://doi.org/10.1002/pen.23543

Shen, L., Haufe, J., & Patel, M. K. (2009). Product overview and market projection of emerging bio-based plastics. PRO-BIP; Final Report, Report No: NWS-E-2009-32. 243.

Süfer, Ö. (2017). Poly (lactic acid) films in food packaging systems. *Food Science & Nutrition Technology*, 2(4). https://doi.org/10.23880/fsnt-16000131

Sullivan, E. M., Oh, Y. J., Gerhardt, R. A., Wang, B., & Kalaitzidou, K. (2014). Understanding the effect of polymer crystallinity on the electrical conductivity of exfoliated graphite nanoplatelet/polylactic acid composite films. *Journal of Polymer Research*, 21(10), 1–9. https://doi.org/10.1007/s10965-014-0563-8

Tao, Y., Liu, M., Han, W., & Li, P. (2021). Waste office paper filled polylactic acid composite filaments for 3D printing. *Composites Part B*, 221(July 2020), 108998. https://doi.org/10.1016/j.compositesb.2021.108998

Tazibt, N., Kaci, M., Dehouche, N., Ragoubi, M., & Atanase, L. I. (2023). Effect of filler content on the morphology and physical properties of poly(lactic acid)-hydroxyapatite composites. *Materials*, 16, 809. https://doi.org/10.3390/ma16020809

Tümer, E. H., & Erbil, H. Y. (2021). Extrusion-based 3D printing applications of PLA composites: A review. *Coatings*, 11, 1–42. https://doi.org/https://doi.org/10.3390/coatings11040390

Turalija, M., Bischof, S., Budimir, A., & Gaan, S. (2016). Antimicrobial PLA films from environment friendly additives. *Composites Part B: Engineering*, 102, 94–99. https://doi.org/10.1016/j.compositesb.2016.07.017

Tyler, B., Gullotti, D., Mangraviti, A., Utsuki, T., & Brem, H. (2016). Polylactic acid (PLA) controlled delivery carriers for biomedical applications. *Advanced Drug Delivery Reviews*, 107, 163–175. https://doi.org/10.1016/j.addr.2016.06.018

Vroman, I., & Tighzert, L. (2009). Biodegradable polymers. *Materials*, 2(2), 307–344. https://doi.org/10.3390/ma2020307

Wan, Y., Wu, C., Xiong, G., Zuo, G., Jin, J., Ren, K., Zhu, Y., Wang, Z., & Luo, H. (2015). Mechanical properties and cytotoxicity of nanoplate-like hydroxyapatite/polylactide nanocomposites prepared by intercalation technique. *Journal of the Mechanical Behavior of Biomedical Materials*, 47, 29–37. https://doi.org/10.1016/j.jmbbm.2015.03.009

Wang, H., Ji, J., Zhang, W., Zhang, Y., Jiang, J., Wu, Z., Pu, S., & Chu, P. K. (2009). Biocompatibility and bioactivity of plasma-treated biodegradable poly(butylene succinate). *Acta Biomaterialia*, 5(1), 279–287. https://doi.org/10.1016/j.actbio.2008.07.017

Wang, J., Wang, M., Chen, F., Wei, Y., Chen, X., Zhou, Y., Yang, X., Zhu, X., Tu, C., & Zhang, X. (2019). Nano-hydroxyapatite coating promotes porous calcium phosphate ceramic-induced osteogenesis via BMP/SMAD signaling pathway. *International Journal of Nanomedicine*, 14, 7987–8000. https://doi.org/10.2147/IJN.S216182

Wang, X., Jiang, M., Zhou, Z., Gou, J., & Hui, D. (2017). 3D printing of polymer matrix composites: A review and prospective. *Composites Part B Engineering*, 110, 442–458. https://doi.org/10.1016/j.compositesb.2016.11.034

Wilson, O. C., Gugssa, A., Mehl, P., & Anderson, W. (2012). An initial assessment of the biocompatibility of crab shell for bone tissue engineering. *Materials Science and Engineering: C*, 32 (2), 78–82. https://doi.org/10.1016/j.msec.2011.06.012

Wokadala, O. C., Emmambux, N. M., & Ray, S. S. (2014). Inducing PLA/starch compatibility through butyl-etherification of waxy and high amylose starch. *Carbohydrate Polymers*, 112, 216–224. https://doi.org/10.1016/j.carbpol.2014.05.095

Xiao, L., Wang, B., Yang, G., & Gauthier, M. (2006). Poly(lactic acid)-based biomaterials: Synthesis, modification and applications. 248–282.

Xiong, Z., Li, C., Ma, S., Feng, J., Yang, Y., Zhang, R., & Zhu, J. (2013). The properties of poly(lactic acid)/starch blends with a functionalized plant oil: Tung oil anhydride. *Carbohydrate Polymers*, 95(1), 77–84. https://doi.org/10.1016/j.carbpol.2013.02.054

Yan, N., & Chen, X. (2015). Sustainability: Don't waste seafood waste. *Nature*, 524, 155–157. https://doi.org/10.1038/524155a

Yang, F., Ye, X., Zhong, J., Lin, Z., Wu, S., Hu, Y., Zheng, W., Zhou, W., Wei, Y., & Dong, X. (2023). Recycling of waste crab shells into reinforced poly (lactic acid) biocomposites for 3D printing. *International Journal of Biological Macromolecules*, 234, 122974. https://doi.org/10.1016/j.ijbiomac.2022.12.193.

Yang, W., Fortunati, E., Dominici, F., Giovanale, G., Mazzaglia, A., Balestra, G. M., Kenny, J. M., & Puglia, D. (2016). Synergic effect of cellulose and lignin nanostructures in PLA based systems for food antibacterial packaging. *European Polymer Journal*, 79, 1–12. https://doi.org/10.1016/j.eurpolymj.2016.04.003

Yang, Z., Peng, H., Wang, W., & Liu, T. (2010). Crystallization behavior of poly(ε-caprolactone)/layered double hydroxide nanocomposites. *Journal of Applied Polymer Science*, 116(5), 2658–2667. https://doi.org/10.1002/app

6 An Overview of the Compression and Flexural Behaviours of Sandwich Composite Structure with *3D-Printed Core*

Nur Ainin F., Azaman M.D.,
Abdul Majid M.S., and Ridzuan M.J.M.

6.1 INTRODUCTION

Rapid prototyping of structural parts has expanded in recent years, as it involves swiftly producing a prototype to evaluate a part or specific part attributes visually and operationally. Direct model fabrication from CAD model data is accomplished using a variety of manufacturing procedures. A more modern manufacturing technique called additive manufacturing (AM), commonly known as 3D printing, creates prototypes of structural parts using different materials and intricate geometries. Three-dimensional printing has recently become a vital element of the production processes of the automobile industry, architecture, in addition to the medical industry. The biggest visible advancement, however, has probably been in the realm of aviation and space exploration (Kristína et al., 2021). The fabrication of airplane parts is frequently challenging and time-consuming and follows strict standards (Singamneni et al., 2019). By eliminating the need for complex tooling and specialised work-tool motions and the forces involved in between, direct digital manufacturing of complex 3D forms allows for shorter manufacturing lead times and possibly localised production, which aids in the development of more efficient supply chain systems (Guo et al., 2018; Vishnu Prashant Reddy et al., 2018).

AM has garnered much interest in the fabrication of aerospace applications such as unmanned aerial vehicles (UAVs) as a wide range of products in the forms of prototypes and final parts which are bonded to the low-weight and high structural strength necessity (Galatas et al., 2018; Ibadaddin et al., 2018). These lightweight UAVs are more desirable as they have better performance in terms of shorter take-off range and longer flight endurance (Goh et al., 2017). The developments in materials technology

DOI: 10.1201/9781003362128-6

made it easy to access and purchase composite materials. With the advances in composite manufacturing technology very complex shaped parts can be built effortlessly within a few days. The composite materials that are most frequently used to create UAVs' fuselage, wings, and landing gear are those reinforced polymers with carbon fibres (CFRP), reinforced polymers with fibreglass (GFRP), boron, and aramid fibres (Anand & Mishra, 2022). The combination of AM techniques and composite materials will enable the development of more effective structural components for UAVs. Also, the orientation of the reinforcement material and volume fraction of the matrix and reinforcement play an important role in determining the final weight and performance of the UAV; this leads to more research work in the field of composites (Ibadaddin et al., 2018). The mechanical behaviour of composite structures has been extensively studied through the use of 3D-printed samples subjected to a variety of experimental tests, such as an impact test, a compression test, and a flexural test. This is because, for aircraft applications, the desired performance may involve a high strength-to-weight ratio, excellent fatigue resistance, a low coefficient of thermal expansion, and a high resistance to environmental factors like moisture and temperature. Thus, much research focuses on energy absorption (EA) capacities and failure mechanisms since a material's or structure's EA capacity directly affects its ability to deform and absorb energy during loading (Krzyzak et al., 2016; Peng et al., 2021; Soliman et al., 2016; Townsend et al., 2020; Xiao et al., 2022). Deformation of a material absorbs and dissipates energy throughout the structure. This EA contributes to the prevention of unexpected and catastrophic collapse.

The structural performance of sandwich composite structures has been the subject of extensive research during the past few years. This review study seeks to critically analyse and explore the earlier research on sandwich composite structures, notably with 3D-printed core energy absorbers, done by eminent experts. This study provides innovative AM technologies relevant to sandwich composite structures' core architecture. Additionally, an intriguing aspect of core fabrication procedures using AM is the numerous kinds of material feedstocks. The review also describes the fundamental ideas behind 3D-printed hexagonal honeycomb core structure and fabrication of sandwich composite structures. The next section explores more regarding the investigation of the EA and typical failure mode analysis by experimentally identifying specific phenomena on the sandwich composite structure, such as the types of complete quasi-static loading conditions (flatwise compression test, in-plane compression test, and flexural test). An overview of the relevant prior works of literature will then be covered.

6.2   AM TECHNOLOGY

With AM, it is possible to directly convert raw materials into complex 3D forms based on digital data obtained by slicing and rasterising CAD files, which results in considerable time and cost reductions when the technique applies to manufacturing end-use parts using different materials (e.g., thermoplastics, thermoset resins, fibre-reinforced plastics, etc.) (Menegozzo et al., 2022; Sara Black, 2015; Singamneni et al., 2019). Several types of 3D printing techniques are available, including stereolithography (SLA), fused deposition modelling (FDM), laminated object manufacturing

(LOM), digital light processing (DLP), selective laser sintering (SLS) for plastic, and selective laser melting (SLM) for metal powders. FDM is among the most popular processes when it comes to affordability, decreased waste output, the potential for recycling, and user-friendliness (Chacón et al., 2019). Due to its extremely stream-lined manufacturing process, this technology continues to be very appealing for the manufacture of structures (Isaac & Duddeck, 2022). This method involves melting plastic filament at the nozzle into a semi-liquid state before extruding it and deposit-ing it along a path determined by a 3D virtual model and slicing orders (Wang et al., 2017). Figure 6.1 shows an outline for how the filament is heated and melted at the hot end of the nozzle before being deposited in layers.

The developments in AM manufacturing have made it possible to create archi-tected cellular cores with free-form 2D and 3D topologies, which were previously impossible to create using traditional sandwich structure manufacturing techniques. Recently, AM methods rapidly gained interest from potential applications in dif-ferent fields, including the aerospace industry. The ability of 3D printing when it comes to reducing aircraft weight, increasing the level of adaptation, and overall construction efficiency poses new challenges for the further potential development of air transport (Kloski et al., 2017). The production and preliminary structural analysis of sandwich configurations are widely used as these new lamination schemes could lead to an important weight reduction without significant decreases in mechanical properties. Therefore, it could be possible, for the designed application (e.g., a mul-tifunctional small UAV produced via FDM), to have stiffener and lighter structures easy to be manufactured with a low-cost 3D printer (Brischetto et al., 2018).

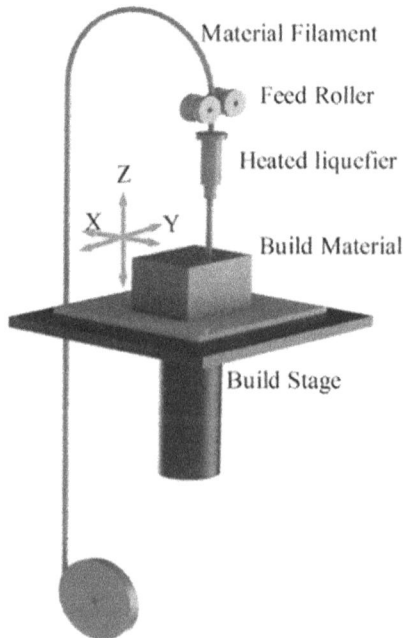

FIGURE 6.1   The mechanism of FDM (Klippstein et al., 2017).

6.3 MATERIALS FOR 3D-PRINTED CORE STRUCTURES

Recently, new materials have been developed to enhance core performance. The categorisation of 3D printing materials covered in this work comprises metals and alloys, polymeric and composite materials, and their hybrids. The AM technique is driving the broad adoption of these core materials (Stocchi et al., 2014), owing to their density and versatility. However, many materials may be used for industrial 3D printing. All these components are unique in their characteristics, advantages, and drawbacks. Furthermore, essential factors like material selection, texture, cost, etc., must be considered to prevent mistakes in 3D printing. In some cases, choosing an appropriate material for the task can be extremely challenging (A. Chen, 2020). Before selecting the appropriate material for their duties, designers and engineers must be aware of the qualities of the materials used for application. In this part, Table 6.1 provides a summary of the most recent previous research on the various materials used for 3D-printed core structures.

6.4 3D-PRINTED HONEYCOMB CORE STRUCTURE

A unique architectural core with periodic geometries that are lightweight and 3D printable has recently been revealed. There are four distinct categories of core structure: corrugated, cellular foams, honeycomb, and balsa wood. Despite the fact that many different types of core structures have been used, the honeycomb core is currently the most popular choice for secondary structural applications in the aerospace industry. These applications include rudders, ailerons, spoilers, and flaps (Chantarapanich et al., 2014; Pollard et al., 2017). The characteristics of honeycomb cores have been thoroughly studied in recent years (Brischetto et al., 2018; Naidu & Kumar, 2018; Pollard et al., 2017; Tanjung et al., 2018; Yan et al., 2020, 2022). Inspired by the honeycombs of bees, honeycomb structures have found widespread use in many different areas (Naidu & Kumar, 2018; Wang, 2019; Zhang et al., 2015) due to their high specific strength and toughness as well as efficient EA. Honeycombs are an example of a specialised cellular structure in which a grid of hollow cells with a specified geometry (usually hexagonal) is produced between thin walls (Araújo et al., 2019; Bitzer, 1997). As can be seen in Figure 6.2(a), there are a variety of honeycomb core configurations that can be used. These include the hexagon, reinforce hexagon, rectangle, flex-core, and square cell. A real honeycomb often has the strongest hexagonal shape out of all of them (Allardyce et al., n.d.). Figure 6.2(b and c) exhibits the recent research into honeycomb designs such as hierarchical and graded, respectively.

One of the most important qualities of the honeycomb sandwich core is its resistance to compression. As a result of its efficient hexagonal shape, where walls support each other, honeycomb cores often have a better compressive strength (at the same weight) than alternative sandwich core constructions like foam or corrugated cores. In comparison to the traditional subtractive approach, AM offers significant promise for the fabrication of honeycomb cores. Using this technique, a working prototype can be fabricated without resorting to adhesive binding, and complex structures can be mass-produced with relative ease. This allows the 3D-printed honeycomb to be widely implemented (Chantarapanich et al., 2014).

TABLE 6.1

Common materials used in designing 3D-printed core structures from the latest research

Researcher and year	AM	3D-printed core structure	Material
Garrido Silva et al. (2022)	FDM	1. Hexagonal honeycomb 2. Lotus 3. Hexagonal honeycomb with plateau borders	1. Polylactic acid (PLA) 2. Aluminium
Wang et al. (2022)	SLS	Re-entrant auxetic	Polyamide 12 + 30% glass fibre
Zhang et al. (2022)	SLM	Truncated-square honeycomb reinforced Hollow Pyramidal lattice sandwich structure (TSH-HP)	Nickel-based superalloys
Yan et al. (2022)	SLM	Honeycomb	Aluminium AlSi10Mg
Xiao et al. (2022)	SLM	Lattice	316 L stainless steel
Li et al. (2022)	SLS	Corrugated lattice	Nylon PA220
Subramaniyan et al. (2022)	FDM	1. Solid rectangular 2. Solid circular cross section	PLA
Gebrehiwot et al. (2021)	FDM	1. Triangular 2. Honeycomb 3. Wiggle 4. Diamond 5. Square	PLA
Smardzewski et al. (2021)	FDM	Auxetic lattice	Wood-based
Acanfora et al. (2021)	SLM	Lattice	Aluminium AlSi10Mg
Usta et al. (2021)	FDM	Honeycomb	PLA
Santos et al. (2021)	FDM	1. Honeycomb 2. Auxetic	1. PLA 2. Polyethylene terephthalate glycol (PETg)
Peng et al. (2021)	FDM	Triply periodic minimal surface (TPMOS)	Acrylonitrile butadiene styrene (ABS)
Meng et al. (2020)	SLA	Lattice	Epoxy resin SPR600B

(Continued)

TABLE 6.1 (*Continued*)
Common materials used in designing 3D-printed core structures from the latest research

Researcher and year	AM	3D-printed core structure	Material
Antony et al. (2020)	FDM	Honeycomb	PLA + 20–25% Hemp fibre
Ayrilmis et al. (2021)	FDM	Honeycomb	PLA + 30–40% Wood
Özen et al. (2020)	FDM	1. Hexagonal honeycomb 2. Re-entrant honeycomb	ABS
Taghvaei (2020)	DLP	1. Honeycomb 2. Rectangular	DA-2 + 0.7 wt% PPO + 0.025 wt% Curcumin
H. Chen (2020)	SLA	Lattice	Aluminium alloy (AA) 1060-O
Dong et al. (2020)	FDM	Diamond	PLA
Essassi et al. (2020)	FDM	Re-entrant honeycomb	PLA + flax fibre
Wang et al. (2020)	SLA	Lattice	Photopolymer DSM Somos 14120
Yan et al. (2020)	FDM	Honeycomb	PLA
Kabir et al. (2020)	FDM	1. Solid 2. Triangular 3. Honeycomb Rectangular	Nylon
Chacón et al. (2019)	FDM	Solid rectangular	Nylon
Tao et al. (2019)	SLA	Honeycomb	Photopolymer VeroWhitePlus
Araújo et al. (2019)	FDM	1. Hexagonal honeycomb 2. Lotus hexagonal honeycomb with plateau borders	1. PLA Aluminium

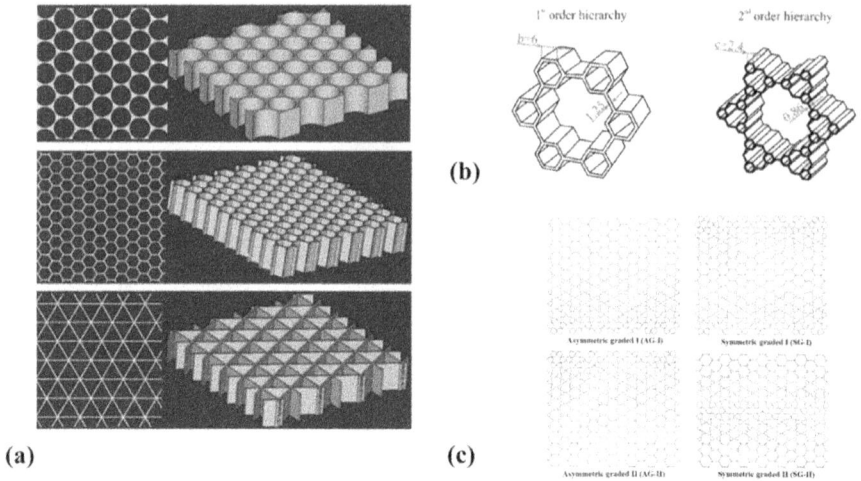

FIGURE 6.2 Honeycomb designs. (a) Typical (Yap & Yeong, 2015); (b) hierarchical (Mansour et al., 2019); (c) graded (Liu et al., 2021).

6.5 SANDWICH COMPOSITE STRUCTURE

In recent decades, modern engineering has embraced AM to use sandwich materials in lightweight constructions used in aviation components. In addition to its high mechanical strength, the sandwich composite construction's low weight is a crucial factor, as is frequently examined in earlier studies (Bharath et al., 2021a; Dikshit et al., 2016; Hou et al., 2018; Sarvestani, Akbarzadeh, Niknam, et al., 2018; Wang et al., 2019). Sandwich composite structures, also called sandwich panels, are commonly used in these applications because of their presence in secondary components including spoilers, floor panels, nacelles, and fairings (Foo, 2009). In the case of anisotropic materials, this means that their strength exhibits unique qualities in desirable directions, depending on the application (Krzyzak et al., 2016). As a result, the important requirements of these structures have encountered major challenges that must be solved, such as stiffness, strength, ability to absorb energy, and damage tolerance. Aeronautics material advancements can serve as a case study for the benefits of innovative sandwich composite structure research through in-depth field experiments into the reasons why some sandwich structure applications are successful while others are not, as well as the identification of specific problem structures and proposed fixes (Russell, 2018).

Figure 6.3 depicts a typical sandwich composite structure, which consists of two thin, stiff, and strong face sheets of metallic or fibre-reinforced composite material sandwiching a thick layer of light core material (such as honeycomb, foam, corrugated, etc.). The face sheets support the weight of the full sandwich constructions because of their ability to withstand in-plane and bending loads, while the core separates the face sheets and supports the transverse loads. Several composite material manufacturing procedures, such as vacuum bag processing, autoclave processing, compression moulding, filament winding, pultrusion, and braiding (Ratwani, 2010),

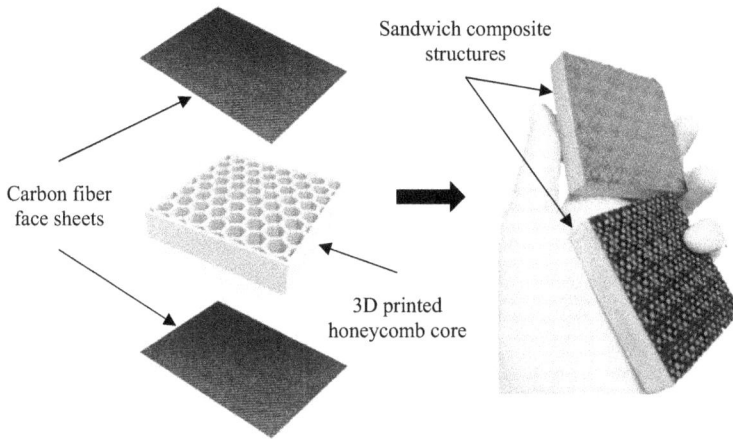

FIGURE 6.3 Typical construction of the sandwich composite structure.

can be used to apply a specific adhesive such as epoxy resin to the core and face sheets to create a bond between them. This combines all the components into a full sandwich composite structure. This leads to an effective, lightweight structure with high specific bending strengths, buckling resistances, and stiffness (Foo, 2009; Gibson & Ashby, 1997). Most of the time, these structures are made with several types of cores, face sheet materials, core materials, and adhesives. Each of these aspects significantly affects a sandwich structure's capabilities. The mechanical strengths of sandwich constructions are directly dependent on sandwich component properties and manufacturing procedures (Krzyzak et al., 2016).

6.6 TYPES OF QUASI-STATIC LOADINGS

A thorough understanding of several tests concentrating on the mechanical performance under various loading circumstances is necessary for the effective design of these structures. Compressive strength tests are the most popular quasi-static loadings used to evaluate the mechanical characteristics of sandwich composites (al Rifaie et al., 2018; Chen et al., 2021; Mansour et al., 2019; Subramaniyan et al., 2022) and flexural tests (Araújo et al., 2019; Bharath et al., 2021b; Gebrehiwot et al., 2021; Paczos et al., 2018), while the most common dynamic loadings are impact tests (Özen et al., 2020; Wang et al., 2019, 2022). From the testings, understanding the damage modes through damage identification, evaluation, and analysis is essential for the effective implementation of such structural components. Due to their resistance to impact loadings as well as compressive and flexural loadings, sandwich structures are beneficial for a range of structural applications in aviation. For a comprehensive quasi-static mechanical characterisation of the sandwich construction, tests were run on both the individual components and the complex structures as a whole to collect crucial comparing characteristics. According to the available literature, the American Society for Testing and Materials (ASTM) standards are the most widely used mechanical testing standard that specifies how to evaluate attributes of composite materials (Antony et al., 2020; Fotsing et al., 2016; Hou et al., 2014;

Lascano et al., 2021). AM as defined by the ISO/ASTM 52900 terminology standard is the process of joining materials to make parts from 3D model data. For quasi-static experiments, all the testing procedures will adhere to ASTM standards for consistency across the article.

6.6.1 FLATWISE COMPRESSION TEST

One of the most crucial experimental tests for elucidating the behaviour of composite materials is the compressive strength test, also known as flatwise or out-of-plane compression testing. Specifically, ASTM C365/C365 M-05 (*Standard Test Method for Flatwise Compressive Properties of Sandwich Cores*, n.d.) is followed for mechanical testing. A standard speed rate is 0.50 mm/min (0.020 in/min) (Intertek Group plc, n.d.-b). Flatwise compressive loading is measured in the same direction that the sandwich core will face during usage, as indicated in Figure 6.4(a), and sample dimensions are provided in Figure 6.4(b). Results for the ultimate flatwise compressive strength, stress at 2% deflection, and compressive chord modulus are computed. It is crucial to the design of sandwich panels once this is determined. Here is equation 1 of compressive modulus of elasticity (E_{fc}) for which the measurement taken from the elastic region and ultimate flatwise compressive strength (σ_{max}) in MPa can be calculated by equation 2 (Toygar et al., 2019).

$$\textit{Slope of proportional, } E_{fc} = \frac{\sigma_c}{\varepsilon_c} \tag{1}$$

where σ_c is the compressive stress in MPa and ε_c is the compressive strain (mm/mm).

$$\sigma_{max} = \frac{P_{max}}{A} \tag{2}$$

where P_max is the ultimate force prior to failure in N, and A is the cross-sectional area in mm².

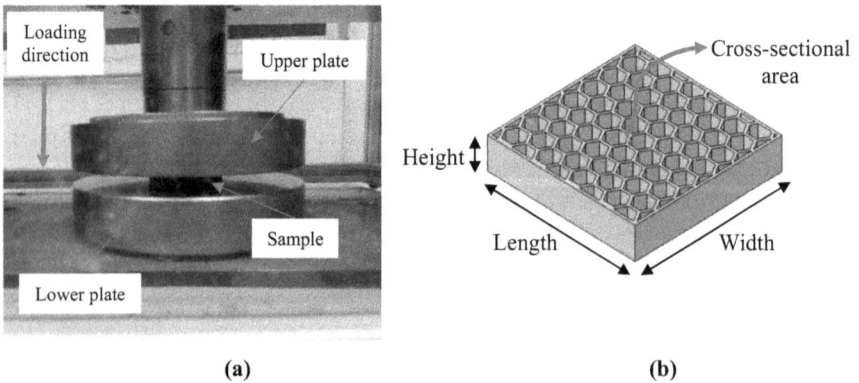

(a)   (b)

FIGURE 6.4   (a) Flatwise compression test and loading direction; (b) sample dimensions of 60 × 60 × 12 mm.

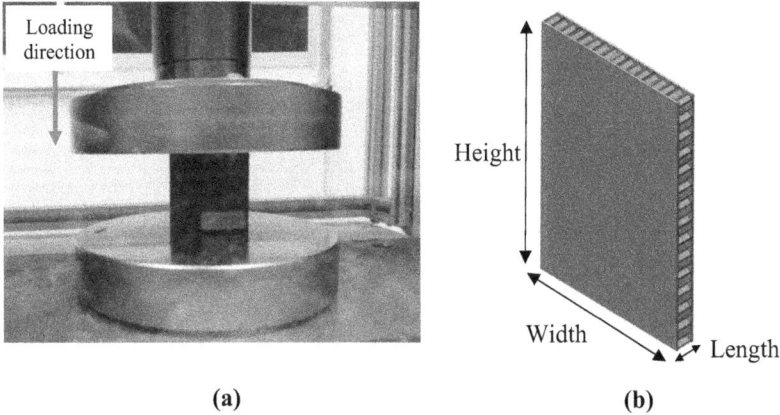

FIGURE 6.5   (a) In-plane compression test and loading direction; (b) sample dimensions of $12 \times 50 \times 50$ mm.

6.6.2 IN-PLANE COMPRESSION TEST

When performing in-plane or edgewise compression tests, ASTM C364/C364-16 specifies a speed of 0.50 mm/min [0.020 in/min] (ASTM International, 2016). Figure 6.5 depicts the results of a compressive strength test performed on a flat structural sandwich construction parallel to the sandwich facing plane and shows the sample dimensions. Compressive stresses are applied to the face sheets and the system's overall strength is evaluated by connecting the sheets. The job of the core is to keep them together to maximise the buckling capacity and lessen the bending effect. This direction of stresses on the sandwich helps to determine modulus data where the finding of compressive modulus of elasticity (E_{fc}) and the ultimate in-plane compressive strength, σ_{max} in MPa, where both formulas are the same as for previous flatwise testing.

6.6.3 FLEXURAL TEST

The procedure for flexural or three-point bending testing sandwich materials is described in ASTM C393/C393-00 (ASTM International, 2017). The standard speed is 6 mm/min [0.25 in/min]. A standard sample with dimensions of 76 mm in width by 200 mm in length and a support span of 150 mm [6.0 in] is required for the three-point arrangement of the standard loading (Intertek Group plc, n.d.-a). The sandwich panel's loading structure is shown in Figure 6.6, while the flexural testing conditions are shown in Figure 6.7 along with an illustration of the sample dimensions. Understanding various elements of material behaviour under straightforward beam loading is the goal of this testing. A complex interplay of forces, including tension, compression, and shear, is applied to the sample when it bends or flexes. It serves the purpose of assessing how materials respond to actual loading scenarios. The span of loading can be altered to measure various attributes. For instance, data for core shear

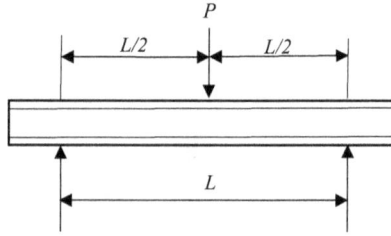

FIGURE 6.6   Flexural/three-point bending test and loading configuration.

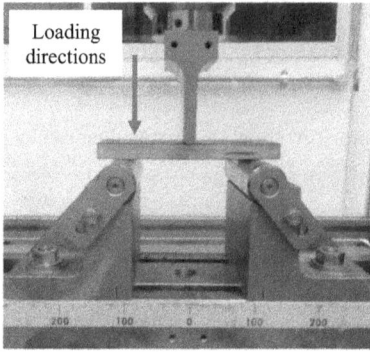

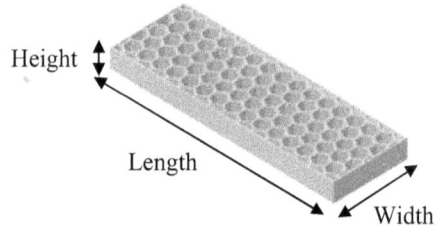

(a)   (b)

FIGURE 6.7   (a) Flexural test and loading direction; (b) sample dimensions of 150 × 50 × 12 mm.

characteristics are provided by a shorter loading span, whereas data for skin proper-ties are provided by a longer loading span.

Here P is the load at a given point on the load-deflection curve in N and L is the support span in mm.

The stresses are normally consistent across the entire component under com-pressive loading parallel to the axis of some homogeneous load-bearing part. Axial stress, or stress in the L direction, varies from one position in the beam to another when stresses are applied in bending normal to the component's main axis. The sur-face of the beam will be compressed on one side and under strain on the other. When compared to the surface compression and tension stress values, the stress through the thickness of the beam varies linearly, with zero stress at the neutral axis (Herrmann & Bucksch, 2014). Equation 3 (Toygar et al., 2019) can be used to determine this stress (σ) for any point along the load-deflection curve.

$$\sigma = \frac{3PL}{2wh^2} \tag{3}$$

where P is the load at a given point on the load-deflection curve in N, L is the support span in mm, w is the width of the sample in mm, and h is the height of sample in mm.

6.7 SIGNIFICANT DATA OBTAINED UNDER QUASI-STATIC LOADINGS

In order to determine structural performance and failure modes, researchers typically examine the mechanical responses of sandwich composites with 3D-printed core structures under quasi-static loading conditions. This section provides a thorough understanding of the general procedures used by most researchers. Numerous test performance indicators can be analysed before, during, and after the deformation process. The elastic limit, proportional limit, yield point, yield strength, and, for some materials, the compressive strength, and the elastic modulus (*What Is Compression Testing? – Instron*, n.d.) are all determined by calculating and plotting the material's fluctuating properties as a stress-strain diagram during testing. Since the area under a curve may be used to determine how much energy a structure has absorbed, these plots made it possible to describe the relationship between the structures' relative configurations in terms of their EA capacity. These stress-strain curves can be generated from the data collected throughout any mechanical test in which a force is applied to a material and prolonged measurements of stress and strain are taken simultaneously. The curve may change, though, depending on the type of structure, the rate of deformation, and the environment (Stanczak et al., 2019). If the structure is fully compressed, the curve's major components are typically visible. According to Figure 6.8, three typical stages can occur: the elastic, plateau, and densification stages.

The registered peak stress, which comes after the structure's linear elastic range of deformation, corresponds to the structure's maximum allowable load or plastic collapse stress. In a force-displacement diagram, this point, known as the yield point,

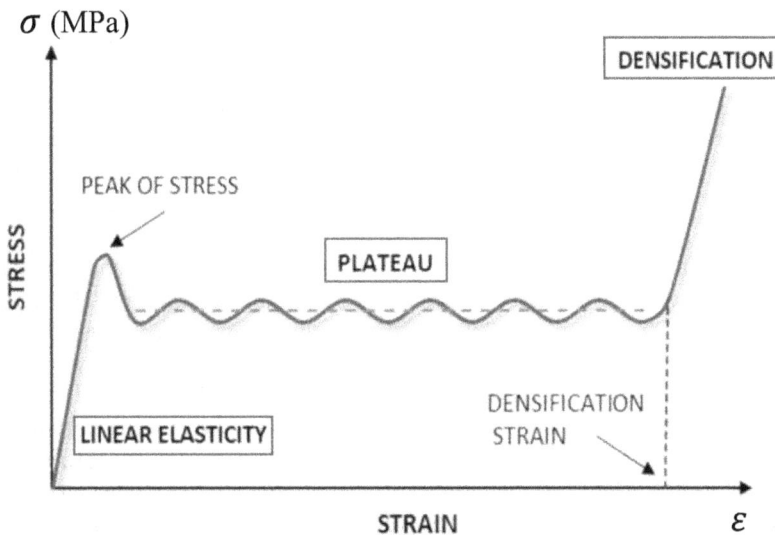

FIGURE 6.8 The schematic of stress-strain curve representing the response of a sandwich composite structure to a compressive load (Stanczak et al., 2019).

also provides the initial peak force or load, abbreviated F_p. The modulus of elasticity is a measure of a material's resistance to elastic, or temporary, deformation. It was used to characterise the stress-strain curve's slope along the straight line. In the field of elastics, the modulus is calculated by dividing the slope between two stress-strain points by the difference in stress and strain at those points. Up until the structure's densification, the next phase, also known as the plateau zone, takes place. The protected generalised load delivered to the structure, represented by the constant mean stress, can be used to approach the plateau regime. The primary process of energy dissipation in honeycombs is the formation of many flexible folds together with the structure height during the plateau regime (Stanczak et al., 2019). There is a large amount of EA at the plateau stress level because the underlying structure is not susceptible to additional compressive stress until the energy provided is adequate to enable the structure to densify. The sandwich element may fracture during this phase if the face sheet or core exceeds its maximum shear or compressive strength (Njim et al., 2021). The plateau's length dictates when the structure enters the densification phase, at which point additional substantial plastic deformation is no longer possible.

Several researchers were interested in the mechanical response of sandwich composites, particularly those having 3D-printed core structures. Prior research by Zhang et al. (2022) examined the impact of temperature on the out-of-plane compressive performance of a novel SLM-built truncated-square honeycomb reinforced hollow pyramidal lattice sandwich structure (TSH-HP). Compressive modulus, initial failure strength, compressive peak strength, and EA performance all decline by 46.2%, 46.9%, 42.3%, and 46.3%, respectively, as the temperature is increased from 25°C to 450°C. TSH-HP outperforms rival core topologies in compression performance at high temperatures. Gyroid-shaped green biocomposite panels with different face layer thicknesses (0.5, 1, 2, and 2.5 mm) were additively constructed and put through mechanical tests (bending properties, edgewise compressive strength, etc.) by Ayrilmis et al. (2020). Comparing composite samples with different surface layer ratios, it was shown that samples with a higher surface layer ratio had better bending strength, bending modulus, and compressive strength. Bates et al. (2019) investigated the effects of various grading methodologies on the compressive behaviour of 3D-printed flexible TPU honeycomb structures and found that the graded structures were 11 ± 3% more energy-absorbing overall than equivalent uniform structures at a compressive stress of 1 MPa.

Dikshit et al. (2016) developed and evaluated innovative core structures of vertical pillared sine wave corrugated (VPSC) and vertical pillared trapezoidal corrugated (VPTC) using 3D printing with flatwise compressive strength. According to the data, the compressive strength and modulus of VPSC structures are higher than those of VPTC structures by 16.6%, or 4.9 MPa. Li and Wang (2017) conducted three-point bending tests on 3D-printed truss, conventional honeycomb, and re-entrant honeycomb core materials. Flexural stiffness and maximum loading pressures are maximum for truss core sandwich composite structures, while bending deflection and minimum stiffness are greatest for re-entrant honeycomb core sandwich composite structures. Both Young's modulus and the stress level are reduced for the re-entrant honeycomb at the same deformation rate. Using SLM technology, Chantarapanich et al. (2014) investigated the possibility of developing and constructing a 3D honeycomb for use in aviation applications. The strength-to-weight ratio of the honeycomb is about 5.0×10^3 Nm/kg, and its elastic modulus is 63.18 MPa and its compressive strength is 1.1 MPa. An increase in beam thickness results in a structure that is notably more rigid since stiffness is a function of geometric parameters, in this case the 3D honeycomb structure's beam thickness.

6.8 ENERGY ABSORPTION

Discovering the EA capacity of sandwich composites with 3D-printed core structures allowed for an assessment of their performance, as presented here. Gibson and Ashby (1997) described this methodology for the first time to describe how cellular honeycomb structures absorb energy. EA is described by Nurul Fazita et al. (2018) as the region under the load-displacement curve, whereas Bates et al. (2016) state that it is the area under the stress-strain curve for a cellular structure under compression. The energy-absorbing capability of sandwich composite structures can be determined using both curves. Figure 6.9 illustrates, for instance, plots of the load versus displacement graph based on the conducted experiments. The slope of the first section of the historical plot was created by the local buckling effect and the friction between the sample's surfaces and the testing machine's grasp. After the consequences of buckling, a failure mechanism was immediately attained. The walls of the building cracked and sheared, which led to the collapse (Antolak-Dudka et al., 2019). The average load is acknowledged as one metric for measuring the absorbed energy capability based on the breath of area under the graph and the peak load value acquired in the initial portion of a quasi-static test. Moreover, the plateau stage

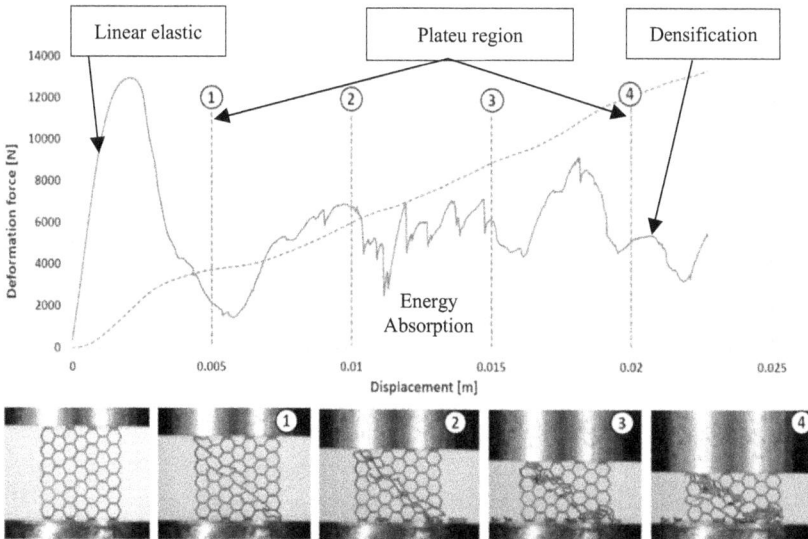

FIGURE 6.9 Schematic of the force-displacement curve of honeycomb core structure (Antolak-Dudka et al., 2019).

is extended and distinguished by brittle crushing, which makes it a highly efficient energy absorber design.

Numerous parameters, such as fibre and matrix characteristics, core architecture, types of materials utilised, and strain rate, control the EA capacity of composites, as demonstrated by the considerable research conducted by quasi-static testing (Habib et al., 2017; Townsend et al., 2020; Xiao et al., 2022). Xiao et al. (2022) constructed a hybrid lattice cell configuration comprising an octet cell and a rhombic dodecahedron (RD) cell that was fabricated using SLM to study the mechanical performance and EA during quasi-static compression trials. The hybrid structure has a smoother post-yielding response than the octet lattice and comparatively higher strength than the RD lattice, which contributes to its enhanced EA capacity. It is observed that the EA capabilities of hybrid lattice structures can be enhanced without sacrificing their load-bearing ability by adjusting their mesoscopic topologies, such as the ratio of each component. Tao et al. (2019) studied the in-plane mechanical characteristics and EA of square hierarchical honeycombs (SHHs) and a regular square honeycomb (RSH) manufactured using a commercial 3D printer. The results demonstrated that the cells of SHHs and RSH ruptured layer by layer along the loading direction, whereas SHHs exhibited more localised bands and damage. Compared to RSH of the same mass, SHHs exhibit superior compressive strength, specific EA, and crushing force efficiency.

In addition, Wang et al. (2020) analysed the compressive behaviour and EA of lattices such as uniform body-centred-cubic (U-BCC), graded body-centred-cubic (G-BCC), uniform body-centred-cubic with z-axis reinforcement (U-BCCz), and graded body-centred-cubic with z-axis reinforcement (G-BCCz) produced by the SLA process. The outcomes demonstrate that BCCz has a greater modulus and strength than BCC. Moreover, uniform lattices exhibit superior EA capabilities at small compression distances, whereas graded lattices absorb more energy at vast compression distances. Townsend et al. (2020) examined the effect of origami fold patterns on the EA behaviour of square 3D-printed TPU honeycombs. Increasing the severity of the fold angle was found to decrease the overall stiffness of the structure and produce quasi-rectangular absorption profiles, whereas increasing the number of folds was found to diminish the width of the stress plateau and produce quasi-linear absorption profiles. Experimentally obtained parameters of absorption efficiency as high as 0.49 rival those of rigid polyurethane foams. Applications requiring customised EA profiles have a great deal of potential with the structures discussed here. Chen et al. (2021) researched and characterised the negative Poisson's ratio (NPR) or enhanced effective elastic modulus (EEEM) by constructing the continuous carbon-reinforced PA (CCF/PA) and carbon fibre-reinforced PA (SCF/PA) produced by AM and subjected to in-plane compression. For metamaterials with NPR and composites with EEEM, the stiffness, peak force, EA, and specific energy absorption (SEA) of CCF/PA are greatly enhanced compared to SCF/PA.

6.9 FAILURE MODES OF THE 3D-PRINTED CORE STRUCTURE

The failure mode of a sandwich structure is the manner in which it fails to bear a specifically applied load. General failure mechanisms may arise based on the design and material of the face sheet and core structure (Ratwani, 2010). Composites entail catastrophic and progressive failure modes when the structure surpasses the yield strength. Thus, it is critical to understand how a sandwich construction can collapse in order to improve mechanical strength and prevent it from probable massive failure. On the other hand, it contributes to the development of a better structural design that can withstand higher loads while being lightweight. Table 6.2 summarises the typical failure mechanisms seen on 3D-printed sandwich composite structures subjected to static loadings in previous research.

TABLE 6.2

Failure mechanisms on 3D-printed sandwich composite structures

Types of failure modes	Researcher and year	Condition loadings	Images
Delamination	Harland et al. (2019)	Flexural	
Core shear	Brischetto et al. (2018)	Flexural	
Core cracking	Hou et al. (2013), Essassi et al. (2020)	Flexural	
Buckling	Sarvestani, Akbarzadeh, Mirbolghasemi, et al. (2018)	Flexural	

(*Continued*)

TABLE 6.2 (*Continued*)

Failure mechanisms on 3D-printed sandwich composite structures

Types of failure modes	Researcher and year	Condition loadings	Images
	Li et al. (2022), Tao et al. (2019), Bates et al. (2016)	In-plane	
	Lascano et al. (2021), Antony et al. (2020)	Flatwise	
Core fracture	Gautam et al. (2020)	Flatwise	
	Azzouz et al. (2019)	Flexural	

6.10   CONCLUSION

In conclusion, AM provides numerous benefits for the construction of sandwich composite structures. It is a versatile method for making lightweight and high-performance components due to its ability to construct complicated geometries and add functional features. Using AM in sandwich composite constructions has yielded positive results in several mechanical tests, including compression and flexural testing. One of the primary advantages of AM in sandwich composites is the ability to customise the EA capacities of the structure. By optimising the design and material composition, it is possible to generate structures with better EA properties, making them appropriate for applications requiring impact resistance or crashworthiness.

The research of failure modes in sandwich composite structures made by AM is essential for comprehending their mechanical behaviour and ensuring their dependability. There are numerous mechanisms of failure possible, including delamination between layers, core crushing, and buckling. Understanding these failure types enables the identification of crucial design elements and the creation of mitigation or prevention techniques. Overall, AM in sandwich composite constructions has the potential to produce components that are lightweight, robust, and energy-absorbing. Continued research and development in this sector will increase the understanding of the mechanical behaviour of these structures and allow for their successful implementation in industries such as UAVs.

ACKNOWLEDGMENT

This research did not receive any specific grant from funding agencies in the public, commercial, or not-for-profit sectors. The authors acknowledge the facilities support provided by Advanced Material Processing & Design, Faculty of Mechanical Engineering Technology, Universiti Malaysia Perlis.

REFERENCES

Acanfora, V., Saputo, S., Russo, A., & Riccio, A. (2021). A feasibility study on additive manufactured hybrid metal/composite shock absorbers. *Composite Structures*, *268*(March), 113958. https://doi.org/10.1016/j.compstruct.2021.113958

al Rifaie, M., Mian, A., & Srinivasan, R. (2018). Compression behavior of three-dimensional printed polymer lattice structures. *Proceedings of the Institution of Mechanical Engineers, Part L: Journal of Materials: Design and Applications*, *233*(8), 1574–1584. https://doi.org/10.1177/1464420718770475

Allardyce, B., De Souza, M., Jennings, M., & Naebe, M. (n.d.). *Honeycomb Structures | Research and Development in Modern Materials*. Retrieved August 6, 2022, from https://blogs.deakin.edu.au/remstep/materials-activities/honeycomb-structures/

Anand, S., & Mishra, A. K. (2022). Computational and theoretical investigation of de Laval Nozzle for various Mach number range view project high-performance materials used for UAV manufacturing: Classified review. *International Journal of All Research Education and Scientific Methods (IJARESM)*, *10*(7).

Antolak-Dudka, A., Płatek, P., Durejko, T., Baranowski, P., Małachowski, J., Sarzyński, M., & Czujko, T. (2019). Static and dynamic loading behavior of Ti6Al4V honeycomb structures manufactured by Laser Engineered Net Shaping (LENSTM) technology. *Materials*, *12*(8). https://doi.org/10.3390/ma12081225

Antony, S., Cherouat, A., & Montay, G. (2020). Fabrication and characterization of hemp fibre based 3D printed honeycomb sandwich structure by FDM process. *Applied Composite Materials*, *27*(6), 935–953. https://doi.org/10.1007/s10443-020-09837-z

Araújo, H., Leite, M., Ribeiro, A. R., Deus, A. M., Reis, L., & Vaz, M. F. (2019). The effect of geometry on the flexural properties of cellular core structures. *Proceedings of the Institution of Mechanical Engineers, Part L: Journal of Materials: Design and Applications*, *233*(3), 338–347. https://doi.org/10.1177/1464420718805511

ASTM International. (2016). *Standard Test Method for Edgewise Compressive Strength of Sandwich Constructions*. https://www.astm.org/c0364_c0364m-16.html

ASTM International. (2017). *Standard Test Method for Flexural Properties of Sandwich Constructions*. https://www.astm.org/c0393-00.html

Ayrilmis, N., Kariz, M., Šernek, M., & Kuzman, M. K. (2021). Effects of sandwich core structure and infill rate on mechanical properties of 3D-printed wood/PLA composites. *International Journal of Advanced Manufacturing Technology, 115*(9–10), 3233–3242. https://doi.org/10.1007/s00170-021-07382-y

Ayrilmis, N., Nagarajan, R., & Kuzman, M. K. (2020). Effects of the face/core layer ratio on the mechanical properties of 3d printed wood/polylactic acid (Pla) green biocomposite panels with a gyroid core. *Polymers, 12*(12), 1–8. https://doi.org/10.3390/polym12122929

Azzouz, L., Chen, Y., Zarrelli, M., Pearce, J. M., Mitchell, L., & Ren, G. (2019). Mechanical properties of 3-D printed truss-like lattice biopolymer non- stochastic structures for sandwich panels with natural fibre composite skins. *Composite Structures, 213*(January), 220–230. https://doi.org/10.1016/j.compstruct.2019.01.103

Bates, S. R. G., Farrow, I. R., & Trask, R. S. (2016). 3D printed polyurethane honeycombs for repeated tailored energy absorption. *Materials and Design, 112*, 172–183. https://doi.org/10.1016/j.matdes.2016.08.062

Bates, S. R. G., Farrow, I. R., & Trask, R. S. (2019). Compressive behaviour of 3D printed thermoplastic polyurethane honeycombs with graded densities. *Materials and Design, 162*, 130–142. https://doi.org/10.1016/j.matdes.2018.11.019

Bharath, H. S., Bonthu, D., Gururaja, S., Prabhakar, P., & Doddamani, M. (2021a). Flexural response of 3D printed sandwich composite. *Composite Structures, 263*. https://doi.org/10.1016/j.compstruct.2021.113732

Bharath, H. S., Bonthu, D., Gururaja, S., Prabhakar, P., & Doddamani, M. (2021b). Flexural response of 3D printed sandwich composite. *Composite Structures, 263*(September 2020), 113732. https://doi.org/10.1016/j.compstruct.2021.113732

Bitzer, T. (1997). *Honeycomb Technology*. Springer-Science+Business Media, B. V. https://doi.org/10.1007/978-94-011-5856-5

Brischetto, S., Ferro, C. G., Torre, R., & Maggiore, P. (2018). 3D FDM production and mechanical behavior of polymeric sandwich specimens embedding classical and honeycomb cores. *Curved and Layered Structures, 5*(1), 80–94. https://doi.org/10.1515/cls-2018-0007

Chacón, J. M., Caminero, M. A., Núñez, P. J., García-Plaza, E., García-Moreno, I., & Reverte, J. M. (2019). Additive manufacturing of continuous fibre reinforced thermoplastic composites using fused deposition modelling: Effect of process parameters on mechanical properties. *Composites Science and Technology, 181*, 107688. https://doi.org/10.1016/J.COMPSCITECH.2019.107688

Chantarapanich, N., Laohaprapanon, A., Wisutmethangoon, S., Jiamwatthanachai, P., Chalermkarnnon, P., Sucharitpwatskul, S., Puttawibul, P., & Sitthiseripratip, K. (2014). Fabrication of three-dimensional honeycomb structure for aeronautical applications using selective laser melting: A preliminary investigation. *Rapid Prototyping Journal, 20*(6), 551–558. https://doi.org/10.1108/RPJ-08-2011-0086

Chen, A. (2020). *Top 10 Materials Used for Industrial 3D Printing*. https://www.cmac.com.au/blog/top-10-materials-used-industrial-3d-printing

Chen, H. (2020). Based on the lattice structure of the sandwich structure of 3D printing. *Journal of Physics: Conference Series, 1549*. https://doi.org/10.1088/1742-6596/1549/3/032121

Chen, Y., Ye, L., Zhang, Y. X., & Fu, K. (2021). Compression behaviours of 3D-printed CF/PA metamaterials: Experiment and modelling. *International Journal of Mechanical Sciences, 206*(July), 106634. https://doi.org/10.1016/j.ijmecsci.2021.106634

Dikshit, V., Yap, Y. L., Goh, G. D., Yang, H., Lim, J. C., Qi, X., Yeong, W. Y., & Wei, J. (2016). Investigation of out of plane compressive strength of 3D printed sandwich composites. *IOP Conference Series: Materials Science and Engineering, 139*(1). https://doi.org/10.1088/1757-899X/139/1/012017

Dong, K., Liu, L., Huang, X., & Xiao, X. (2020). 3D printing of continuous fiber reinforced diamond cellular structural composites and tensile properties. *Composite Structures, 250*(December 2019), 112610. https://doi.org/10.1016/j.compstruct.2020.112610

Essassi, K., Rebiere, J.-L., El Mahi, A., Ben Souf, M. A., Bouguecha, A., & Haddar, M. (2020). Experimental and analytical investigation of the bending behaviour of 3D-printed bio-based sandwich structures composites with auxetic core under cyclic fatigue tests. *Composites Part A: Applied Science and Manufacturing*, *131*(October 2019), 105775. https://doi.org/10.1016/j.compositesa.2020.105775

Foo, K. C. C. (2009). Energy absorption characteristics of sandwich structures subjected to low-velocity impact. *Dissertation, NanYang Technological University*, 215.

Fotsing, E. R., Leclerc, C., Sola, M., Ross, A., & Ruiz, E. (2016). Mechanical properties of composite sandwich structures with core or face sheet discontinuities. *Composites Part B: Engineering*, *88*, 229–239. https://doi.org/10.1016/j.compositesb.2015.10.037

Galatas, A., Hassanin, H., Zweiri, Y., & Seneviratne, L. (2018). Additive manufactured sandwich composite/ABS parts for unmanned aerial vehicle applications. *Polymers*, *10*(11). https://doi.org/10.3390/polym10111262

Garrido Silva, B., Alves, F., Sardinha, M., Reis, L., Leite, M., Deus, A. M., & Vaz, M. F. (2022). Functionally graded cellular cores of sandwich panels fabricated by additive manufacturing. *Proceedings of the Institution of Mechanical Engineers, Part L: Journal of Materials: Design and Applications*, April. https://doi.org/10.1177/14644207221084611

Gautam, R., Sridharan, V. S., & Idapalapati, S. (2020). Flatwise compression and local indentation response of 3D-printed strut-reinforced kagome with polyurethane filling. *Jom*, *72*(3), 1324–1331. https://doi.org/10.1007/s11837-019-03968-w

Gebrehiwot, S. Z., Espinosa Leal, L., Eickhoff, J. N., & Rechenberg, L. (2021). The influence of stiffener geometry on flexural properties of 3D printed polylactic acid (PLA) beams. *Progress in Additive Manufacturing*, *6*(1), 71–81. https://doi.org/10.1007/s40964-020-00146-2

Gibson, L. J., & Ashby, M. F. (1997). *Cellular Solids Structure and Properties.pdf*. Cambridge Solid State Science Series.

Goh, G. D., Agarwala, S., Goh, G. L., Dikshit, V., Sing, S. L., & Yeong, W. Y. (2017). Additive manufacturing in unmanned aerial vehicles (UAVs): Challenges and potential. In *Aerospace Science and Technology* (Vol. 63, pp. 140–151). Elsevier Masson SAS. https://doi.org/10.1016/j.ast.2016.12.019

Guo, X., Cheng, G., & Liu, W.-K. (2018). Report of the workshop predictive theoretical, computational and experimental approaches for additive manufacturing (WAM 2016). In *SpringerBriefs in Applied Sciences and Technology*. Springer.

Habib, F. N., Iovenitti, P., Masood, S. H., & Nikzad, M. (2017). In-plane energy absorption evaluation of 3D printed polymeric honeycombs. *Virtual and Physical Prototyping*, *0*(0), 1–15. https://doi.org/10.1080/17452759.2017.1291354

Harland, D., Alshaer, A. W., & Brooks, H. (2019). An experimental and numerical investigation of a novel 3D printed sandwich material for motorsport applications. *Procedia Manufacturing*, *36*, 11–18. https://doi.org/10.1016/j.promfg.2019.08.003

Herrmann, H., & Bucksch, H. (2014). Compression test. *Dictionary Geotechnical Engineering/ Wörterbuch GeoTechnik*, 269–269. https://doi.org/10.1007/978-3-642-41714-6_33779

Hou, Y., Neville, R., Scarpa, F., Remillat, C., Gu, B., & Ruzzene, M. (2014). Graded conventional-auxetic Kirigami sandwich structures: Flatwise compression and edgewise loading. *Composites Part B: Engineering*, *59*, 33–42. https://doi.org/10.1016/j.compositesb.2013.10.084

Hou, Y., Tai, Y. H., Lira, C., Scarpa, F., Yates, J. R., & Gu, B. (2013). The bending and failure of sandwich structures with auxetic gradient cellular cores. *Composites: Part Aes: Part A*, *49*, 119–131. https://doi.org/10.1016/j.compositesa.2013.02.007

Hou, Z., Tian, X., Zhang, J., & Li, D. (2018). 3D printed continuous fibre reinforced composite corrugated structure. *Composite Structures*, *184*(July 2017), 1005–1010. https://doi.org/10.1016/j.compstruct.2017.10.080

Ibadaddin, S., Paleshwar, D. V., & Sainath, K. (2018). Sandwich composite for UAV wing design and fabrication. *International Journal of Research in Engineering, Science and Management, 1*(11), 2581–5792.

Intertek Group plc. (n.d.-a). *Core Shear Properties of Sandwich Constructions by ASTM C393.* Retrieved August 19, 2022, from https://www.intertek.com/polymers/composites/astm-c393/

Intertek Group plc. (n.d.-b). *Flatwise Compressive Properties by ASTM C365.* Retrieved August 19, 2022, from https://www.intertek.com/polymers/composites/astm-c365/

Isaac, C. W., & Duddeck, F. (2022). Current trends in additively manufactured (3D printed) energy absorbing structures for crashworthiness application – A review. *Virtual and Physical Prototyping, 0*(0), 1–44. https://doi.org/10.1080/17452759.2022.2074698

Kabir, S. M. F., Mathur, K., & Seyam, A. M. (2020). Impact resistance and failure mechanism of 3D printed continuous fiber-reinforced cellular composites. *The Journal of the Textile Institute, 0*(0), 1–15. https://doi.org/10.1080/00405000.2020.1778223

Klippstein, H., Diaz, A., Cerio Sanchez, D., Hassanin, H., Zweiri, Y., & Seneviratne, L. (2017). Fused deposition modelling for unmanned aerial vehicles (UAVs): A review. *Advanced Engineering Materials, 20*(2), 1700552. https://doi.org/10.1002/adem.201700552

Kloski, L. W., Kloski, N., & Goner, J. (2017). *Getting Started with 3D Printing.* Albatros Media a.s.

Kristína, Š., Jozef, Č., & Michal, J. (2021). Possibilities of using 3D printing technology in production of aircraft component. *University of Zilina*, 10–15.

Krzyzak, A., Mazur, M., Gajewski, M., Drozd, K., Komorek, A., & Przybyłek, P. (2016). Sandwich structured composites for aeronautics: Methods of manufacturing affecting some mechanical properties. *International Journal of Aerospace Engineering, 2016.* https://doi.org/10.1155/2016/7816912

Lascano, D., Guillen-Pineda, R., Quiles-Carrillo, L., Ivorra-Martínez, J., Balart, R., Montanes, N., & Boronat, T. (2021). Manufacturing and characterization of highly environmentally friendly sandwich composites from polylactide cores and flax-polylactide faces. *Polymers, 13*(3), 1–14. https://doi.org/10.3390/polym13030342

Li, B., Liu, H., Zhang, Q., Yang, X., & Yang, J. (2022). Crushing behavior and energy absorption of a bio-inspired bi-directional corrugated lattice under quasi-static compression load. *Composite Structures, 286*(January). https://doi.org/10.1016/j.compstruct.2022.115315

Li, T., & Wang, L. (2017). Bending behavior of sandwich composite structures with tunable 3D-printed core materials. *Composite Structures, 175*, 46–57. https://doi.org/10.1016/j.compstruct.2017.05.001

Liu, H., Zhang, E. T., & Ng, B. F. (2021). In-plane dynamic crushing of a novel honeycomb with functionally graded fractal self-similarity. *Composite Structures, 270*(May), 114106. https://doi.org/10.1016/j.compstruct.2021.114106

Mansour, M. T., Tsongas, K., Tzetzis, D., & Antoniadis, A. (2019). The in-plane compression performance of hierarchical honeycomb additive manufactured structures. *IOP Conference Series: Materials Science and Engineering, 564*(1). https://doi.org/10.1088/1757-899X/564/1/012015

Menegozzo, M., Cecchini, A., Just-Agosto, F. A., Serrano Acevedo, D., Flores Velez, O. J., Acevedo-Figueroa, I., & de Jesús Ruiz, J. (2022). A 3D-printed honeycomb cell geometry design with enhanced energy absorption under axial and lateral quasi-static compression loads. *Applied Mechanics, 3*(1), 296–312. https://doi.org/10.3390/applmech3010019

Meng, L., Qiu, X., Gao, T., Li, Z., & Zhang, W. (2020). An inverse approach to the accurate modelling of 3D-printed sandwich panels with lattice core using beams of variable cross-section. *Composite Structures, 247*(April), 112363. https://doi.org/10.1016/j.compstruct.2020.112363

Naidu, B. V. V., & Kumar, G. D. (2018). Additive manufacturing of honeycomb structure and analysis of infill and material characteristics. *International Journal of Scientific Research and Review*, *7*(3), 226–234.

Njim, E. K., Al-Waily, M., & Bakhy, S. H. (2021). A review of the recent research on the experimental tests of functionally graded sandwich panels. *Journal of Mechanical Engineering Research and Developments*, *44*(3), 420–441.

Nurul Fazita, M. R., Abdul Khalil, H. P. S., Nor Amira Izzati, A., & Rizal, S. (2018). Effects of strain rate on failure mechanisms and energy absorption in polymer composites. In *Failure Analysis in Biocomposites, Fibre-Reinforced Composites and Hybrid Composites*. Elsevier Ltd. https://doi.org/10.1016/B978-0-08-102293-1.00003-6

Özen, İ., Çava, K., Gedikli, H., Alver, Ü., & Aslan, M. (2020). Low-energy impact response of composite sandwich panels with thermoplastic honeycomb and reentrant cores. *Thin-Walled Structures*, *156*(April). https://doi.org/10.1016/j.tws.2020.106989

Paczos, P., Wichniarek, R., & Magnucki, K. (2018). Three-point bending of sandwich beam with special structure of the core. *Composite Structures*, *201*(June), 676–682. https://doi.org/10.1016/j.compstruct.2018.06.077

Peng, C., Fox, K., Qian, M., Nguyen-xuan, H., & Tran, P. (2021). 3D printed sandwich beams with bioinspired cores: Mechanical performance and modelling. *Thin-Walled Structures*, *161*(September 2020), 107471. https://doi.org/10.1016/j.tws.2021.107471

Pollard, D., Ward, C., Herrmann, G., & Etches, J. (2017). The manufacture of honeycomb cores using fused deposition modeling. *Advanced Manufacturing: Polymer and Composites Science*, *3*(1), 21–31. https://doi.org/10.1080/20550340.2017.1306337

Ratwani, M. M. (2010). Composite materials and sandwich structures – A primer. *RTO-EN-AVT*, *156*, 1–16.

Russell, J. D. (2018). 3.3 The impact of large integrated and bonded composite structures on future military transport aircraft. *Comprehensive Composite Materials II*, 91–130. https://doi.org/10.1016/B978-0-12-803581-8.09925-2

Santos, F. A., Rebelo, H., Coutinho, M., Sutherland, L. S., Cismasiu, C., Farina, I., & Fraternali, F. (2021). Low velocity impact response of 3D printed structures formed by cellular metamaterials and stiffening plates: PLA vs. PETg. *Composite Structures*, *256*(August 2020), 113128. https://doi.org/10.1016/j.compstruct.2020.113128

Sara Black. (2015). *3D Printing Moves Into Tooling Components*. CompositesWorld. https://www.compositesworld.com/articles/3d-printing-moves-into-tooling-components

Sarvestani, H. Y., Akbarzadeh, A. H., Mirbolghasemi, A., & Hermenean, K. (2018). 3D printed meta-sandwich structures: Failure mechanism, energy absorption and multi-hit capability Face-sheets. *Materials & Design*, *160*, 179–193. https://doi.org/10.1016/j.matdes.2018.08.061

Sarvestani, H. Y., Akbarzadeh, A. H., Niknam, H., & Hermenean, K. (2018). 3D printed architected polymeric sandwich panels: Energy absorption and structural performance. *Composite Structures*, *200*(March), 886–909. https://doi.org/10.1016/j.compstruct.2018.04.002

Singamneni, S., Yifan, L. V., Hewitt, A., Chalk, R., Thomas, W., & Jordison, D. (2019). Additive manufacturing for the aircraft industry: A review. *Journal of Aeronautics & Aerospace Engineering*, *8*(1). https://doi.org/10.35248/2168-9792.19.8.215

Smardzewski, J., Maslej, M., & Wojciechowski, K. W. (2021). Compression and low velocity impact response of wood-based sandwich panels with auxetic lattice core. *European Journal of Wood and Wood Products*, *79*(4), 797–810. https://doi.org/10.1007/s00107-021-01677-3

Soliman, H. E., Black, J. T., & Brown, A. J. (2016). *Mechanical Properties of Cellular Core Structures Mechanical Properties of Cellular Core Structures*.

Stanczak, M., Fras, T., Blanc, L., Pawlowski, P., & Rusinek, A. (2019). Blast-induced compression of a thin-walled and modeling. *Metals*, *9*(12), 1350.

Standard Test Method for Flatwise Compressive Properties of Sandwich Cores. (n.d.). Retrieved August 17, 2022, from https://www.astm.org/c0365_c0365m-05.html

Stocchi, A., Colabella, L., Cisilino, A., & Álvarez, V. (2014). Manufacturing and testing of a sandwich panel honeycomb core reinforced with natural-fiber fabrics. *Materials & Design, 55*, 394–403. https://doi.org/10.1016/J.MATDES.2013.09.054

Subramaniyan, M., Karuppan, S., Prabanjan, P., Anand A. P., & Vasanthan A. P. (2022). Survey on compression property of sandwich 3D printed PLA components. *Materials Today: Proceedings, xxxx.* https://doi.org/10.1016/j.matpr.2022.04.749

Taghvaei, M. (2020). *Optimization of Additively Manufactured Sandwich Panels with Composite Face Sheets.*

Tanjung, R. A., Hidayat, M. I. P., & Wicaksono, S. T. (2018). Stress analysis on tensile loaded honeycomb sandwich structured material of poly lactic acid filament. *International Conference on Science and Applied Science,* 020107.

Tao, Y., Li, W., Wei, K., Duan, S., Wen, W., Chen, L., Pei, Y., & Fang, D. (2019). Mechanical properties and energy absorption of 3D printed square hierarchical honeycombs under in-plane axial compression. *Composites Part B: Engineering, 176*(January), 107219. https://doi.org/10.1016/j.compositesb.2019.107219

Townsend, S., Adams, R., Robinson, M., Hanna, B., & Theobald, P. (2020). 3D printed origami honeycombs with tailored out-of-plane energy absorption behavior. *Materials & Design,* 135577. https://doi.org/10.1016/j.scitotenv.2019.135577

Toygar, M. E., Tee, K. F., Maleki, F. K., & Balaban, A. C. (2019). Experimental, analytical and numerical study of mechanical properties and fracture energy for composite sandwich beams. *Journal of Sandwich Structures and Materials, 21*(3), 1167–1189. https://doi.org/10.1177/1099636217710003

Usta, F., Türkmen, H. S., & Scarpa, F. (2021). Low-velocity impact resistance of composite sandwich panels with various types of auxetic and non-auxetic core structures. *Thin-Walled Structures, 163*(November 2020), 1–13. https://doi.org/10.1016/j.tws.2021.107738

Vishnu Prashant Reddy, K., Meera Mirzana, I., & Koti Reddy, A. (2018). Application of Additive Manufacturing technology to an Aerospace component for better trade-off's. *Materials Today: Proceedings, 5*(2), 3895–3902. https://doi.org/10.1016/j.matpr.2017.11.644

Wang, L., Sun, J., Ding, T., Liang, Y., Ho, J. C. M., & Lai, M. H. (2022). Manufacture and behaviour of innovative 3D printed auxetic composite panels subjected to low-velocity impact load. *Structures, 38*(January), 910–933. https://doi.org/10.1016/j.istruc.2022.02.033

Wang, S., Wang, J., Xu, Y., Zhang, W., & Zhu, J. (2020). Compressive behavior and energy absorption of polymeric lattice structures made by additive manufacturing. *Frontiers of Mechanical Engineering, 15*(2), 319–327. https://doi.org/10.1007/s11465-019-0549-7

Wang, S., Xu, Y., & Zhang, W. (2019). Low-velocity impact response of 3D-printed lattice sandwich panels. *IOP Conference Series: Materials Science and Engineering, 531,* 012056. https://doi.org/10.1088/1757-899X/531/1/012056

Wang, X., Jiang, M., Zhou, Z., Gou, J., & Hui, D. (2017). 3D printing of polymer matrix composites: A review and prospective. *Composites Part B: Engineering, 110*, 442–458. https://doi.org/10.1016/J.COMPOSITESB.2016.11.034

Wang, Z. (2019). Recent advances in novel metallic honeycomb structure. *Composites Part B: Engineering, 166*, 731–741. https://doi.org/10.1016/J.COMPOSITESB.2019.02.011

What Is Compression Testing? – Instron. (n.d.). Retrieved August 20, 2022, from https://www.instron.com/en/resources/test-types/compression-test

Xiao, L., Xu, X., Feng, G., Li, S., Song, W., & Jiang, Z. (2022). Compressive performance and energy absorption of additively manufactured metallic hybrid lattice structures. *International Journal of Mechanical Sciences, 219*(January), 107093. https://doi.org/10.1016/j.ijmecsci.2022.107093

Yan, J., Liu, Y., Yan, Z., Bai, F., Shi, Z., Si, P., & Huang, F. (2022). Ballistic characteristics of 3D-printed auxetic honeycomb sandwich panel using CFRP face sheet. *International Journal of Impact Engineering*, *164*(January), 104186. https://doi.org/10.1016/j.ijimpeng.2022.104186

Yan, L., Zhu, K., Zhang, Y., Zhang, C., & Zheng, X. (2020). Effect of absorbent foam filling on mechanical behaviors of 3D-printed honeycombs. *Polymers*, *12*(9). https://doi.org/10.3390/POLYM12092059

Yap, Y. L., & Yeong, W. Y. (2015). Shape recovery effect of 3D printed polymeric honeycomb. *Virtual and Physical Prototyping*, *10*(2), 91–99. https://doi.org/10.1080/17452759.2015.1060350

Zhang, Q., Yang, X., Li, P., Huang, G., Feng, S., Shen, C., Han, B., Zhang, X., Jin, F., Xu, F., & Lu, T. J. (2015). Bioinspired engineering of honeycomb structure – Using nature to inspire human innovation. *Progress in Materials Science*, *74*, 332–400. https://doi.org/10.1016/J.PMATSCI.2015.05.001

Zhang, Z., Li, B., Wang, Y., Zhang, W., Yue, C., Zhang, Q., & Jin, F. (2022). Elevated temperature compression behaviors of 3D printed hollow pyramidal lattice sandwich structure reinforced by truncated square honeycomb. *Composite Structures*, *286*(July 2021), 115307. https://doi.org/10.1016/j.compstruct.2022.115307

7 A Brief Review of the Structure Designed Using Metallic 3D Printing for Biomechanics Applications

*Rayappa Shrinivas Mahale, Shika Shaygan,
Ali Attaeyan, Shamanth Vasanth, Hemanth Krishna,
Sharath P.C., Shashanka R., Malekipour M.H.,
Atefeh Ghorbani, Harsha Prasad, and Ghaffari Y.*

NOMENCLATURE

Abbreviation	Full Form
SLS	Selective Laser Sintering
MRI	Magnetic Resonance Imaging
CT	Computed Tomography
RP	Rapid Prototyping
CAD	Computer-Aided Design
SLM	Selective Laser Melting
FDM	Fused Deposition Modeling
SLA	Stereolithography
DSLS	Direct Selective Laser Sintering
DDM	Direct Digital Manufacturing
FGM	Functionally Graded Material
DED	Direct Energy Deposition
PBF	Powder Bed Fusion
SPD	Supersonic Particle Deposition
PEEK	Polyetheretherketone
PVA	Poly Vinyl Alcohol
PCL	Polycaprolactone
PLLA	poly-L-lactic acid
UAV	Unmanned Aerial Vehicle

DOI: 10.1201/9781003362128-7

7.1 INTRODUCTION

Today, shortening the time of development and development of a product from design to production is the key to the success of a production organization in the competitive world. To achieve this, new technologies known as rapid prototyping (RP) and new production methods have been introduced [1–4]. Prototyping a piece or a designed product is traditionally done through physical modeling in the modeling workshop with hand tools and a lot of trial and error [5–7]. This process is a difficult, time-consuming, and costly task. By using RP methods, a 3D physical model of a complex piece, albeit a complex one, can be made in a short time (about a few hours) at a low cost with high accuracy and used in evaluating design, product, or other uses [8–11]. The superiority and capability of this technology becomes apparent when, first, the parameter of short prototyping time is an important priority for us, and, second, the part has a complex geometric shape. As a result, RP can be defined as a set of techniques for a 3D model made by CAD or computer design. This part is usually made by 3D printers or through an "additive manufacturing process". Incremental fabrication is a technique in which objects are created using a digital model under a layered process [12–15]. In other words, incremental fabrication is a method in which parts are made in layers by fusion or deposition of materials such as plastic, ceramic, metal, powders, liquids, and living cells. This process is called incremental fabrication, or prototyping [16–19]. It is also called 3D printing, or shapeless solid manufacturing technology. In this method, 3D parts are created using the instructions of a computer-designed file [20–23]. The basis of this method is specified in the form of layering according to the instructions, so the printer nozzle on the x-y page moves according to the program and creates a layer; in the next step the nozzle is moved along the z-axis and the next layer on the previously formed layer is created, which is repeated by repeating this process to create the desired 3D shape [24–28].

Figure 7.1 illustrates the classification of RP methods based on the type of material used. RP methods are techniques used to produce 3D objects with high speed and accuracy [29–33]. These methods are commonly used in industrial manufacturing, product design, and even in the production of medical components. Some of the benefits are speed, accuracy, the possibility of using a wide range of different materials

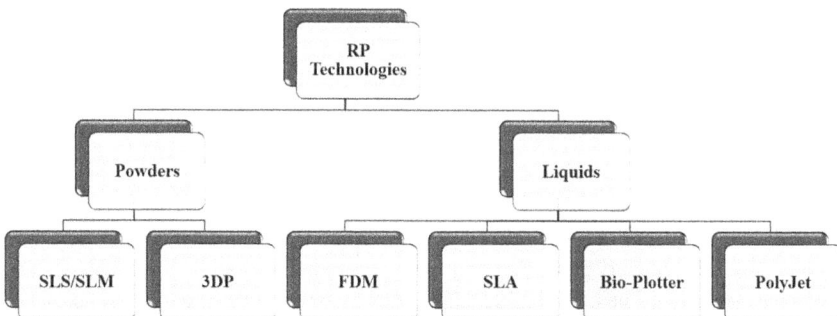

FIGURE 7.1 Classification of RP methods according to the type of material used: SLS, SLM, fused deposition modeling (FDM), and stereolithography (SLA).

and printing machines with different capabilities as well as economic efficiency, the application of this method in various industrial, research, and educational fields. By using these manufacturing techniques it is possible to build anatomically shaped scaffolds with complex interior architecture and to allow precise control of porosity (including pore size, shape, and connection) and mechanical properties [34–37]. RP methods are diverse due to the nature of their operation, the process that takes place within a particular method, and the materials used in them. The classification of rapid sampling methods according to the type of material used is as follows [38–41].

7.2   SELECTIVE LASER SINTERING (SLS)

SLS was first proposed by Descardes and Bimen. In this method, powder particles are joined together by a laser as an energy source. During the process, the laser is guided to create a specific pattern on the surface of the powder. When the first layer is completed, a roller puts a new layer of powder on top of the first layer, and this process is repeated until the piece is complete, which finally restores the piece made from under the powder bed. However, this method usually produces powders from plastics, ceramics, and metal alloys that require high temperatures and high laser energy. The use of this method in tissue engineering is well known and is effective in some non-medical industries. However, there are concerns in pharmaceutical applications because the high energy production of the laser beam can lead to the destruction of drugs [42–45]. Direct selective laser sintering (DSLS), a high-power laser beam that acts on a metal powder substrate, results in direct curing. This type of SLS eliminates costly and time-consuming heat treatment. DSLS technology allows the application of a mixture of different powders with different properties, which makes it possible to make specialized materials with good and unique properties [46–48]. Fermentation is a process of stabilization at high temperature (thermal aggregation) in which the powder is freely dispersed and, as a result, as the temperature increases, the free enthalpy of the system decreases, reducing the porosity and increasing the density of the material. In the sintering process, the mechanical connections between the powder particles are transformed into stronger metal bonds. In analyzing the theoretical aspects of the sintering process, the driving forces and mechanisms of material transfer must also be considered [49–51].

7.2.1   SELECTIVE LASER SINTERING

The cylinder of the piece is placed at the height necessary to form the first layer. The powder material is spread by a roller from the powder chamber on the surface of the cylinder of the part and the thickness of the layer is adjusted by the same roller. The CO_2 laser beam draws the first section on the surface of the powder. The powder particles are then heated by the laser powder and sintered together. In this way, the first layer is formed. The cylinder of the piece goes down to the thickness of one layer and the next layer of powder is spread. The laser beam scans the new section and creates the next layer so that this layer connects to the previous layer. The above steps continue until the formation of all layers. In some cases, depending on the type of material used in the construction of the part, the final operation is required.

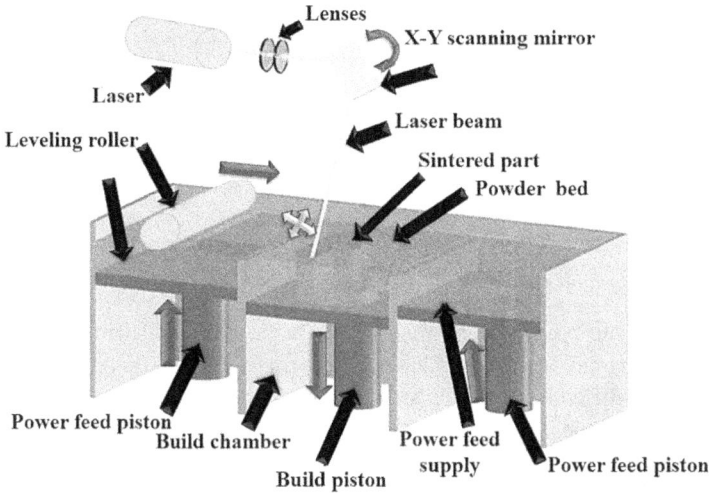

FIGURE 7.2 SLS process.

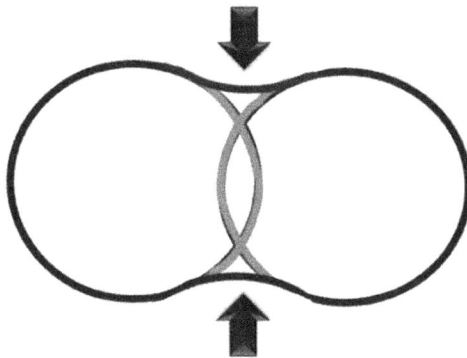

FIGURE 7.3 Formation of the neck in two synthesized powder particles. The main shapes are shown in red.

Figure 7.2 illustrates the SLS process, one of the most commonly used RP methods. The SLS process begins with a layer of powder material, which is spread over a build platform. A high-powered laser then selectively heats the powder to fuse it together, forming a solid layer. This process is repeated layer by layer until the final object is created.

7.2.2 Sintering Mechanisms

Baking in SLS first takes place in the liquid state, where the powder particles form a micro-melted layer on the surface, thereby reducing the viscosity and forming a concave radius between the particles, known as the necklace [52–55].

Figure 7.3 illustrates the formation of a neck between two synthesized powder particles. The neck is the small region of contact between the two particles that occurs during the sintering process. The sintering process involves heating the particles until they are just below their melting point, at which point they begin to fuse together. As shown in the figure, the two particles are initially separate and have distinct shapes, shown in red. As the particles are heated and begin to fuse together, the neck forms, connecting the two particles. The neck typically has a smaller cross-sectional area than the rest of the particle, due to the localized heating and melting that occurs in this region.

7.2.2   DIFFERENCE BETWEEN SLM AND OTHER AM METHODS

Different parts can be placed on the border. In the manufacturing process, selective laser melting (SLM) is like SLA except that in this method powder particles (weldable) with different materials such as polymers (hard and soft), sand, metals, ceramics, polystyrene, and wax replace the light-sensitive resin. Compared to the SLM technique, the two methods are briefly the same concept but differ in technical details. In SLM, the material is completely melted instead of sintered, creating different properties such as crystal structure, porosity, and so on [56–59].

7.3   ADVANTAGES AND DISADVANTAGES OF THE SLM METHOD

7.3.1   ADVANTAGES

Good Stability of Parts: Parts are created in an environment with precise control. A wide range of materials including nylon, polycarbons, metals, and ceramics can be used in this system. Therefore, the flexibility of the system for different applications is very high. In this technique, there is no need for support; it requires low post-processing, and there is no need for final baking: The laser-sintered piece is hard enough and does not require final baking. The synthesized powder substrate is fully self-supporting and allows for high angles (0–45° from the horizontal plane).

7.3.2   DISADVANTAGES

Components have a porous surface, which can be sealed by several different post-processing methods such as cyanoacrylate coatings, or by hot isostatic pressing. Some other disadvantages are large device dimensions, high energy consumption, and relatively poor level of payment [60–63].

7.4   APPLICATIONS

As mentioned, SLM technology has a great ability to build complex and geometric geometries, so it is widely used in many industries around the world. The range of users of the product and industries using this technology are military industries, aerospace industries, automotive industries, medical engineering, instrumentation industries, and oil, gas, petrochemical, and power plant industries. In this chapter, the aim was to discuss the application of 3D print technology in the aerospace and medical sciences industries [64–66].

7.4.1 Industrial Applications

The defense and aerospace industries were among the first to operate the next printer. The first use of 3D printers in these industries dates to 1989. Three decades later, the defense and aerospace industries account for 16.8% of the $10.4 billion incremental manufacturing market budget, and industry professionals are constantly expanding the frontiers of aerospace technology with 3D printing. As expected, the use of 3D printers in the aerospace industry is not limited to prototyping. The main and functional parts are also made by 3D printers and are used in airplanes and other flying devices [67–69]. The components made by this method include air ducts, wall panels, and even metal structural components. In industries such as aerospace and defense, where the production of complex parts in low circulation is required, 3D printing is very ideal, and with the help of this technology, very complex geometric shapes are possible without the need to invest in tooling and molding. Three-dimensional printers allow manufacturers to produce the required parts in small numbers, such as aerodynamics and engine performance, with less time and cost. Weight is one of the most important factors in aircraft design. Reducing the weight of any aircraft can significantly affect carbon dioxide emissions, fuel consumption, and transport capacity [70–72]. The invention was patented in 1989 under the name DTM, which was later acquired by 3D systems. SLM is primarily used for RP of non-functional components and direct digital fabrication (DDM) of non-critical components [73–78]. According to studies, the crystallinity percentage of nylon-12 SLS parts depends on the degree of melt of the particle, to the extent that the different crystallinity percentages of the volume of the parts can be considered as different materials. This opens the possibility to produce FGM components with SLS. SLS parts made of acrylic styrene and polyamide (nylon) show almost the same mechanical properties as plastic injection parts [26–38].

Other companies using this technology include the Spanish company Aero ITP, which recently designed one of the main structures of its new RULTRAFAN engine using SLS 3D printer technology. This component, called the Tail Bearing Housing (TBH), can be used as a connecting element between the aircraft and the engine [79–82]. Advanced Aerials also uses 3D systems SLS technology to build hard components for its unmanned vehicle systems. Honeywell also used SLS to make control sheath covers on its RQ-16 T-Hawk drone [39–48]. Table 7.1 provides an overview of the additive non-metal technologies that are being used in the aerospace industry. These technologies allow for the production of complex, high-performance parts that can withstand the harsh conditions of aerospace applications.

7.4.2 Applications of SLS in Biomedicine

The applications of 3D printing in this field are so wide that these applications cross the boundaries of research and development and fall into other sub-disciplines such as preparation for surgery, artificial limbs, dental and maxillofacial implants, and scaffolds, 3D printing of living tissues and organs, and prescription drugs. The application of SLS in medicine has developed with astonishing speed from its earliest stages. This technology is used in surgery, orthopedics, tooth restoration, fabrication

TABLE 7.1

Additive non-metal technologies in the aerospace industry [43–55]

Application	Example part	Requirements	Recommended process	Recommended material
Engine compartment	Tarmac nozzle bezel	Heat-resistant functional parts	SLS	Glass-filled nylon
Air ducts	Airflow ducting	Flexible ducts and below directors	SLS	Nylon 12
Full-size panels	Seat backs and entry doors	Large parts with smooth surface finish	SLA	Standard resin
Casted metal parts	Brackets and door handles	Cast metal parts using 3D-printed patterns	SLA and PolyJet	Castable resin or wax
Lights	Headlight prototypes	Fully transparent, high-detail models	SLA and PolyJet	Clear resin
Bezels	Dashboard interface	End-use custom screen bezels	PolyJet	Digital ABS
UAV	Wings and fuselage	High strength, lightweight, durability	FDM	ULTEM
Prototypes and tools	Camera case prototype and tool to install wiring	Rigidity, dimensional stability	FDM	ULTEM
Cabin accessories	Door handle covers	Customizable and robust handles	FDM	ULTEM

of skeletal prostheses in tissue engineering, and even for educational or research purposes [49–53]. PVA has been selected because of its similar tensile strength to human articular cartilage, its ability to form complex shapes, and its good adhesion properties. Hydroxyapatite (HA) is a bone-conducting bioactive substance that can bond to bones as strong as that of a 3- to 6-month-old and can also increase collagen production. Experiments on HA bioactivity have shown that the laser tempering process does not change anything, and HA is still active in the environment. Therefore, by using a mixture of PVA and HA, suitable scaffolds can be produced [54–60]. The use of SLS technology to build models used in bioengineering allows medical products to be produced according to the expected dimensions and properties. The long-term stability of laser-sintered elements ensures that their geometry is maintained for a long time [83–85].

Figure 7.4 shows a skull implant that was produced using SLS technology. The implant was created by American manufacturer Oxford Performance Materials, using a material called polyether ketone ketone (PEKK). This implant was successfully implanted in a patient in March 2013 to replace 75% of the patient's skull, making it a significant breakthrough in the field of medical implants. PEKK is a high-performance polymer that is known for its excellent mechanical and thermal properties, making it an ideal material for use in medical implants. PEKK has been used in a variety of medical applications, including spinal implants, dental implants, and orthopedic implants.

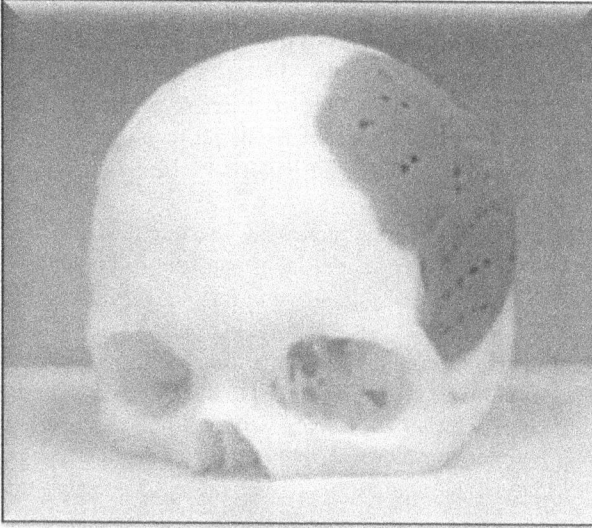

FIGURE 7.4 A skull implant made by SLS (the implant was printed by American manufacturer Oxford Performance Materials using polyethylene ketone ketone, PEKK, which resulted in its first successful implantation for 75% of the skull in March 2013).

7.4.3 SLS IN DENTISTRY

SLS is a great alternative to traditional techniques, such as casting, to make dental crowns. For this purpose, Co-Cr powder alloy with a particle diameter of 3–14 μm is used. In the case of SLS, there is nothing wrong with producing a large series of products. This allows the production of dental implants with high quality and 20 times more density. Also, unlike casting, improved geometric control is more predictable in terms of deformation [86–89].

7.4.4 USE OF 3D PRINTING MODELS IN LIVER SURGERIES

Three-dimensional printing technology with remarkable prospects is being rapidly developed for liver surgeries. Liver surgery can be a pivotal role to create 3D printing models. However, the two main applications of 3D printing technology in the field of liver surgery include training or design necessary for surgery and printing functional liver cells through bioprinting technology that can be used in the study of liver disease and pharmaceutical research. Complex surgeries, such as liver transplants (with a living donor), require accurate knowledge of each patient's unique anatomy. Because the cost of printing 3D models is highly dependent on the complexity of the process and the accuracy of the model (in terms of medical training or pre-practice training), printing 3D models for liver surgeries is one of the costliest surgeries. But what studies have shown is that reducing the dimensions of 3D models and printing them in smaller sizes may reduce costs, which in some cases seems reasonable. However, one of the limitations of reducing the dimensions of 3D models is the lack of accurate reflection of a patient's anatomy [90–93]. The use of SLS and PolyJet printing technologies in the creation of 3D liver printing models makes printing relatively easy and fast. However, the availability and high cost of printing relative to FDM technology are the main limitations in their use. Although using 3D printing models to educate patients and students has not yet received much attention in the field of liver surgery, it can be useful in relation to therapeutic relationships when advising patients [94–96].

The model presented had a very high accuracy so that through direct comparison with validation protocols, the average dimensional error obtained was less than 4 mm for the whole model and less than 1.3 mm for the printed vessel diameter. The results of this study showed that printing different parts by 3D printing technology is a valuable tool for understanding the spatial relationship of vascular and biliary anatomical structures, which may ultimately facilitate surgery and minimize complications during surgery [97–99]. Because printing 3D models can sometimes be time-consuming, it is not common to use these 3D models in emergencies such as liver rupture due to the time required to produce these 3D models [100–103].

7.4.5 SLS IN PHARMACY

Production of pharmaceutical pills is done in large factories with the same properties and with a large production volume. This method is not suitable to produce pharmaceutical pills with the desired formulation and properties because it is mass-produced, and the pills are made in high production volume. One of the methods that can solve this problem and be used to produce pills according to the conditions and needs of each patient is the use of additive manufacturing technology and 3D printing in pharmacy. Using 3D printing technology, it is possible to prepare pills with the shape, formulation, dose, and medicinal properties according to the condition of each patient and move toward personalization and uniqueness of medicines [100–104]. To produce pills that enter the body through the gastrointestinal tract (oral entry), the FDM 3D printing method was first used for polyvinyl alcohol (PVA) base pills in the molten sediment modeling method. Pre-prepared filaments heat and flow through the nozzles and by moving the nozzle in certain directions on the pill, the device creates the final shape layer by layer [88–94]. To produce pills by fused sediment modeling, many pharmaceutical polymers whose formulations are suitable for use in fused sediment modeling 3D printing machines have been identified and reported [105–108]. Many researchers succeeded in producing pills by melting sediment modeling. They printed pills with different geometric shapes and controlled release in this way and performed various drug tests on the produced pills [109–113]. Fina et al. [4] studied the feasibility of producing SLS pills and printed the pills using a 3D laser printer by sintering various drugs. Fina et al. [4] showed that 3D printing by SLS is a method that is fully adaptable to the pharmaceutical industry. In this method, unlike the melted sediment modeling method, there is no need to produce the filament of any drug beforehand. Of course, there are many variables involved in the production of pills that affect the properties of the final pill. According to the results, with increasing laser advancement speed, the breaking force of the pill decreases, but increasing the filling percentage and the amount of added color increases the breaking force of the pills [114–117].

7.5 CONCLUSIONS

SLS is an advanced global RP technology that can be used to create metal, plastic, and ceramic objects which produce very durable and very complex shapes with relatively high strength along with a wide range of available materials that can be applied. This is related to the fact that the SLS process is performed in an inert gas shield space. The degree of complexity is limited only by the accuracy of the laser and the size of the powder particles, which affects the ability to create highly accurate parts. The strength, porosity, and elastic modulus of the alloy made by these methods enhance compatibility with the damaged tissue. In addition, SLS and SLM are the most cost-effective methods for producing large-scale additives. Therefore, this technology is used for hard custom parts such as aerospace and medical guilds, especially patient-specific surgery, prostheses, and pharmaceuticals. Another advantage of using SLS and SLM is that all types of biocompatible polymers can be sintered together, which allows the production of ineffective or absorbable implants for the patient. Making very accurate models is an important challenge with this technique; other additive manufacturing techniques such as PolyJet and stereolithography offer higher accuracy that can be solved by obtaining a laser with appropriate parameters. Application software also plays an important role. This is because the software allows advanced settings of process technology parameters to be applied and can therefore offer more versatile applications. The application of 3D printing in medicine is unlimited and represents a revolutionary future in the whole field of modern medicine as well as in other disciplines.

ACKNOWLEDGMENT

We would like to express our sincere gratitude to the School of Mechanical Engineering at REVA University in Bengaluru, Karnataka, India, and the Department of Pharmacy at Cyprus Health and Social Science for their outstanding contributions and support in our research endeavors.

REFERENCES

1. Aimar, A., Palermo, A., & Innocenti, B. (2019). The role of 3D printing in medical applications: A state of the art. *Journal of Healthcare Engineering*, 2019.
2. Lou, A., & Grosvenor, C. (2012). *Selective Laser Sintering, Birth of an Industry* (p. 8). Austin, TX: Department of Mechanical Engineering, University of Texas.
3. Olakanmi, E. O., Cochrane, R. F., & Dalgarno, K. W. (2015). A review on selective laser sintering/melting (SLS/SLM) of aluminium alloy powders: Processing, microstructure, and properties. *Progress in Materials Science*, 74, 401–477.
4. Fina, F., Goyanes, A., Gaisford, S., & Basit, A. W. (2017). Selective laser sintering (SLS) 3D printing of medicines. *International Journal of Pharmaceutics*, 529(1–2), 285–293.
5. Ventola, C. L. (2014). Medical applications for 3D printing: Current and projected uses. *Pharmacy and Therapeutics*, 39(10), 704.
6. Sun, W. (2013, July). Bio-3D printing. In *NSF Workshop on Frontiers of Additive Manufacturing Research and Education*. Filadelfia: Drexel University.
7. Maulvi, F. A., Shah, M. J., Solanki, B. S., Patel, A. S., Soni, T. G., & Shah, D. O. (2017). Application of 3D printing technology in the development of novel drug delivery systems. *International Journal of Drug Development and Research*, 9(1), 44–49.

8. Jeon, B., Lee, C., Kim, M., Choi, T. H., Kim, S., & Kim, S. (2016). Fabrication of three-dimensional scan-to-print ear model for microtia reconstruction. *Journal of Surgical Research*, 206(2), 490–497.

9. Kruth, J. P., Mercelis, P., Van Vaerenbergh, J., Froyen, L., & Rombouts, M. (2005). Binding mechanisms in selective laser sintering and selective laser melting. *Rapid Prototyping Journal*,11(1), 26–36.

10. Fina, F., Goyanes, A., Gaisford, S., & Basit, A. W. (2017). Selective laser sintering (SLS) 3D printing of medicines. *International Journal of Pharmaceutics*, 529(1–2), 285–293.

11. Das, S. (2003). Physical aspects of process control in selective laser sintering of metals. *Advanced Engineering Materials*, 5(10), 701–711.

12. Awad, A., Fina, F., Goyanes, A., Gaisford, S., & Basit, A. W. (2020). 3D printing: Principles and pharmaceutical applications of selective laser sintering. *International Journal of Pharmaceutics*, 586, 119594.

13. Msallem, B., Sharma, N., Cao, S., Halbeisen, F. S., Zeilhofer, H. F., & Thieringer, F. M. (2020). Evaluation of the dimensional accuracy of 3D-printed anatomical mandibular models using FFF, SLA, SLS, MJ, and BJ printing technology. *Journal of Clinical Medicine*, 9(3), 817.

14. Zhansitov, A. A., Slonov, A. L., Shetov, R. A., Baikaziev, A. E., Shakhmurzova, K. T., Kurdanova, Z. I., & Khashirova, S. Y. (2018). Synthesis and properties of polyetheretherketones for 3D printing. *Fibre Chemistry*, 49(6), 414–419.

15. Török, J., Kaščak, J., Kočiško, M., Teliškova, M., & Dobránsky, J. (2018). Orientation of the model in SLS printing and its influence on mechanical properities. *TEM Journal*, 7(4), 723.

16. Moshayedi, A. J., & Gharpure, D. C. (2012, May). Development of position monitoring system for studying performance of wind tracking algorithms. In *ROBOTIK 2012; 7th German Conference on Robotics* (pp. 1–4). VDE.

17. Foroutan, S., Hashemian, M., Khosravi, M., Nejad, M. G, Asefnejad, A., Saber-Samandari, S., & Khandan, A. (2021). A porous sodium alginate-CaSiO₃ polymer reinforced with graphene nanosheet: Fabrication and optimality analysis. *Fibers and Polymers*, 22, 540–549.

18. Moshayedi, A. J., Abbasi, A., Liao, L., & Li, S. (2019). Path planning and trajectroy tracking of a mobile robot using bio-inspired optimization algorithms and PID control. In *2019 IEEE International Conference on Computational Intelligence and Virtual Environments for Measurement Systems and Applications (CIVEMSA)* (pp. 1–6). IEEE.

19. Raisi, A., Asefnejad, A., Shahali, M., Kazerouni, Z. A. S., Kolooshani, A., Saber-Samandari, S., Moghadas, B. K., & Khandan, A. (2020). Preparation, characterization, and antibacterial studies of N, O-carboxymethyl chitosan as a wound dressing for bedsore application. *Archives of Trauma Research*, 9(4), 181–188.

20. Esfahani, O. T., & Moshayedi, A. J. (2014). Accuracy of the positioning systems for the tracking of Alzheimer's patients—A review. *International Journal of Applied Electronics in Physics & Robotics*, 2(2), 10–16.

21. Heidari, A., Forouzan, M. R., & Akbarzadeh, S. (2014). Effect of friction on tandem cold rolling mills chattering. *ISIJ International*, 54(10), 2349–2356.

22. Dong, X, Heidari, A., Mansouri, A., Hao, W. S., Dehghani, M., Saber-Samandari, S., Toghraie, D., & Khandan, A. (2021). Investigation of the mechanical properties of a bony scaffold for comminuted distal radial fractures: Addition of akermanite nanoparticles and using a freeze-drying technique. *Journal of the Mechanical Behavior of Biomedical Materials*, 121, 104643.

23. Hajshirmohammadi, B., Forouzan, M. R., & Heidari, A. (2019). Effect of interstand tensions on lubrication regime in cold strip rolling with O/W emulsion. *Tribology Transactions*, 62(4), 548–556.

24. Moradi, A., Heidari, Amini, K., Aghadavoudi, F., & Abedinzadeh, R. (2021). Molecular modeling of Ti-6Al-4V alloy shot peening: The effects of diameter and velocity of shot particles and force field on mechanical properties and residual stress. *Modelling and Simulation in Materials Science and Engineering*, 29(6), 065001.
25. Moradi, A., Heidari, A., Amini, K., Aghadavoudi, F., & Abedinzadeh, R. (2022). The effect of shot peening time on mechanical properties and residual stress in Ti-6Al-4V alloy. *Metallurgical Research & Technology*, 119(4), 401.
26. Li, X., Heidari, A., Nourbakhsh, S. M., Mohammadi, R., & Semiromi, D. (2022). Design and fabrication of elastic two-component polymer-metal disks using a 3D printer under different loads for the lumbar spine. *Polymer Testing*, 112, 107633.
27. Rajabi, H., & Heidari, A. (2018). Analysis and presenting an optimum post weld heat treatment cycle to maximum reduction of residual stresses of electron beam welding. *Journal of Mechanical Engineering and Vibration*, 9(2), 55–65.
28. Moshayedi, A. J., Roy, A. S., Sambo, S. K., Zhong, Y., & Liao, L. (2022). Review on: The service robot mathematical model. *EAI Endorsed Transactions on AI and Robotics*, 1(1), 1–19.
29. Moshayedi, A. J., & Gharpure, D. (2014). Implementing breath to improve response of gas sensors for leak detection in plume tracker robots. In *Proceedings of the Third International Conference on Soft Computing for Problem Solving: SocProS 2013* (Vol. 2, pp. 337–348). Springer India.
30. Fereidooni, B., Morovvati, M. R., & Sadough-Vanini, S. A. (2018). Influence of severe plastic deformation on fatigue life applied by ultrasonic peening in welded pipe 316 Stainless Steel joints in corrosive environment. *Ultrasonics*, 88, 137–147.
31. Asadian-Ardakani, M. H., Morovvati, M. R., Mirnia, M. J., & Dariani, B. M. (2017). Theoretical and experimental investigation of deep drawing of tailor-welded IF steel blanks with non-uniform blank holder forces. *Proceedings of the Institution of Mechanical Engineers, Part B: Journal of Engineering Manufacture*, 231(2), 286–300.
32. Haddadzadeh, M., Razfar, M. R., & Mamaghani, M. R. M. (2009). Novel approach to initial blank design in deep drawing using artificial neural network. *Proceedings of the Institution of Mechanical Engineers, Part B: Journal of Engineering Manufacture*, 223(10), 1323–1330.
33. Yarahmadi, A., Hashemian, M., Toghraie, D., Abedinzadeh, R., & Ali Eftekhari, S. (2022). Investigation of mechanical properties of epoxy-containing Detda and Degba and graphene oxide nanosheet using molecular dynamics simulation. *Journal of Molecular Liquids*, 347, 118392.
34. Morovvati, M. R., Mollaei-Dariani, B., Lalehpour, A., & Toghraie, D. (2023). Fabrication and finite element simulation of aluminum/carbon nanotubes sheet reinforced with Thermal Chemical Vapor Deposition (TCVD). *Journal of Materials Research and Technology*, 23, 1887–1902.
35. Wu, J., Ling, C., Ge, A., Jiang, W., Baghaei, S., & Kolooshani, A. (2022). Investigating the performance of tricalcium phosphate bioceramic reinforced with titanium nanoparticles in friction stir welding for coating of orthopedic prostheses application. *Journal of Materials Research and Technology*, 20, 1685–1698.
36. Chen, X., Kolooshani, A., Heidarshenas, B., Mortezagholi, B., Yuan, Y., & Semiruomi, D. T. (2023). Effects of tricalcium phosphate-titanium nanoparticles on mechanical performance after friction stir processing on titanium alloys for dental applications. *Materials Science and Engineering: B*, 293, 116492.
37. Khandan, A., Karamian, E., Mehdikhani-Nahrkhalaji, M., Mirmohammadi, H., Farzadi, A., Ozada, N., Heidarshenas, B., & Zamani, K. (2015). Influence of spark plasma sintering and baghdadite powder on mechanical properties of hydroxyapatite. *Procedia Materials Science*, 11, 183–189.

38. Abedinzadeh, R., & Nejad, M. F. (2021). Effect of embedded shape memory alloy wires on the mechanical behavior of self-healing graphene-glass fiber-reinforced polymer nanocomposites. *Polymer Bulletin*, 78(6), 3009–3022.

39. Afshary, K., Chamanara, M., Talari, B., Rezaei, P., & Nassireslami, E. (2020). Therapeutic effects of minocycline pretreatment in the locomotor and sensory complications of spinal cord injury in an animal model. *Journal of Molecular Neuroscience*, 70, 1064–1072.

40. Zarei, M. H., Pourahmad, J., Aghvami, M., Soodi, M., & Nassireslami, E. (2017). Lead acetate toxicity on human lymphocytes at non-cytotoxic concentrations detected in human blood. *Main Group Metal Chemistry*, 40(5–6), 105–112.

41. Jones, R., Matthews, N., Green, R., & Peng, D. (2015). On the potential of supersonic particle deposition to repair simulated corrosion damage. *Engineering Fracture Mechanics*, 137, 26–33.

42. Do, A. K., Wright, P. K., & Sequin, C. H. (2000). Latest generation SLA resins enable direct tooling for injection molding. *Society of Manufacturing Engineers*, 5, 1–15.

43. Goh, G. D., Agarwala, S., Goh, G. L., Dikshit, V., Sing, S. L., & Yeong, W. Y. (2017). Additive manufacturing in unmanned aerial vehicles (UAVs): Challenges and potential. *Aerospace Science and Technology*, 63, 140–151.

44. Quincieu, J., Robinson, C., Stucker, B., & Mosher, T. (2005). Case study: Selective laser sintering of the USUSat II small satellite structure. *Assembly Automation*, 25(4), 267–272.

45. Hauser, C. (2014). Case Study: Laser powder metal deposition manufacturing of complex real Parts. TWI. https://docplayer. net/48458850-Case-study-laser-powder-metal-depositionmanufacturing-of-complex-real-parts. html.

46. Najmon, J. C., Raeisi, S., & Tovar, A. (2019). Review of additive manufacturing technologies and applications in the aerospace industry. *Additive Manufacturing for the Aerospace Industry*, 7–31.

47. Harbaugh, J. (2017). Space station 3-D printer builds ratchet wrench to complete first phase of operations. NASA, available at: http://www. nasa. gov/mission_pages/station/research/news/3Dratchet_wrench (accessed 28 September 2018).

48. Saffarzadeh, M., Gillispie, G. J., & Brown, P. (2016). Selective Laser Sintering (SLS) rapid protytping technology: A review of medical applications. In *Proceedings of the 53rd Annual Rocky Mountain Bioengineering Symposium, RMBS 2016 and 53rd International ISA Biomedical Sciences Instrumentation Symposium* (pp. 8–10). Denver, CO.

49. Sljivic, M., Mirjanic, D., Sljivic, N., Fragassa, C., & Pavlovic, A. (2019). 3D printing and 3D bioprinting to use for medical applications. *Contemporary Materials*, 10(1), 82–92.

50. Zhao, Y., Yao, R., Ouyang, L., Ding, H., Zhang, T., Zhang, K., & Sun, W. (2014). Three-dimensional printing of Hela cells for cervical tumor model in vitro. *Biofabrication*, 6(3), 035001.

51. Goyanes, A., Fina, F., Martorana, A., Sedough, D., Gaisford, S., & Basit, A. W. (2017). Development of modified release 3D printed tablets (printlets) with pharmaceutical excipients using additive manufacturing. *International Journal of Pharmaceutics*, 527(1–2), 21–30.

52. Choonara, Y. E., du Toit, L. C., Kumar, P., Kondiah, P. P., & Pillay, V. (2016). 3D-printing and the effect on medical costs: A new era. *Expert Review of Pharmacoeconomics & Outcomes Research*, 16(1), 23–32.

53. Mohamed, O. A., Masood, S. H., & Bhowmik, J. L. (2015). Optimization of fused deposition modeling process parameters: A review of current research and future prospects. *Advances in Manufacturing*, 3(1), 42–53.

54. Melocchi, A., Parietti, F., Maroni, A., Foppoli, A., Gazzaniga, A., & Zema, L. (2016). Hot-melt extruded filaments based on pharmaceutical grade polymers for 3D printing by fused deposition modeling. *International Journal of Pharmaceutics*, 509(1–2), 255–263.

55. Saffarzadeh, M., Gillispie, G. J., & Brown, P. (2016, April). Selective Laser Sintering (SLS) rapid protytping technology: A review of medical applications. In *53rd Annual Rocky Mountain Bioengineering Symposium, RMBS 2016 and 53rd International ISA Biomedical Sciences Instrumentation Symposium* (pp. 142–149).

56. Khandan, A., Abdellahi, M., Ozada, N., & Ghayour, H. (2016). Study of the bioactivity, wettability and hardness behaviour of the bovine hydroxyapatite-diopside bio-nanocomposite coating. *Journal of the Taiwan Institute of Chemical Engineers*, 60, 538–546.

57. Karamian, E., Motamedi, M. R. K., Khandan, A., Soltani, P., & Maghsoudi, S. (2014). An in vitro evaluation of novel NHA/zircon plasma coating on 316L stainless steel dental implant. *Progress in Natural Science: Materials International*, 24(2), 150–156.

58. Karamian, E., Abdellahi, M., Khandan, A., & Abdellah, S. (2016). Introducing the fluorine doped natural hydroxyapatite-titania nanobiocomposite ceramic. *Journal of Alloys and Compounds*, 679, 375–383.

59. Najafinezhad, A., Abdellahi, M., Ghayour, H., Soheily, A., Chami, A., & Khandan, A. (2017). A comparative study on the synthesis mechanism, bioactivity and mechanical properties of three silicate bioceramics. *Materials Science and Engineering: C*, 72, 259–267.

60. Ghayour, H., Abdellahi, M., Ozada, N., Jabbrzare, S., & Khandan, A. (2017). Hyperthermia application of zinc doped nickel ferrite nanoparticles. *Journal of Physics and Chemistry of Solids*, 111, 464–472.

61. Kazemi, A., Abdellahi, M., Khajeh-Sharafabadi, A., Khandan, A., & Ozada, N. (2017). Study of in vitro bioactivity and mechanical properties of diopside nano-bioceramic synthesized by a facile method using eggshell as raw material. *Materials Science and Engineering: C*, 71, 604–610.

62. Khandan, A., & Ozada, N. (2017). Bredigite-Magnetite (Ca7MgSi4O16-Fe3O4) nanoparticles: A study on their magnetic properties. *Journal of Alloys and Compounds*, 726, 729–736.

63. Khandan, A., Jazayeri, H., Fahmy, M. D., & Razavi, M. (2017). Hydrogels: Types, structure, properties, and applications. *Biomaterials and Tissue Engineering*, 4(27), 143–69.

64. Sharafabadi, A. K., Abdellahi, M., Kazemi, A., Khandan, A., & Ozada, N. (2017). A novel and economical route for synthesizing akermanite (Ca2MgSi2O7) nano-bioceramic. *Materials Science and Engineering: C*, 71, 1072–1078.

65. Khandan, A., Abdellahi, M., Ozada, N., & Ghayour, H. (2016). Study of the bioactivity, wettability and hardness behaviour of the bovine hydroxyapatite-diopside bio-nanocomposite coating. *Journal of the Taiwan Institute of Chemical Engineers*, 60, 538–546.

66. Shayan, A., Abdellahi, M., Shahmohammadian, F., Jabbarzare, S., Khandan, A., & Ghayour, H. (2017). Mechanochemically aided sintering process for the synthesis of barium ferrite: Effect of aluminum substitution on microstructure, magnetic properties and microwave absorption. *Journal of Alloys and Compounds*, 708, 538–546.

67. Heydary, H. A., Karamian, E., Poorazizi, E., Khandan, A., & Heydaripour, J. (2015). A novel nanofiber of Iranian gum tragacanth-polyvinyl alcohol/nanoclay composite for wound healing applications. *Procedia Materials Science*, 11, 176–182.

68. Khandan, A., Karamian, E., & Bonakdarchian, M. (2014). Mechanochemical synthesis evaluation of nanocrystalline bone-derived bioceramic powder using for bone tissue engineering. *Dental Hypotheses*, 5(4), 155.

69. Karamian, E., Khandan, A., Kalantar Motamedi, M. R., & Mirmohammadi, H. (2014). Surface characteristics and bioactivity of a novel natural HA/zircon nanocomposite coated on dental implants. *BioMed Research International*, 2014, 1–10.

70. Jabbarzare, S., Abdellahi, M., Ghayour, H., Arpanahi, A., & Khandan, A. (2017). A study on the synthesis and magnetic properties of the cerium ferrite ceramic. *Journal of Alloys and Compounds*, 694, 800–807.

71. Razavi, M., & Khandan, A. (2017). Safety, regulatory issues, long-term biotoxicity, and the processing environment. In *Nanobiomaterials Science, Development and Evaluation* (pp. 261–279). Woodhead Publishing.

72. Khandan, A., Ozada, N., & Karamian, E. (2015). Novel microstructure mechanical activated nano composites for tissue engineering applications. *Journal of Bioengineering and Biomedical Science*, 5(1), 1.

73. Ghayour, H., Abdellahi, M., Bahmanpour, M., & Khandan, A. (2016). Simulation of dielectric behavior in RFeO $$ _ {3} $$3 orthoferrite ceramics (R= rare earth metals). *Journal of Computational Electronics*, 15(4), 1275–1283.

74. Saeedi, M., Abdellahi, M., Rahimi, A., & Khandan, A. (2016). Preparation and characterization of nanocrystalline barium ferrite ceramic. *Functional Materials Letters*, 9(05), 1650068.

75. Khandan, A., Karamian, E., Faghih, M., & Bataille, A. (2014). Formation of AlN nano particles precipitated in St-14 low carbon steel by micro and nanoscopic observations. *Journal of Iron and Steel Research International*, 21(9), 886–890.

76. Karamian, E. B., Motamedi, M. R., Mirmohammadi, K., Soltani, P. A., & Khandan, A. M. (2014). Correlation between crystallographic parameters and biodegradation rate of natural hydroxyapatite in physiological solutions. *Indian Journal of Scientific Research*, 4(3), 92–99.

77. Khandan, A., & Esmaeili, S. (2019). Fabrication of polycaprolactone and polylactic acid shapeless scaffolds via fused deposition modelling technology. *Journal of Advanced Materials and Processing*, 7(4), 16–29.

78. Mahale, R. S., Vasanth, S., Krishna, H., Shashanka, R., Sharath, P. C., & Sreekanth, N. V. (2022). Electrochemical sensor applications of nanoparticle modified carbon paste electrodes to detect various neurotransmitters: A review. *Applied Mechanics and Materials*, 908, 69–88.

79. Mahale, R. S., Shamanth, V., Hemanth, K., Nithin, S. K., Sharath, P. C., Shashanka, R., ... & Shetty, D. (2022). Processes and applications of metal additive manufacturing. *Materials Today: Proceedings*, 54, 228–233.

80. Safaei, M., Abedinzadeh, R., Khandan, A., Barbaz-Isfahani, R., & Toghraie, D. (2023). Synergistic effect of graphene nanosheets and copper oxide nanoparticles on mechanical and thermal properties of composites: Experimental and simulation investigations. *Materials Science and Engineering: B*, 289, 116248.

81. Kamble, P., & Mahale, R. (2016). Simulation and parametric study of clinched joint. *International Research Journal of Engineering and Technology*, 3(5), 2730–2734.

82. Asgari, F., Minooei, A., Abdolahi, S., Shokrani Foroushani, R., & Ghorbani, A. (2021). A new approach using Machine Learning and Deep Learning for the prediction of cancer tumor. *Journal of Simulation and Analysis of Novel Technologies in Mechanical Engineering*, 13(4), 41–51.

83. Maghsoudlou, M. A., Nassireslami, E., Saber-Samandari, S., & Khandan, A. (2020). Bone regeneration using bio-nanocomposite tissue reinforced with bioactive nanoparticles for femoral defect applications in medicine. *Avicenna Journal of Medical Biotechnology*, 12(2), 68.

84. Karimianmanesh, M., Azizifard, E., Javidanbashiz, N., Latifi, M., Ghorbani, A., & Shahriari, S. (2021). Feasibility study of mechanical properties of alginates for neuroscience application using finite element method. *Journal of Simulation and Analysis of Novel Technologies in Mechanical Engineering*, 13(3), 53–62.

85. Ehsani, A., Mahale, R., Shayegan, S., Attaeyan, A., Ghorbani, A., Vasanth, S., P C, S., Shahriari, S., Asefnejad, A. (2022). A review of the treatment of bone tumours by hyperthermia using magnetic nanoparticles. *Journal of Nanoanalysis*, doi: 10.22034/jna.2022.1944876.1278

86. Ghorbani, A., Shahriari, S., & Gholami, A. M. (2021). Investigation of cell biomechanics and the effect of biomechanical stimuli on cancer and their characteristics. *Journal of Simulation and Analysis of Novel Technologies in Mechanical Engineering*, 13(4), 67–79.

87. Malekipour Esfahani, M. H., Sharifinezhad, N., Hemati, M., & Gholami, A. M. (2021). Evaluation of mechanical properties of bioglass materials for dentistry application. *Journal of Simulation and Analysis of Novel Technologies in Mechanical Engineering*, 13(4), 19–29.

88. Ghomi, F., Asefnejad, A., Daliri, M., Godarzi, V., & Hemati, M. (2022). Fabrication and characterization of chitosan/gelatin scaffold with bioactive glass reinforcement using PRP to regenerate bone tissue. *Nanomedicine Research Journal*, 7(2), 205–213.

89. Ravishankar, S., & Mahale, R. (2015). A study on magneto rheological fluids and their applications. *International Research Journal of Engineering and Technology*, 2, 2023–2028.

90. Patil, A., Banapurmath, N., Hunashyal, A. M., Meti, V., & Mahale, R. (2022). Development and performance analysis of novel cast AA7076-graphene amine-carbon fiber hybrid nanocomposites for structural applications. *Biointerface Research in Applied Chemistry*, 12(2), 1480–1489.

91. Mahale, R. S., & Shashanka, R. (2022, March). Mechanical testing of spark plasma sintered materials: A review. In *AIP Conference Proceedings* (Vol. 2469, No. 1, p. 020026). AIP Publishing LLC.

92. Abedpour, M., Kamyab Moghadas, B., & Tamjidi, S. (2020). Equilibrium and kinetic study of simultaneous removal of Cd (II) and Ni (II) by acrylamide-based polymer as effective adsorbent: Optimisation by response surface methodology (RSM). *International Journal of Environmental Analytical Chemistry*, 102(15), 1–18.

93. Tamjidi, S., Esmaeili, H., & Moghadas, B. K. (2021). Performance of functionalized magnetic nanocatalysts and feedstocks on biodiesel production: A review study. *Journal of Cleaner Production*, 305, 127200.

94. Heydari, S., Attaeyan, A., Bitaraf, P., Gholami, A. M., & Kamyab Moghadas, B. (2021). Investigation of modern ceramics in bioelectrical engineering with proper thermal and mechanical properties. *Journal of Simulation and Analysis of Novel Technologies in Mechanical Engineering*, 13(3), 43–52.

95. Ravi, G. R., & Subramanyam, R. V. (2012). Calcium hydroxide-induced resorption of deciduous teeth: A possible explanation. *Dental Hypotheses*, 3(3), 90.

96. Farazin, A., Aghadavoudi, F., Motififard, M., Saber-Samandari, S., & Khandan, A. (2021). Nanostructure, molecular dynamics simulation and mechanical performance of PCL membranes reinforced with antibacterial nanoparticles. *Journal of Applied and Computational Mechanics* 7(4), 1907–1915.

97. Gupta, R., Thakur, N., Thakur, S., Gupta, B., & Gupta, M. (2013). Talon cusp: A case report with management guidelines for practicing dentists. *Dental Hypotheses*, 4(2), 67.

98. Monfared, R. M., Ayatollahi, M. R., & Isfahani, R. B. (2018). Synergistic effects of hybrid MWCNT/nanosilica on the tensile and tribological properties of woven carbon fabric epoxy composites. *Theoretical and Applied Fracture Mechanics*, 96, 272–284.

99. Lucchini, R., Carnelli, D., Gastaldi, D., Shahgholi, M., Contro, R., & Vena, P. (2012). A damage model to simulate nanoindentation tests of lamellar bone at multiple penetration depth. In *6th European Congress on Computational Methods in Applied Sciences and Engineering, ECCOMAS 2012* (pp. 5919–5924).

100. Mahjoory, M., Shahgholi, M., & Karimipour, A. (2022). The effects of initial temperature and pressure on the mechanical properties of reinforced calcium phosphate cement with magnesium nanoparticles: A molecular dynamics approach. *International Communications in Heat and Mass Transfer*, 135, 106067.

101. Talebi, M., Abbasi-Rad, S., Malekzadeh, M., Shahgholi, M., Ardakani, A. A., Foudeh, K., & Rad, H. S. (2021). Cortical bone mechanical assessment via free water relaxometry at 3 T. *Journal of Magnetic Resonance Imaging*, 54(6), 1744–1751.

102. Shahgholi, M., Oliviero, S., Baino, F., Vitale-Brovarone, C., Gastaldi, D., & Vena, P. (2016). Mechanical characterization of glass-ceramic scaffolds at multiple characteristic lengths through nanoindentation. *Journal of the European Ceramic Society*, 36(9), 2403–2409.

103. Fada, R., Farhadi Babadi, N., Azimi, R., Karimian, M., & Shahgholi, M. (2021). Mechanical properties improvement and bone regeneration of calcium phosphate bone cement, polymethyl methacrylate and glass ionomer. *Journal of Nanoanalysis*, 8(1), 60–79.

104. Kamarian, S., Bodaghi, M., Isfahani, R. B., & Song, J. I. (2021). Thermal buckling analysis of sandwich plates with soft core and CNT-Reinforced composite face sheets. *Journal of Sandwich Structures & Materials*, 23(8), 3606–3644.

105. Kamarian, S., Bodaghi, M., Isfahani, R. B., & Song, J. I. (2022). A comparison between the effects of shape memory alloys and carbon nanotubes on the thermal buckling of laminated composite beams. *Mechanics Based Design of Structures and Machines*, 50(7), 2250–2273.

106. Barbaz-I, R. (2014). *Experimental determining of the elastic modulus and strength of composites reinforced with two nanoparticles* (Doctoral dissertation, Doctoral dissertation, MSc Thesis, School of Mechanical Engineering Iran University of Science and Technology, Tehran, Iran).

107. Morovvati, M. R., Niazi Angili, S., Saber-Samandari, S., Ghadiri Nejad, M., Toghraie, D., & Khandan, A. (2023). Global criterion optimization method for improving the porosity of porous scaffolds containing magnetic nanoparticles: Fabrication and finite element analysis. *Materials Science and Engineering: B* 292, 116414.

108. Safaei, M., Abedinzadeh, R., Khandan, A., Barbaz-Isfahani, R., & Toghraie, D. (2023). Synergistic effect of graphene nanosheets and copper oxide nanoparticles on mechanical and thermal properties of composites: Experimental and simulation investigations. *Materials Science and Engineering: B*, 289, 116248.

109. Moarrefzadeh, A., Morovvati, M. R., Angili, S. N., Smaisim, G. F., Khandan, A., & Toghraie, D. (2022). Fabrication and finite element simulation of 3D printed poly L-lactic acid scaffolds coated with alginate/carbon nanotubes for bone engineering applications. *International Journal of Biological Macromolecules*, 224, 1496–1508.

110. Esmaeili, S., Khandan, A., & Saber-Samandari, S. (2018). Mechanical performance of three-dimensional bio-nanocomposite scaffolds designed with digital light processing for biomedical applications. *Iranian Journal of Medical Physics*, 15(Special Issue-12th. Iranian Congress of Medical Physics), 328–328.

111. Rajaei, A., Kazemian, M., & Khandan, A. (2022). Investigation of mechanical stability of lithium disilicate ceramic reinforced with titanium nanoparticles. *Nanomedicine Research Journal*, 7(4) 350–359.

112. Soleimani, M., Salmasi, A. A., Asghari, S., Yekta, H. J., Moghadas, B. K., Shahriari, S., Saber-Samandari, S., & Khandan, A. (2021). Optimization and fabrication of alginate scaffold for alveolar bone regeneration with sufficient drug release. *International Nano Letters*, 11(3), 295–305.

113. Kjaer, I. (2013). External root resorption: Different etiologies explained from the composition of the human root-close periodontal membrane. *Dental Hypotheses*, 4(3), 75.

114. Motamedi, M. R. K., Behzadi, A., Khodadad, N., Zadeh, A. K., & Nilchian, F. (2014). Oral health and quality of life in children: A cross-sectional study. *Dental Hypotheses*, 5(2), 53.
115. Narayanan, N., & Thangavelu, L. (2015). Salvia officinalis in dentistry. *Dental Hypotheses*, 6(1), 27.
116. Heydary, H. A., Karamian, E., Poorazizi, E., Heydaripour, J., & Khandan, A. (2015). Electrospun of polymer/bioceramic nanocomposite as a new soft tissue for biomedical applications. *Journal of Asian Ceramic Societies*, 3(4), 417–425.
117. Qian, W.-M., Vahid, M. H., Sun, Y.-L., Heidari, A., Barbaz-Isfahani, R., Saber-Samandari, S., Khandan, A., & Toghraie, D. (2021). Investigation on the effect of functionalization of single-walled carbon nanotubes on the mechanical properties of epoxy glass composites: Experimental and molecular dynamics simulation. *Journal of Materials Research and Technology*, 12, 1931–1945.

8 Investigation on the Effect of Different Joint-Based Topology of PLA Core Structure Using 3D Printing Technology

Zichen W., Zuhri M.Y.M., Hang Tuah B.T., As'arry A., and Lalegani Dezaki M.

8.1 INTRODUCTION

A lightweight structure is highly desirable for engineering applications as it can deliver excellent performance, particularly in specific energy absorption (EA) (Dong et al. 2017; Tamburrino et al. 2018). These structures are renowned for their impressive capacity to absorb energy from impacts. Among different types of sandwich structures, the hexagonal shape structure is recognized as being particularly effective in terms of EA under compression loading (Sun et al. 2022; Tatlier 2021; Wang et al. 2009). The use of hexagonal shapes in the honeycomb structure of commercial products is a well-known practice, as it is employed in various industries ranging from aircraft and automobile manufacturing to building construction. Hexagonal honeycomb structures are utilized to enhance properties such as EA, air directionalization, acoustic panels, and light diffusion (Qi et al. 2021; Thomas and Tiwari 2019a; Wang 2019; Zhang et al. 2015). Three-dimensional printing, in this case, fused deposition modeling (FDM), is an optimal method for creating honeycomb structures, which are achieved by designing specific structures consisting of periodically arranged unit cells (Palaniyappan et al. 2022; Płaczek et al. 2021). This technology allows for the manufacturing of novel and intricate structures in a relatively short amount of time compared to conventional methods. FDM offers the opportunity for precise, cost-effective, and time-saving manufacturing of honeycomb structures with a wide range of materials, especially polymers, due to their ease of printing, cost-effectiveness, lightness, and flexibility (Azmi et al. 2018; Dong et al. 2018; Lalegani et al. 2021). In recent studies, new lattice structures have been introduced and their topological optimization has been discussed to achieve effective Young's modulus and higher EA capability (Yousefi et al. 2022). The geometric parameters of honeycombs, a common type of auxetic structure, such as the ratio of base length and connection angle, have been evaluated for their effect

DOI: 10.1201/9781003362128-8

on the hardness, strength, and EA properties (Choudhry et al. 2022; Gohar et al. 2021). Mehrpouya et al. (2020) conducted a study on the impact of printing parameters, such as printing speed and nozzle diameter, on the reversibility of polylactic acid (PLA) sandwich structures. Their findings revealed that decreasing printing speed and increasing nozzle size resulted in improved shape recovery capabilities. Serjouei et al. (2022) investigated the EA and shape recovery potential of two bio-inspired horseshoe-shaped structures with negative Poisson's ratios, analyzing the impact of various manufacturing parameters. The results showed that increasing structure thickness led to an increase in EA rate and specific EA. Other researchers have also explored the design and development of meta-structures from natural or biological structures, such as 3D-printed bio-inspired honeycomb structures reinforced by starfish shape elements, as a means of improving EA and thermal conductivity (Saufi et al. 2021).

Moreover, Yap et al. (2015) investigated the capability of honeycomb structures using PolyJet technology and found that the compression strength increased when the loading rate increased accordingly. An et al. (2017) observed that introduction of hierarchy into the honeycomb structure can enhance its strength and EA. They also suggested that the selection of good hierarchical parameters was required to gain the optimum properties. One has developed a second-order level of hierarchical hexagonal structure and found that the EA increased up to 56.45% as compared to that regular honeycomb unit cell (Korupolu et al. 2022). Additionally, thickness parameters of sub-regions for self-similar hierarchical structures should also be taken into consideration as they can affect the failure deformation under compression loading (Liang et al. 2021). Also, Xu et al. (2019) studied the EA capacity between the hexagonal standard shape with a combination of hybrid structures consisting of auxetic and hexagonal honeycomb cells under in-plane compression. The hybrid structure improved 38% compared with the original hexagonal honeycomb structure, which indicated that the internal geometry of the structure had influence on the EA performance.

Modifying the EA of honeycomb structures is an effective way to enhance their crashworthiness capabilities. Research works demonstrated that bio-inspired structures could increase EA (Audibert et al. 2018; Tasdemirci et al. 2018), and introducing a secondary structure and increasing the stiffness value can also enhance the honeycomb structure's performance (Wang et al. 2015). In addition, the failure behavior of sandwich structures can be influenced by the wall thickness and cell size of the core structure (Shan et al. 2019). Compared to rectangular honeycomb structures, the multilayer hexagonal shape can absorb 31%–60% more energy (Ali et al. 2015). A comparative study has shown that the hexagonal shape honeycomb exhibits better stiffness than foam sandwiches, mainly due to their cell size and buckling. Therefore, controlling the crashworthiness of honeycomb structures is more manageable (Crupi et al. 2013). Varying cell size, material, and wall thickness yields different results in EA. By optimizing these parameters, researchers can increase EA, as suggested by Tao et al. (2017), which showed that changing the wall thickness resulted in a different crushing strength and EA. Furthermore, enhancing the material at the intersecting area can improve EA by promoting plastic deformation near the cell wall (Thomas and Tiwari 2019b).

In this work, an investigation of the effect of different joint-based designs, which is at the nodal part of the base core structure, is conducted under edgewise compression testing. The purpose of this research is to enhance the compression strength and EA capabilities of the conventional honeycomb structure shape. The experiment will involve testing the structure under edgewise compression loading, with a specific emphasis on the impact of varying joint-based shapes and sizes. The study includes understanding of their maximum strength, energy-absorbing capacity, as well as failure deformation.

8.2 METHODOLOGY

8.2.1 DESIGN

The primary objective of this study is to explore the impact of various additional structures of different shapes and sizes on the nodal points of the standard hexagonal honeycomb structure. Initially, circular, triangular, and hexagonal specimens with 2 mm, 3 mm, and 4 mm contacts, both hollow and solid, were designed, in addition to the original hexagonal honeycomb structure. AutoCAD 2016 software was used to create the sample structure to get a better visual of the solid and hollow joint-based nodal. The samples are coded for easy identification, such as original hexagonal honeycomb as a benchmark (OG), hexagonal hollow (HH), hexagonal solid (HS), circular hollow (CH), circular solid (CS), triangular hollow (TH), and triangular solid (TS). The size of the lattice designs is shown in Figure 8.1(a–g).

8.2.2 FABRICATION OF SAMPLE

Here, the samples are printed using the 3D printing machine model Ultimaker 2+ with a volume build-up capability of 223 × 223 × 205 mm. The diameter of the nozzle is approximately 0.25 mm and the temperature ranges from 190°C to 230°C. The material used in this study is polylactic acid (PLA) filament from Polymaker. The density is between 1.17 g/cm^3 and 1.24 g/cm^3 with a melting temperature of 149°C. The printing parameters for all samples are 200°C nozzle temperature, 60°C bed temperature, 100% infill density, 0.2 mm layer height, 70 mm/s printing speed, and 100% material flow. The base core design is the hexagonal shape which is like the honeycomb structure. Three different joint-based topologies are selected, these being hexagonal, circle, and triangular. The size of the joint-based topologies used is 2 mm, 3 mm, and 4 mm. While using the Ultimaker 2+ machine for 3D printing, it was discovered that the node radius was too small, and the 3D printer required a minimum thickness of 0.4 mm. Upon examination, it was found that the hollow specimens with radii of 2 mm and 3 mm were almost indistinguishable from the solid specimens since the hollow part was filled with printing material. Therefore, the hollow specimens with radii of 2 mm and 3 mm were not used, and only the specimen with a 4 mm radius and a hollow joint-based nodal was created. The specimens with radii of 2 mm and 3 mm were only evaluated using the solid nodal. Overall, 39 samples are printed, and examples are shown in Figure 8.1(h–j).

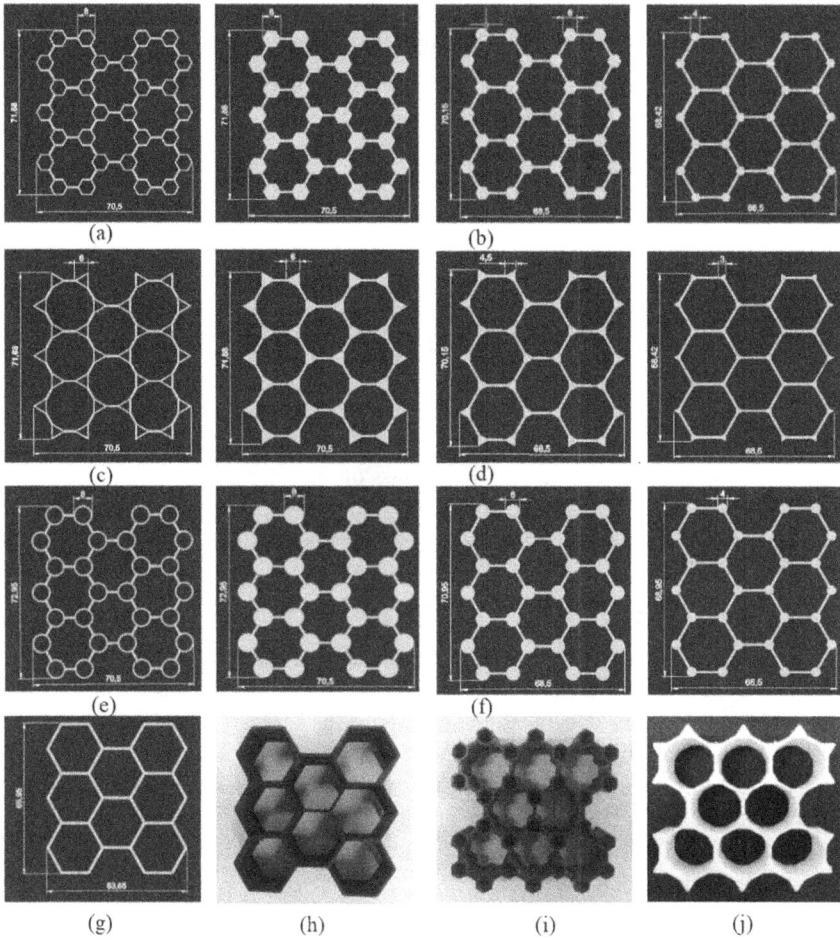

FIGURE 8.1 Design of (a) hollow hexagonal joint-based topology, (b) solid hexagonal joint-based topology, (c) hollow triangular joint-based topology, (d) solid triangular joint-based topology (e) hollow circular joint-based topology (f) solid circular joint-based topology, and (g) original honeycomb. Examples of 3D-printed samples of (h) original honeycomb, (i) hexagonal joint-based topology and (j) solid triangular joint-based topology.

8.3.3 COMPRESSION TESTING

The samples are tested under compression loading, which is edgewise compression testing following the ASTM C364 standard. An Instron machine model 3382 is used to investigate the compression properties of the specimen. The samples are placed between the bottom and the top plate of the machine. Initially, the top plate is adjusted to give an ample distance between the top sample surface and the top plate. A crosshead displacement of 2 mm/min is used with a load cell of 10 kN. During the testing, the failure deformations are captured for every 2 mm displacement. The testing is conducted until the sample is fully crushed. All samples are weighed before the testing begins, which will later be used for the determination of their specific energy absorption (SEA) value. The EA and SEA are calculated from the area under the load-displacement curve by using the trapezoidal rule (Haghdan and Smith 2015) (see Equations 1 and 2).

$$EA = \int_0^d F(x)\, dx \qquad (1)$$

$F(x)$ is the function of displacement (x) and d is the deformation. Total EA is the accumulative load-displacement curve from zero to the maximum deformation.

$$SEA = \frac{W_f}{M} \qquad (2)$$

where Wf is work of fracture and M is the mass of the sample.

8.3 RESULTS AND DISCUSSION

8.3.1 FAILURE DEFORMATION

Figure 8.2(a–d) shows examples of load-displacement curves for few samples. The failure was taken at four different stages, these being the elastic, elastic-plastic, plateau, and finally at their densification area where the samples have been completely crushed. Upon examining each stage, it becomes apparent that core failure in most cases is caused by buckling under bending stress. However, many joint-based topologies delay the onset of bending failure and instead transition into a secondary structure to support the primary structure. Hollow joint-based topologies, despite their main structure being deformed, show little to no changes until stage 3, likely due to their smaller diameter. In contrast, most solid joint-based topologies fill the gap area once the structure reaches stages 3 and 4.

The load-displacement diagram in Figure 8.2 indicates that most specimens experienced one or several fluctuations before the stress vanished. During the experiment, it was observed that the specimen was compressed between pressure plates, and the middle part broke, resulting in immediate stress. As the pressure plate continued to push down until the specimen lost its elastic deformation, the elastic stress disappeared, leaving only plastic deformation which could not be recovered. As the volume of the specimen was continuously compressed, it became denser. The larger the radius of the joint-based solid, the greater was the fluctuation because the larger the joint-based core (shorter side length of the original honeycomb structure), the shorter the object and the more prone it was to irreversible deformation. When non-elastic deformation, direct fracture, and specimen stress disappeared, the load-displacement diagram showed a sudden decrease. The solid joint caused stress to arise during a decrease in the pressure plate. The side length of the original honeycomb structure of a specimen with a smaller joint-based core will be longer, making it more likely to undergo elastic deformation. Therefore, the load-displacement diagram of a specimen with a smaller radius of a solid joint exhibited significant undulation but less intensity.

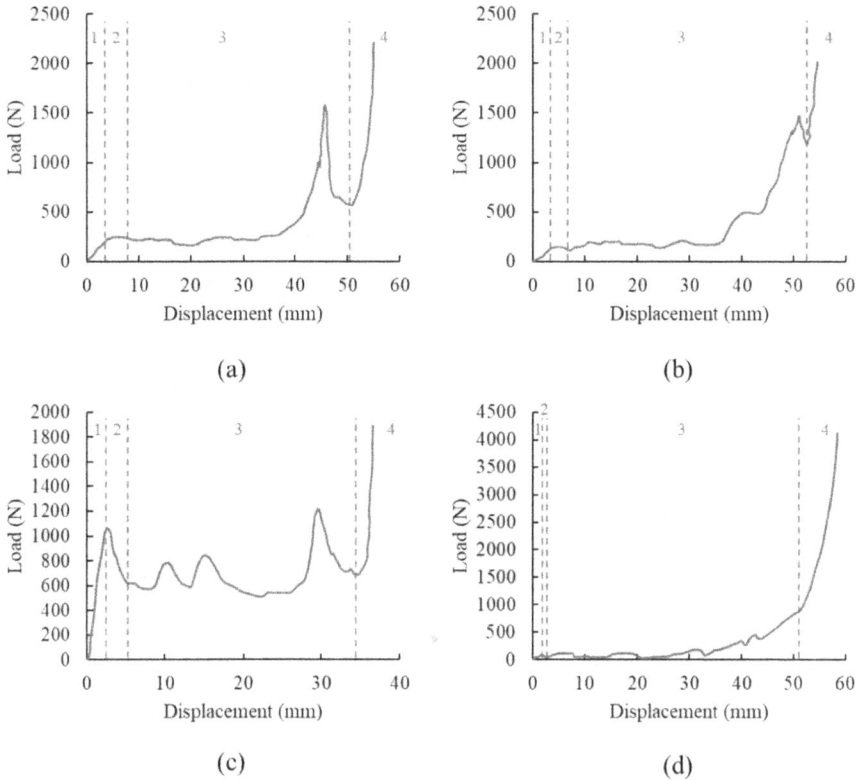

FIGURE 8.2 Load-displacement graph of (a) original honeycomb, (b) hollow hexagonal joint-based core, (c) solid hexagonal joint-based core, and (d) hollow circular joint-based core.

After reaching the maximum peak value, the structure began collapsing. As the load continued, the structure expanded to the sides and the main structure, compressing its joint-based topology. For the solid joint-based samples, only the main honeycomb structure was compressed, and the joint-based core remained uncompressed. The stage of failure of all different joint-based topologies is presented in Table 8.1. Most of the specimens fail in some layers when they encounter overlap, and then fail in other layers. When compressing a hollow specimen, the test piece with a hollow triangle contact differs from a hollow circle or hexagon. Due to the stability of triangles, the sample generally retains its original shape during compression and is not easily deformed. When the entire structure fails completely, most of the hollow triangles remain intact and undamaged, whereas the hollow hexagons and circular specimens collapse along with the processor and main structure during compression.

8.3.2 COMPRESSION STRENGTH

Figure 8.3 presents the compression strength value tested under edgewise compression. Here, the highest value lies at the core structure at the solid circular with diameter of 4 mm (CS4), which is around 9.11 MPa, while the lowest value is CH at

TABLE 8.1

Stage of failures comparison of different joint-based topologies

	Stage 1	Stage 2	Stage 3	Stage 4
Original honeycomb				
Hexagonal hollow				
Hexagonal solid (2 mm)				
Hexagonal solid (3 mm)				
Hexagonal solid (4 mm)				
Triangular Hollow				
Triangular Solid (2 mm)				

(Continued)

TABLE 8.1 (*Continued*)
Stage of failures comparison of different joint-based topologies

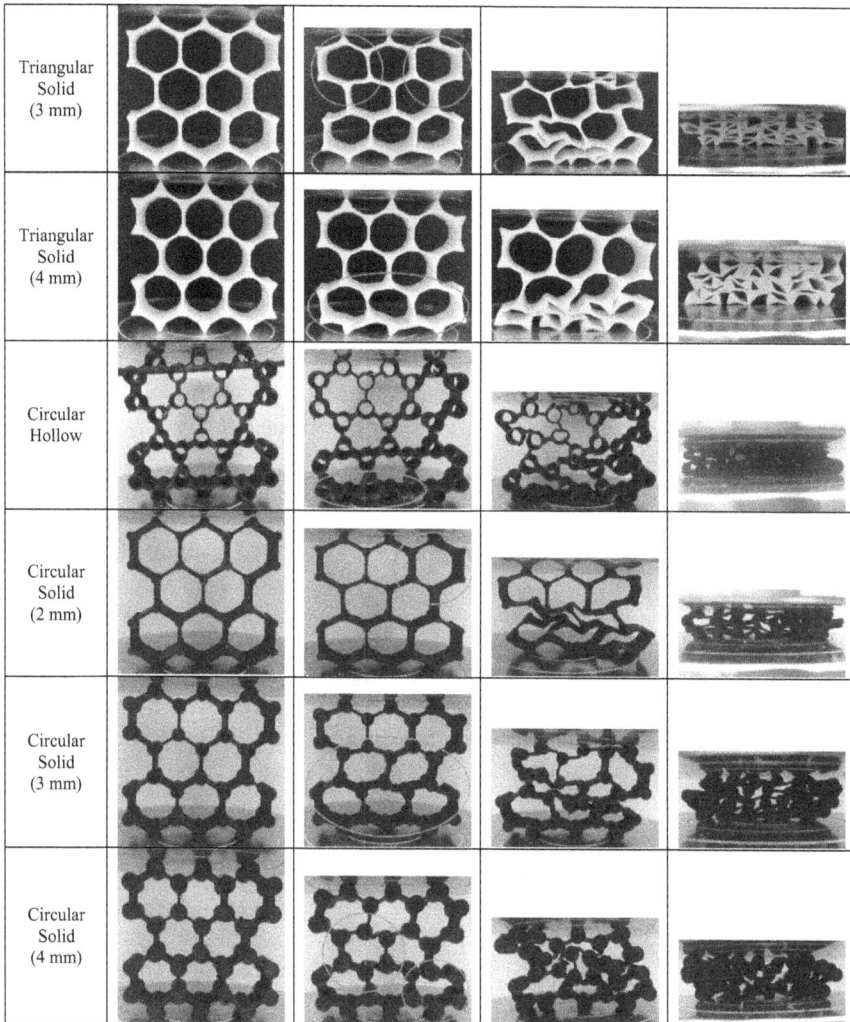

Triangular Solid (3 mm)				
Triangular Solid (4 mm)				
Circular Hollow				
Circular Solid (2 mm)				
Circular Solid (3 mm)				
Circular Solid (4 mm)				

approximately 0.9 MPa. The core with HS is seen to offer comparable strength to that CS4 with less than 10% differences. It is observed that the core with CS offers the higher value compared to the other joint-based shape. Close observation showed that as the size of the joint-based shape decreased, the strength of the core also decreased in all cases. By comparing at each individual parameter, for example, at 2 mm with a different joint-based shape, one can see no significant difference in terms of its strength value. The hollow shape of the joint-based shape indeed gives lower value of strength when compared to the original core; however, when the joint-based solid is

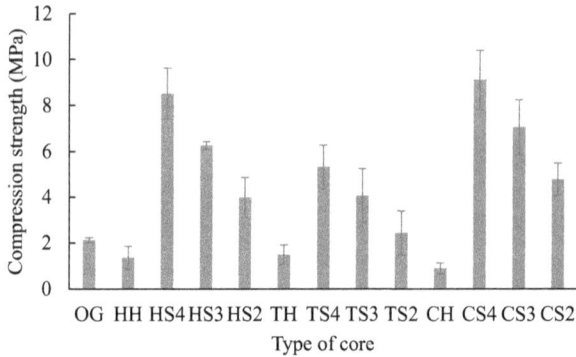

FIGURE 8.3 Comparison of compression strength of different joint-based topologies.

introduced, the core strength increases rapidly. Furthermore, the circular joint-based shape can be seen to give close value to the hexagonal joint-based shape, while the triangular joint-based shape provides lower strength than those two shapes at their individual parameters, e.g. TS4 against HS4 and CS4.

Compared to the original honeycomb structure specimen, the hexagon joint-based core specimens showed significant improvement in compression strength, especially in the solid specimens. The improvement in compression strength was found to be much greater than the improvement in quality and increased even more with larger radii. Also, in triangular samples, it was observed that the load-bearing capacity of the hollow sample with a radius of 4 mm was lower than that of the original honeycomb structure sample and had decreased. On the other hand, the compression strength of the other three solid samples had increased. The solid sample with a 4 mm radius still had the highest compression strength, which was 2.5 times that of the original honeycomb structure sample. Meanwhile, for circular joint-based shapes, the compression strength of a solid specimen with a joint-based core radius of 4 mm was found to have increased by 328% compared to the original honeycomb structure specimen. The solid circular specimen with a 3 mm radius also had a compression strength that was 3.3 times greater than the original honeycomb structure specimen. Additionally, even the solid circular specimen with a 2 mm radius had a compression strength that was 2.25 times greater than the original honeycomb structure specimen. This suggests that the compression strength of specimens with a solid circular joint-based core was significantly improved.

8.3.3 ENERGY ABSORPTION CAPABILITY

The result on the EA capability can be seen in that there is some improvement offered by the core with hollow joint-based shape, especially a triangular shape, which gives the highest SEA value, about 30% than the original core design (see Figure 8.4 (a and b)). The SEA value showed a different trend as compared to the result in Figure 8.3. Here, most of the core gives values ranging from 1.8 to 2 kJ/kg; however, only the core of HS4, HS3, HS2, and TS2 has lower than 1.8 kJ/kg. On the other hand, the SEA of hexagonal shape shows an increasing trend when the size

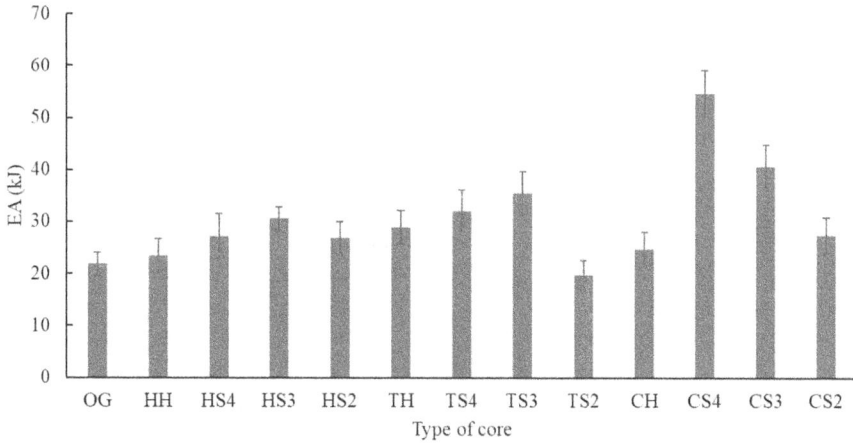

(a)

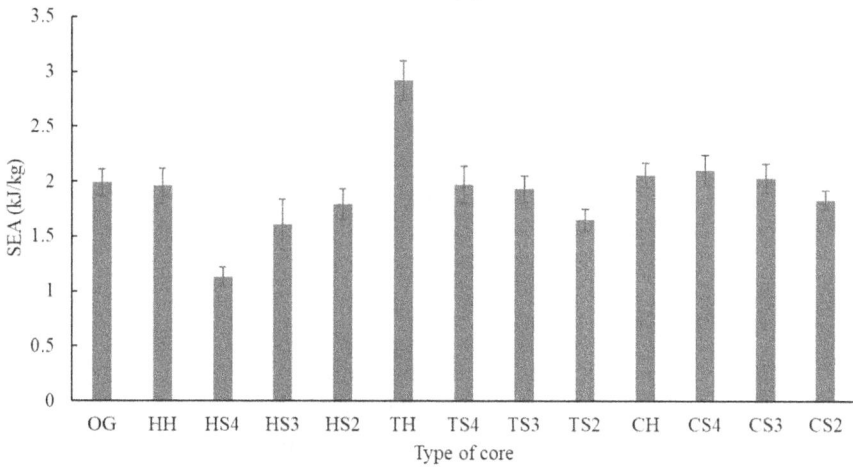

(b)

FIGURE 8.4 Comparison of (a) EA and (b) SEA values of different joint-based topologies.

of the joint-based shape decreases, while the SEA of circle shape is vice versa. The joint-based core hollow specimens showed varying SEA values, with the triangular specimen exhibiting the highest increase. It was found that the joint-based core of the round and hexagonal specimens was deformed or broken first. However, the compression process of the triangular specimen was different. This is because the triangle has higher stability than circles and hexagons, which are more prone to deformation under load. As a result, the triangle is less likely to deform during compression and almost the entire specimen is crushed before the triangles begin to deform.

By comparing the solid joint-based core specimens with a radius of 2 mm, it is evident that the EA of the triangle specimens has the least increase. On the other hand, the hexagons and circles have almost the same growth, with an increase of 23% and 25%, respectively, as compared to the original honeycomb structure. Due to the insufficient precision of the machine, it is difficult to distinguish between the circle and hexagon shapes when the radius of the contact is small. The SEA value of the three specimens is the smallest in the triangle, and the largest in the circle. From the results of the specimen with a 3 mm radius, it is evident that only the solid specimen with a circular contact has a higher SEA value than the original honeycomb structure. However, all the specimens have improved EA values, with the circular contact specimen showing the most significant improvement. The core specimen with a 4 mm radius, based on solid joints, showed the highest increase in EA when a round-shaped contact was used. This specimen also had a higher SEA value compared to the original honeycomb structure while it had the lowest mass. When comparing contacts with different shapes but the same radius, the benefit of using joint-based core circular blocks was evident, with their SEA values being higher than the original honeycomb structure. However, the most effective addition among all specimens was a joint-based core with a hollow triangle shape and a 4 mm radius. This specimen had a much higher SEA value than the original honeycomb structure.

The improvement ratio showed that the hollow joint-based core with triangular nodes had the most significant improvement, while the other specimens showed less noticeable improvements. Upon analysis, it was found that triangles differ from circles and hexagons in their deformation behavior under pressure. When circles and hexagons are deformed, the deformation occurs along the sides of the shapes, causing stress to decrease until an unrecoverable deformation is formed. However, triangles are more stable and difficult to deform, allowing them to absorb more energy when compressed. Tension does not disappear during the continuous decline of the pressure plate, and the triangle always plays a role.

8.4   CONCLUSIONS

The study focused on analyzing the compression strength and EA capacity of various honeycomb core structures with different joint-based topologies made from PLA using commercial FDM 3D printing technology. The quasi-static compression testing revealed that all modified samples had increased EA, especially the round structure which showed the biggest improvement. The highest SEA value was observed in the hollow triangular row specimen, which was nearly 50% higher than the original hexagonal honeycomb structure. However, all hollow specimens had lower compression strength values than the original structure, while all solid samples had greater compression strength. Even the sample with the smallest increase in compression strength (solid triangle sample with a radius of 2 mm) showed a 15% improvement over the original honeycomb structure. The compression strength of the sample with a hexagonal and a round joint was not much different from that of the sample with a core based on a triangular joint. The larger the core of the solid joint attached to the original hexagonal honeycomb structure node, the larger the radius, the greater the compression strength, and the more likely that the edges of the honeycomb structure will break during compression. Further research under dynamic loading conditions to better understand the capabilities of the structure is recommended. Additionally, the precision of 3D printers should be improved to ensure printed specimens are consistent with the designed size. Finally, it is suggested that future research investigate the use of different materials to determine if any major differences in results occur.

ACKNOWLEDGMENT

Gratitude to the Ministry of Higher Education Malaysia (MOHE) for supporting this project through the Fundamental Research Grant Scheme (FRGS/1/2021/TK0/UPM/02/21) as well as funding from Universiti Putra Malaysia through the Putra Grant (GP-IPM/2016/9499400 and GP-IPS/2018/9663200).

REFERENCES

Ali, M., Ohioma, E., Kraft, F., and Alam, K. 2015. "Theoretical, numerical, and experimental study of dynamic axial crushing of thin walled pentagon and cross-shape tubes." *Thin-Walled Structures* 94: 253–272. https://doi.org/10.1016/j.tws.2015.04.007

An, L.-Q., Zhang, X.-C., Wu, H.-X., and Jiang, W.-Q. 2017. "In-plane dynamic crushing and energy absorption capacity of self-similar hierarchical honeycombs." *Advances in Mechanical Engineering* 9(6). https://doi.org/10.1177/1687814017703896

Audibert, C., Chaves-Jacob, J., Linares, J.-M., and Lopez, Q.-A. 2018. "Bio-inspired method based on bone architecture to optimize the structure of mechanical workspieces." *Materials & Design* 160: 708–717. https://doi.org/10.1016/j.matdes.2018.10.013

Azmi, M.S., Hasan, R., Ismail, R., Rosli, N.A., and Alkahari, M.R. 2018. "Static and dynamic analysis of FDM printed lattice structures for sustainable lightweight material application." *Progress in Industrial Ecology* 12(3): 247–259. https://doi.org/10.1504/PIE.2018.097063

Choudhry, N.K., Panda, B., and Kumar, S. 2022. "In-plane energy absorption characteristics of a modified re-entrant auxetic structure fabricated via 3D printing." *Composites Part B: Engineering* 228: 109437. https://doi.org/10.1016/j.compositesb.2021.109437

Crupi, V., Epasto, G., and Guglielmino, E. 2013. "Comparison of aluminium sandwiches for lightweight ship structures: Honeycomb vs. foam." *Marine Structures* 30: 74–96. https://doi.org/10.1016/j.marstruc.2012.11.002

Dong, G., Tang, Y., and Zhao, Y.F. 2017. "A survey of modeling of lattice structures fabricated by additive manufacturing." *Journal of Mechanical Design* 139(10): 100906. https://doi.org/10.1115/1.4037305

Dong, G., Wijaya, G., Tang, Y., and Zhao, Y.F. 2018. "Optimizing process parameters of fused deposition modeling by Taguchi method for the fabrication of lattice structures." *Additive Manufacturing* 19: 62–72. https://doi.org/10.1016/j.addma.2017.11.004

Gohar, S., Hussain, G., Ilyas, M., and Ali, A. 2021. "Performance of 3D printed topologically optimized novel auxetic structures under compressive loading: experimental and FE analyses." *Journal of Materials Research and Technology* 15: 394–408. https://doi.org/10.1016/j.jmrt.2021.07.149

Haghdan, S., and Smith, G.D. 2015. "Fracture mechanisms of wood/polyester laminates under quasi-static compression and shear loading." *Composites Part A: Applied Science and Manufacturing* 74: 114–122. https://doi.org/10.1016/j.compositesa.2015.04.006

Korupolu, D.K., Budarapu, P.R., Vusa, V.R., Pandit, M.K., and Reddy, J.N. 2022. "Impact analysis of hierarchical honeycomb core sandwich structures." *Composite Structures* 280: 114827. https://doi.org/10.1016/j.compstruct.2021.114827

Lalegani Dezaki, M., Ariffin, M.K.A., and Hatami, S. 2021. "An overview of fused deposition modelling (FDM): research, development and process optimisation." *Rapid Prototyping Journal* 27(3): 562–582. https://doi.org/10.1108/RPJ-08-2019-0230

Liang, H., Wang, Q., Pu, Y., Zhao, Y., and Ma, F. 2021. "In-plane compressive behavior of a novel self-similar hierarchical honeycomb with design-oriented crashworthiness." *International Journal of Mechanical Sciences* 209: 106723. https://doi.org/10.1016/j.ijmecsci.2021.106723

Mehrpouya, M., Gisario, A., Azizi, A., and Barletta, M. 2020. "Investigation on shape recovery of 3D printed honeycomb sandwich structure." *Polymer for Advanced Technologies* 31: 3361–3365. https://doi.org/10.1002/pat.5020

Palaniyappan, S., Veeman, D., Narain Kumar, S., Surendhar G.J., and Natrayan, L. 2022. "Effect of printing characteristics for the incorporation of hexagonal-shaped lattice structure on the PLA polymeric material." *Journal of Thermoplastic Composite Materials* 089270572210898. https://doi.org/10.1177/08927057221089832

Płaczek, M., Ariffin, M.K.A., Baharudin, B.T.H.T., and Lalegani Dezaki, M. 2021. "The effects of 3D printing structural modelling on compression properties for material jetting and FDM process." In: Kyratsis, P., Davim, J.P. (eds) *Experiments and Simulations in Advanced Manufacturing. Materials Forming, Machining and Tribology.* Springer, Cham. https://doi.org/10.1007/978-3-030-69472-2_7

Qi, C., Jiang, F., and Yang, S. 2021. "Advanced honeycomb designs for improving mechanical properties: A review." *Composites Part B: Engineering* 227: 109393. https://doi.org/10.1016/j.compositesb.2021.109393

Saufi, S.A.S.A., Zuhri, M.Y.M., Dezaki, M.L., Sapuan, S.M., Ilyas, R.A., As'arry, A., Ariffin, M.K.A., and Bodaghi, M. 2021. "Compression behaviour of bio-inspired honeycomb reinforced starfish shape structures using 3D printing technology." *Polymers* 13(24): 4388. https://doi.org/10.3390/polym13244388

Serjouei, A., Yousefi, A., Jenaki, A., Bodaghi, M., and Mehrpouya, M. 2022. "4D printed shape memory sandwich structures: experimental analysis and numerical modeling." *Smart Materials and Structures* 31: 055014. https://doi.org/10.1088/1361-665X/ac60b5

Shan, J., Xu, S., Zhou, L., Wang, D., Liu, Y., Zhang, M., and Wang, P. 2019. "Dynamic fracture of aramid paper honeycomb subjected to impact loading." *Composite Structures* 223: 110962. https://doi.org/10.1016/j.compstruct.2019.110962

Sun, G., Chen, D., Zhu, G., and Li, Q. 2022. "Lightweight hybrid materials and structures for energy absorption: A state-of-the-art review and outlook." *Thin-Walled Structures* 172: 108760. https://doi.org/10.1016/j.tws.2021.108760

Tamburrino, F., Graziosi, S., and Bordegoni, M. 2018. "The design process of additively manufactured mesoscale lattice structures: A review." *Journal of Computing and Information Science in Engineering* 18(4): 040801. https://doi.org/10.1115/1.4040131

Tao, Y., Duan, S., Wen, W., Pei, Y., and Fang, D. 2017. "Enhanced out-of-plane crushing strength and energy absorption of in-plane graded honeycombs." *Composites Part B: Engineering* 118: 33–40. https://doi.org/10.1016/j.compositesb.2017.03.002

Tasdemirci, A., Akbulut, E.F., Guzel, E., Tuzgel, F., Yucesoy, A., Sahin, S., and Guden, M. 2018. "Crushing behavior and energy absorption performance of a bio-inspired metallic structure: Experimental and numerical study." *Thin-Walled Structures* 131: 547–555. https://doi.org/10.1016/j.tws.2018.07.051

Tatlier, M.S. 2021. "A numerical study on energy absorption of re-entrant honeycomb structures with variable alignment." *International Journal of Crashworthiness* 26: 237–245. https://doi.org/10.1080/13588265.2019.1701891

Thomas, T., and Tiwari, G. 2019a. "Crushing behavior of honeycomb structure: A review." *International Journal of Crashworthiness* 24: 555–579. https://doi.org/10.1080/13588265.2018.1480471

Thomas, T., and Tiwari, G. 2019b. "Energy absorption and in-plane crushing behavior of aluminium reinforced honeycomb." *Vacuum* 166: 364–369. https://doi.org/10.1016/j.vacuum.2018.10.057

Wang, D.-M., Wang, Z.-W., and Liao, Q.-H. 2009. "Energy absorption diagrams of paper honeycomb sandwich structures." *Packaging Technology and Science* 22 (2): 63–67. https://doi.org/10.1002/pts.818

Wang, Z. 2019. "Recent advances in novel metallic honeycomb structure." *Composites Part B: Engineering* 166: 731–741. https://doi.org/10.1016/j.compositesb.2019.02.011

Wang, Z., Zhang, Y., and Liu, J. 2015. "Comparison between five typical reinforced honey-comb structures." In: *Proceedings of the 5th International Conference on Advanced Engineering Materials and Technology (AEMT 2015)*. Atlantis Press, Paris, France.

Xu, M., Xu, Z., Zhang, Z., Lei, H., Bai, Y., and Fang, D. 2019. "Mechanical properties and energy absorption capability of AuxHex structure under in-plane compression: Theoretical and experimental studies." *International Journal of Mechanical Sciences* 159: 43–57. https://doi.org/10.1016/j.ijmecsci.2019.05.044

Yap, Y.L., and Yeong, W.Y. 2015. "Shape recovery effect of 3D printed polymeric honeycomb." *Virtual Physical Prototyping* 10: 91–99. https://doi.org/10.1080/17452759.2015.1060350

Yousefi, A., Jolaiy, S., Lalegani Dezaki, M., Zolfagharian, A., Serjouei, A., and Bodaghi, M. 2022. "3D-printed soft and hard meta-structures with supreme energy absorption and dissipation capacities in cyclic loading conditions." *Advanced Engineering Materials* 25: 2201189. https://doi.org/10.1002/adem.202201189

Zhang, Q., Yang, X., Li, P., Huang, G., Feng, S., Shen, C., Han, B., Zhang, X., Jin, F., Xu, F., and Lu, T.J. 2015. "Bioinspired engineering of honeycomb structure – Using nature to inspire human innovation." *Progress in Materials Science* 74: 332–400. https://doi.org/10.1016/j.pmatsci.2015.05.001

9 Application of Artificial Intelligence (Machine Learning) in Additive Manufacturing, Bio-Systems, Bio-Medicine, and Composites

Vahid Monfared

9.1 INTRODUCTION

For millions of people around the world, supervising complicated and challenging bio-systems conditions may be a fight and a conflict. Recent developments and advancements in artificial intelligence (AI) and machine learning (ML) technology have made considerable strides in predicting and identifying engineering, healthcare emergencies, bio-medicine predictions, disease populations (cancer, drug, etc.), bio-systems analysis, and disease status and immune responses, amongst a few others. We are thankful for recent advances in computer science and informatics IT; AI is swiftly becoming an integral part of modern engineering and healthcare.

AI and ML technology in engineering and healthcare have achieved considerable traction throughout the pandemic, from combing scientific articles on coronaviruses to searching CT scans for Covid-19 symptoms. AI/ML has taken on a much bigger part in healthcare through the pandemic period.

Recently, in the field of additive manufacturing (AM) and 3D printing, monitoring anomalies in 3D bioprinting with deep neural networks (DNNs) has been done. AM technologies have advanced in the previous decades, particularly when employed to print bio-system constructions like scaffolds and vessels with living cells for bio-medical and tissue engineering applications. Quality and reliability are vital to maintaining the bio-compatibility and mechanical/physical strength and integrity required for engineered tissue concepts. It is important to distinguish for any anomalies that might happen in the 3D bioprinting process that may cause a mismatch between favorite designs and printed shapes. However, challenges are in detecting the faultiness within frequently clear bio-printed and multifaceted printing features precisely and professionally. An anomaly detection system is established based on layer-by-layer sensor images and ML/AI models are developed to separate

DOI: 10.1201/9781003362128-9

and categorize defects for clear hydrogel-based bio-printed materials. High anomaly detection precision is attained using convolutional neural network (CNN) methods as well as advanced image processing and augmentation techniques on extracted small image patches. Along with the prediction of various anomalies, the group of infill samples and location information on the image patches may be precisely obtained. By employing our detection system to classify and localize printing anomalies, real-time autonomous improvement of process parameters may be realized to attain high-quality tissue constructs in 3D bioprinting processes. Employing ML/AI techniques, we may put together large-scale heterogeneous data like medical, clinical, imaging, and genomic/bio-system data, and allocate researchers to better understand the genomic basis of disease and recognize the optimum therapeutic approaches. Numerous research programs have been designed and created to take advantage of these models (Collins and Varmus 2015; Zheng, Li, and Liang 2019; Jin et al., 2021b). Lately, ML/AI approaches have advanced intensely in terms of proficiency and have seen prospective applications in predictive materials modeling, advanced manufacturing, self-directed vehicles, and bio-mechanics, among many others (Begg and Kamruzzaman 2005; Collins and Varmus 2015; Gobert et al. 2018; Gu, Chen, and Buehler 2018; Dias and Torkamani 2019; Hanakata et al. 2019; Jaganathan et al. 2019; Zheng, Li, and Liang 2019; Chen and Gu 2020; Jin, Zhang, and Gu 2020; Jin et al. 2021a; Zeng and Li 2022; Kosorok and Laber, n.d.). Obviously, two baseline models as well as two advanced CNN models are applied by means of our collected and processed image dataset. For example, an anomaly detection system is developed to distinguish and predict anomalies precisely and efficiently in a layer-by-layer configuration for bio-printed materials. Three main anomalies including discontinuity (broken raster), nonuniformity (unsmoothed surface), and irregularity (improper line width) are discovered for the first layer of the print, which is measured on a significant basis for the whole print. Because of the translucent and multifaceted features of these anomalies, ML/AI methods are dynamically incorporated for this problem to discover and separate the underlying hidden patterns behind real-time printing images (Jin et al. 2021b). It is required to know all things around ML/AI, because some techniques used in healthcare and medicine can be beneficial and applied in the field of application of ML/AI in AM of bio-composites in terms of data science/ analysis and image processing (CNN).

As discussed and based on the state of precision medicine, the models and approaches in ML/AI often contain genomic data to establish the best course of treatment (Dias and Torkamani 2019; Jaganathan et al. 2019; Kosorok and Laber 2019; Wang et al. 2022; Zeng and Li 2022). The merging of AI/ML and precision medicine promises to reform healthcare issues. Precision medicine methods recognize phenotypes of patients with fewer common responses to treatment or single healthcare requirements. Various studies are gradually incorporating AI/ML models and algorithms across a large variety of domains (Rajkomar, Dean, and Kohane 2019; Topol 2019; Dong et al. 2020). ML is an old-fashioned idea and concept which has achieved a lot of attention lately because of the explosion of data generation processes in healthcare. According to one report, around 86% of healthcare organizations employ some form of ML solutions, and more than 80% of healthcare organization leaders have an AI plan (Alanazi 2022). ML and its applications in healthcare have earned a lot of attention. When increased computational power is combined with big data, there is an opportunity to use ML/AI techniques and algorithms to improve healthcare. Supervised learning (SL) is the type of ML that may be implemented to predict labeled data based on algorithms like linear or logistic regression (LR), support vector machine (SVM), decision tree, LASSO regression, K-Nearest-Neighbor (KNN), and naive Bayes classifier with labels (X, Y ~ input, output). Unsupervised ML models may identify data patterns in datasets that do not contain information about the outcome without labels (X~input). Such models can be used for fraud or anomaly detection. Examples of clinical applications of ML contain the formulation of different clinical decision support systems. An essential public health application of ML is the identification and prediction of populations at high risk for developing certain adverse health outcomes and the development of public health interventions targeted toward these populations. Numerous ideas related to ML need to be integrated into the medical curriculum so that health professionals can effectively guide and interpret research in this area (Alanazi 2022).

In this chapter, we have introduced some significant applications of ML/AI in the fields of bio-systems, bio-composites, AM, and related similar applications such as healthcare and bio-medicine as well as presented some examples of these applications along with related codes in Python. Another aim is to introduce some useful websites to learn more about ML/AI. Briefly explaining some famous predictive models is another purpose. In general, the main structure of this chapter is as follows: abstract, introduction, some questions and challenges, ML/AI in healthcare and medicine, ML/AI in bio-systems, illustrations about AI/ML process, solved and analyzed practices, conclusions, and finally references.

Introducing, presenting, and analyzing these topics are necessary to fulfill the abovementioned purposes. That is, we need to understand useful analysis of the main categories of AM, bio-systems/bio-composites, healthcare/bio-medicine, and bio-systems. Note that, in this chapter, because of limitations in available studies in the field of application of ML/AI in AM of bio-composites, some similar related research in healthcare and bio-systems has been presented and added to possible applications in the special ML/AI in AM of bio-composite topics.

9.2   SOME QUESTIONS AND CHALLENGES

Here, it is better to briefly state some important inquiries and questions along with introducing some challenges in the field of application of ML/AI in AM of bio-composites, bio-systems, healthcare/bio-medicine subjects.

9.2.1   GENERALITIES, AMBIGUITIES, AND QUESTIONS

Some highlighted challenges are as follows:

1. What/where was the problem?
2. What are the required materials for solving the problem and how can we solve that?
3. Why do we need ML/AI?

4. Can we employ ML/AI in AM, healthcare, medicine, and bio-systems?
5. What are the challenges and questions?
6. What are the roadmap and flowchart?
7. Where are we?
8. Big Inquiry: Can we stop doing the experiments and analyze and work only with our former available dataset using ML for the rest of our research life?

9.2.2 SHORT ANSWERS TO AMBIGUITIES AND QUESTIONS

Here there are some short answers to the above concerns:

1. What/where was the problem?

     Answer: saving time, money (economical aspects), lack/shortage of space, need for high-accuracy results and quality, tendency/willingness to work remotely or hybrid, easy supervising, and management.

2. What were the required materials for solving the problem and how can we solve that?

     Answer: clean dataset, ML/AI models.

3. Why do we need ML/AI?

     Answer: fast prediction, high accuracy and quality, use of high potential data in hospitals and labs.

4. Can we employ ML/AI in AM, bio-composites, healthcare/medicine, and biosciences?

     Answer: yes, it is nascent in this field. AI and ML methods are powerful methods for decision-making in AM and bio-systems like tissue engineering and other healthcare and bio-medicine realms.

5. What are the challenges and questions?

     Answer: uncertainty, compatibility, transparency, unemployment, bias issues, heterogeneity of data, surviving without them? is there a risk?

6. What are the roadmap and flowchart?

     Answer: Figure 9.1 depicts a comprehensive illustration graphically.

7. Where are we?

     Answer: now, we claim that we can design and build a predictive model to produce a desired and ideal bio-medical product with high quality and targeted ones as well as high-accuracy predictions in the healthcare and medicine fields like cancer predictions. Also, we can predict all parameters directly and indirectly (back/forward) (input to output and vice versa: X (input)$\rightarrow Y$ (output) and Y (output)$\rightarrow X$ (input)).

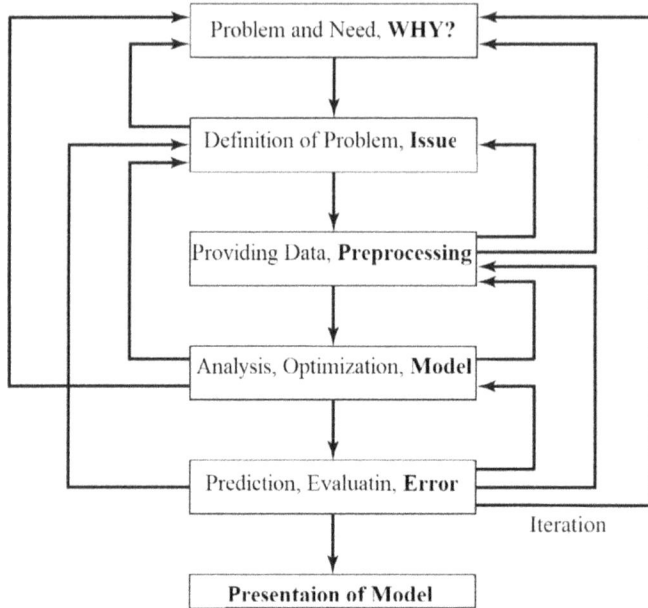

FIGURE 9.1   Road map and flowchart (problem solution steps).

9.2.3   WHERE DID THE PROBLEM (DIFFICULTY/NEED) START?

Some short answers are as follows:

1. Limitation in financial resources (economic aspects).
2. Difficulty and limitation in finding a suitable space/place/lab.
3. Cost of materials and tools/devices.
4. Device maintenance and repairing costs.
5. Salary of personnel and researchers.
6. Time-consuming and costly processes for doing a project with average accuracy.
7. Dangers arising from doing the experiments (safety issues).
8. Managing many students and researchers (research team).
9. How can we solve and remove this problem? Remote and hybrid working rotationally, and using ML/AI in these related fields.

9.2.4   WHY DO WE NEED IT (ML/AI)?

1. Because of available big data.
2. Massive amount of data that must be evaluated and calculated and properly used.
3. It is time to employ this big data instead of repeating the experimental attempts with high and acceptable accuracy.

9.2.5 WHERE ML DON'T WORK?

The next important thing is if you have excessively huge data, and you have zero knowledge of the domain to understand the feature variables then ML won't work well. Also, it does not work when we have many outlier data, high volumes of missing data, small and low features, high errors in collecting data, small and unclean data.

9.2.6 THE FUTURE OF SOME LABS (AUTOMATION)

A clear diagram and figure to draw the future of ML/AI in AM, healthcare, medicine, and bio-systems that would be a revolution in this category is depicted in Figure 9.2. ML techniques could revolutionize how materials science is done!

Briefly, some important challenges, concerns, and inquiries regarding the use of ML/AI in AM, healthcare, medicine, and bio-systems are as given below:

1. Compatibility and uncertainty problems and issues.
2. Clean dataset including variations of all parameters (outliers, missing, balancing the data, normalization, number of data and predictors/attributes, error in collecting data).
3. Heterogeneity of data (the data needs to be adequately structured and cleaned first before ML algorithms can effectively use it).

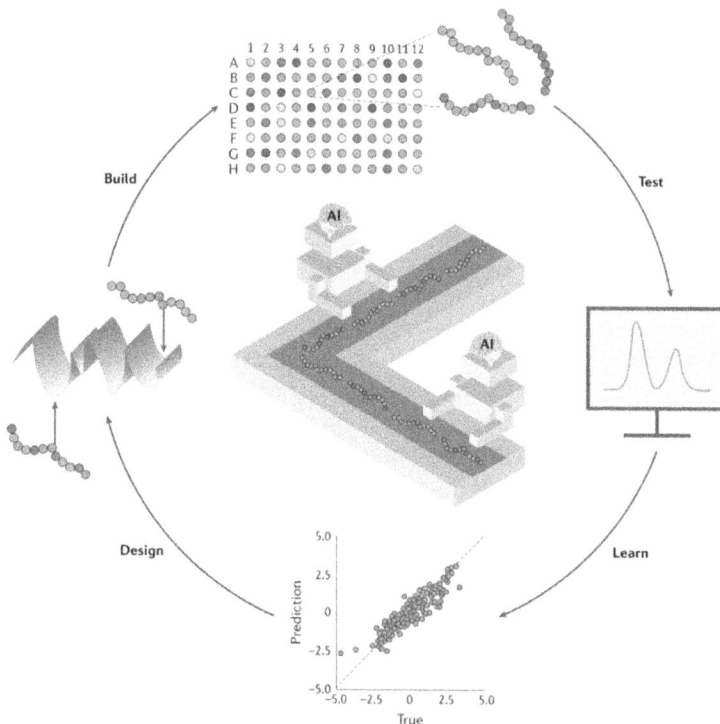

FIGURE 9.2   The future of some labs (revolution and automation).

4. Lack and shortage of qualified resources as well as provider resistance.
5. Technical safety (will AI systems work as they are promised, or will they fail? if and when they fail, what will be the results of those failures? and if we are dependent upon them, will we be able to survive without them?).
6. Transparency and privacy concerns.
7. Bias in data, training sets, etc.
8. Unemployment problems.
9. Automation structures and some related issues to create and design the system with high quality of supervising and by changing the rules.

9.3   A REVIEW ON ML/AI IN HEALTHCARE AND MEDICINE/BIO

Before beginning to state our main chapter topic, it is required to review some previous available research works related/similar to our chapter such as ML/AI in medicine/healthcare as follows: Prediction analysis of healthcare in medicine like cancer checking, Covid-19 screening, and preterm neonate mortality is essential and substantial for benchmarking and evaluating healthcare services in medicine at various hospitals and other medical centers. Application of AI and ML algorithms is a challenging and hot topic in healthcare/medicine; can physicians' skills better predict some issues like preterm neonatal deaths in infants after delivery at hospitals. Also, because of similarity of these solution methods, researchers and engineers can use them to analyze AM problems via AI/ML. For example, after recognizing the critical status of an infant, physicians and other healthcare personnel can help the infant to survive by using special medical NICU care. We have tried to present some appropriate models with high accuracy and compare the results.

AI is a wide term referring to the application of computational techniques that may analyze large datasets to categorize, predict, or achieve helpful conclusions. Underneath the umbrella of AI and its branches is ML. ML is the process of making or learning statistical models applying formerly observed real world data to predict conclusions, or classify observations based on "training" provided by humans. These predictions are also applied to future data, all the while folding in the new data into its continually developing and calibrated statistical model. The future of AI and ML in healthcare research is thrilling and expansive. AI and ML are becoming foundations in the medical and healthcare research fields and are fundamental in our continued processing and capitalization of strong patient EMR data. The use and application of ML in healthcare settings comprise measuring the quality of data inputs and decision-making, which serve as the foundations of the ML model, ensuring the end-product is interpretable, clear, and moral matters are considered during the development process (Rubinger, Gazendam, and Ekhtiari 2022). The present and future applications of ML comprise improving the quality and quantity of data collected from EMRs to progress registry data, employing these robust datasets for better and standardized research protocols and obtained results, clinical decision-making applications, natural language processing, and improving the fundamentals of value-based care, to name only a few.

The eagerness neighboring the possible for human-level presentation and consistency of ML applications for healthcare must be balanced by the requirement to lessen ethical worries, like the potential for these models to aggravate current differences in our health systems (Maxmen, n.d.; Davidben-Israel et al. 2020).

AI/ML models should constantly be checked precisely by PCPs. They could outperform humans in diagnosing a restricted although increasing number of clinical conditions, like prostate and skin cancers, lung and breast cancers, cardiac arrhythmia, and Alzheimer's disease. ML models have the potential of much more improved uniformity than traditional risk models in predicting hospital readmissions and chronic kidney disease progression. PCPs should continually be concerned about these vetting questions when introducing the algorithms in their practice: have several metrics been presented? has the technique been compared well enough to a standard-of-care method? has the procedure been studied in an external pattern? is the algorithm biased toward or opposed to certain groups? Blind introduction of AI/ML technology in the clinical settings with no PCP guidance will not lead to better population health consequences and results (Sendak, Balu, and Schulman 2017; Rajkomar et al. 2018; Rose 2018; Ding et al. 2019; Hannun et al. 2019; Lin, Mahoney, and Sinsky 2019; Liu et al. 2019; Soenksen et al. 2021; Yang et al. 2022).

Cancer fatalities, especially those of skin, breast, brain, and lung cancers, are more sensitive and at higher risk of Covid-19 and associated outcomes as a result of compromised immune systems, which makes them remarkably susceptible. Due to a range of conditions, cancer patients' diagnosis, treatment, and aftercare are very complex and time-consuming throughout an epidemic. In this kind of circumstance, advances in AI and ML suggest the capacity to boost cancer patient diagnosis, therapy, and care through the use of cutting-edge technologies. For instance, employing clinical and imaging data combined with ML methods, researchers may be able to differentiate between lung alterations induced by corona virus and those produced by immunotherapy and radiation. Throughout this epidemic, AI may be used to ensure that proper people are recruited in cancer clinical trials more rapidly and efficiently than in the past, which was done in a traditional and confused manner. To improve care for cancer patients and find novel and more useful therapies, it would be crucial that we move beyond traditional research methods and utilize AI and ML to update our research. AI and ML are being used to aid with various aspects of the Covid-19 epidemic, like epidemiology, molecular research and medication development, medical diagnosis and treatment, and socioeconomics. The use of AI and ML in the diagnosis and treatment of Covid-19 patients is also being studied. The mixture of AI and ML in Covid-19 may well help to recognize positive patients more rapidly. To know the dynamics of an epidemic that is relevant to artificial intelligence, when used in various patient groups, AI-based algorithms can immediately discover CT scans with Covid-19-linked pneumonia, as well as categorize non-COVID-connected pneumonia with high specificity and accuracy. It is possible to understand the current difficulties and future views presented in this study to guide an optimum implementation of AI and ML technologies in an epidemic (Boddu et al. 2022).

AI/ML models may fight Covid-19 on several counts. This includes finding those at high risk with multivariable clinical or populational data, identifying patients, identifying obviously happening interactions among predictors or identifying risk clusters, developing drugs more rapidly, finding medicines that can help, predicting the disease spread, and understanding the virus (see Figures 9.3 and 9.4).

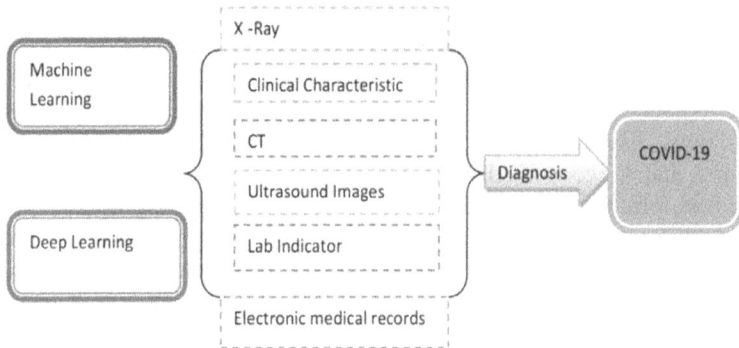

FIGURE 9.3   AI for Covid-19 diagnosis (Boddu et al. 2022).

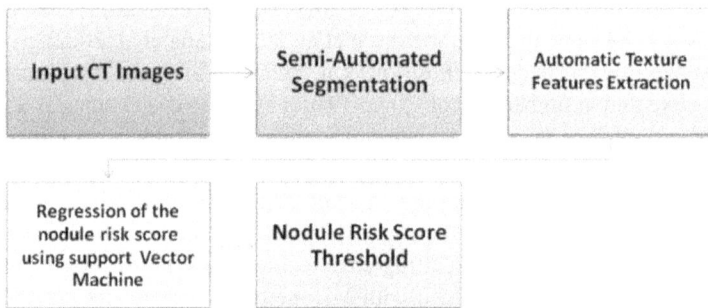

FIGURE 9.4   Computer vision for lung cancer detection and cancer risk estimation (Boddu et al. 2022).

AI models may be as efficient as professional doctors for Covid-19 predictions and diagnosis. The clinical use of AI in Covid-19 diagnosis is encouraging and a more complete study is required. Deep learning (DL) and ML techniques exhibit superb accuracy for differentiating Covid-19 from non-Covid-19 chest pneumonia. These methods have made it simpler to evaluate these pictures automatically. DL techniques nonetheless persist from the absence of openness and applicability, since the exact imagery characteristic employed to make the result cannot be identified. Since no method is capable of distinguishing between all pulmonary diseases on the basing images on chest CT scans, the use of interdisciplinary solutions to defeat diagnostic difficulties is recommended strongly (Boddu et al. 2022).

The role of healthcare AI in boosting abilities and programming tasks raises concerns regarding its capacity to deskill and in certain cases even replace healthcare workers. Deskilling refers to workers experiencing "reduced discretion, autonomy, decision-making quality and knowledge as they perform their jobs" (Aquino et al. 2023).

In healthcare, deskilling may result in deterioration of clinical skills, compromising decision-making across different stages of clinical management, and potentially undermining patient safety. Also, AI-enabled automation raises fears regarding workforce replacement, particularly in medical disciplines that rely on sample recognition. With the development of AI in imaging applications and analysis apparently approaching human capability, the examination of the impact of AI on workforce and professional skills has been exceptionally active in radiology. While some studies flag the risk of AI in deskilling clinicians (Diprose and Buist 2016; Lu 2016; Panch, Szolovits, and Atun 2018; Varga-Szemes, Jacobs, and Schoepf 2018; Becker 2019; Macrae 2019), most studies claim that AI is unlikely to result in workforce replacement given the slow growth and uptake of healthcare AI applications, and the combination of cognitive and emotional skills required to make healthcare work (Langlotz 2019; Kim et al. 2020; Ross and Spates 2020; Saheb, Saheb, and Carpenter 2021; Aquino et al. 2023; Kumar and Munish 2023).

The analysis presented in Aquino et al. (2023) showed participants had different viewpoints on three critical issues concerning AI and deskilling. The first concerned conflicting views around the appropriate extent of AI-enabled automation in healthcare work, and which clinical tasks should or should not be automated. They identified a cluster of types of tasks that were considered more appropriate for automation. The second involved expectations concerning the effect of AI on clinical skills, and whether AI-enabled automation would be an inferior or superior quality of healthcare. The third indirectly contrasted two models of healthcare work: a human-centric model and a technology-centric model. These models assumed various values and priorities for healthcare work and its relationship to AI-enabled automation. Conclusion: Our study shows that a diverse group of professional stakeholders involved in healthcare AI development, gaining, deployment, and regulation are sensitive to the potential impact of healthcare AI on clinical skills, but have dissimilar views about the nature and valence (positive or negative) of this impact. Detailed engagement with numerous types of professional stakeholders allowed us to recognize relevant ideas and values that could guide decisions about AI algorithm development and use (Aquino et al. 2023).

In several surgical fields, ML and AI have begun to be used for various surgeries like cancer, spine, bone, and brain. Prognostic analytics for results after lumbar fusion were used to increase patient counseling and surgical decision-making (Fukuchi et al. 2011; Zaharchuk et al. 2018; Nam et al. 2019; Jin, Schröder, and Staartjes 2023). Additionally, a DL algorithm was employed to create synthetic calculated tomography images from magnetic resonance images to enable improved presurgical planning and near radiationless, robotic-guided spinal instrumentation. Here, a lung cancer-detecting AI algorithm shows impressive results (Figure 9.5).

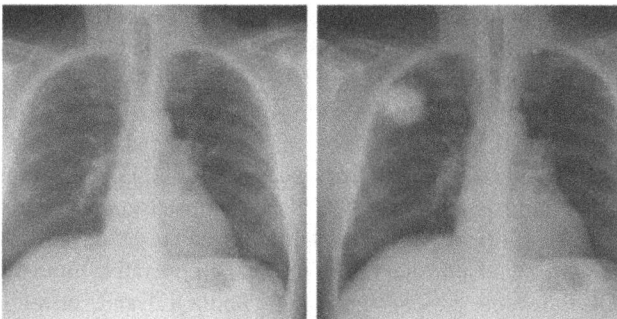

FIGURE 9.5 Lung cancer-detecting AI algorithm (www.onmyowntechnology.com).

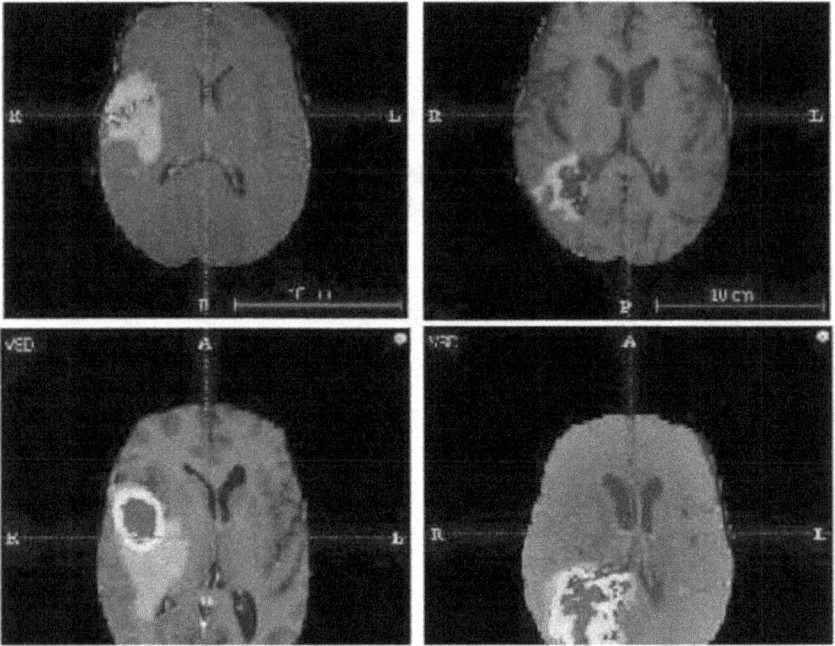

FIGURE 9.6   Brain cancer-detecting AI algorithm (www.devmesh.intel.com).

The techniques of ML and AI along with data mining are being efficiently employed for brain tumor detection and prevention at an early stage (Figure 9.6). It can be very helpful for fast prediction of cancer.

The available AI research shows that it is possible to use a CNN especially for breast cancer classification tasks. However, a general problem in any AI algorithm is its dependence on the dataset and the quality of the data (Figure 9.7). Figure 9.7 shows that AI prediction can be useful for better prognosis of breast cancer in women with proper accuracy.

Bone cancer is assumed a major health risk, and, unfortunately, it mostly causes patient death. Some techniques like CT scan, X-ray, and MRI are employed by physicians and specialized doctors to recognize bone cancer. The manual method is time-consuming and needs expertise in that field. Thus, it is crucial to create a computerized system like AI-/ML-based tools to categorize and detect cancerous bone and healthy bone (Figure 9.8).

Bone mineral density (BMD) is of central importance for fusion surgery (Nam et al. 2019). Though dual X-ray absorptiometry is considered as the gold standard for evaluating BMD, quantitative calculated tomography (QCT) provides more precise data in spine osteoporosis. However, QCT has the disadvantage of extra radiation risk and charge. The utility of AI/ML models has been presented and stated to evaluate osteoporosis utilizing Hounsfield units (HU) of preoperative lumbar CT coupling with data of QCT. Seventy patients undergoing both QCT and conventional lumbar CT for spine surgery were reviewed. The T-scores of 198 lumbar vertebra were measured using QCT and the HU of the vertebral body at a similar level was determined

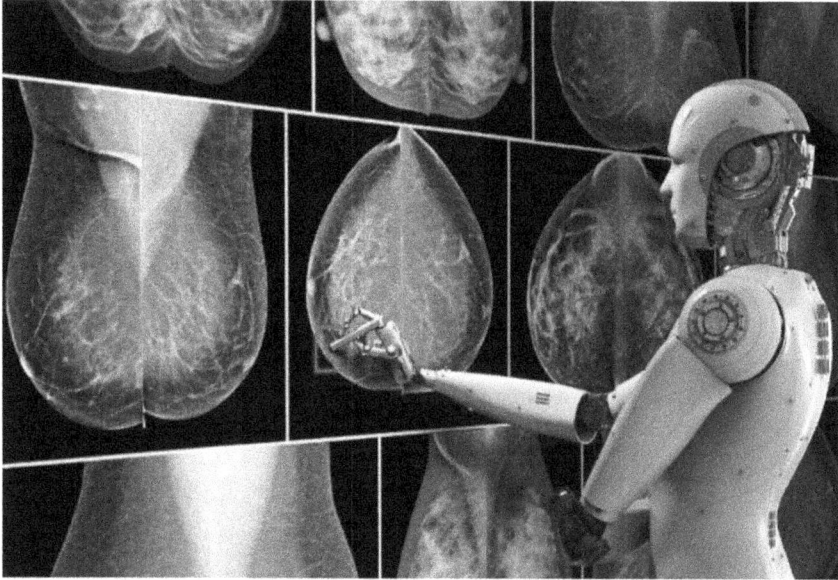

FIGURE 9.7   Breast cancer-detecting AI algorithm (www. medium.com).

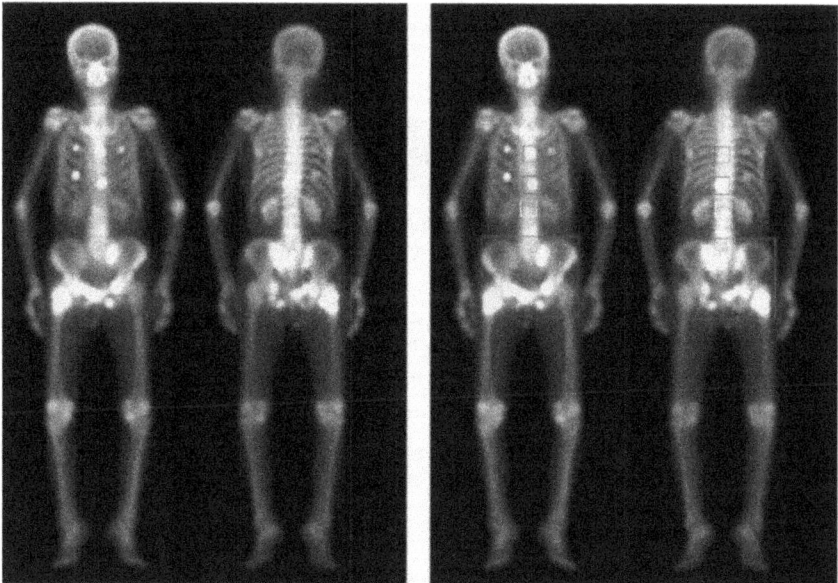

FIGURE 9.8   Bone cancer-detecting AI algorithm (www.expo.taiwan-healthcare.org).

in traditional CT by the picture archiving and communication system (PACS). A varying regression algorithm was employed to forecast the T-score applying three separate variables (age, sex, and HU of vertebral body on conventional CT) coupled with the *T*-score of QCT. Then, a LR algorithm was utilized to predict osteoporotic or non-osteoporotic vertebra. The TensorFlow (tf) and Python codes were applied as the ML tools. The prognostic model with multiple regression algorithm estimated same *T*-scores with data of QCT. HU demonstrated similar results as QCT without the discordance in only one non-osteoporotic vertebra that indicated osteoporosis. From the training set, the prognostic model categorized the lumbar vertebra into two groups (osteoporotic vs. non-osteoporotic spine) with 88.0% accuracy. In a test set of 40 vertebrae, classification accuracy was 92.5% when the learning rate was 0.0001 (precision, 0.939; recall, 0.969; F1_score, 0.954; area under the curve, 0.900). Finally, the mentioned study (Nam et al. 2019) showed that an easy ML model is applicable in the spine research field. The ML model may simply predict the *T*-score and osteoporotic vertebrae solely by measuring the HU of conventional CT, and this would help spine surgeons not to under-estimate the osteoporotic spine preoperatively. If applied to a bigger dataset, we believe that the predictive accuracy of our model will improve. These cancer detections (predictive AI models) may be used for the prediction and recognition of the imperfections and defects in bio-composite products via CNN's models and AI-/ML-based algorithms.

9.4 ML/AI IN BIO-SYSTEMS

ML and AI are involved in interpreting fundamental mechanisms in bioprinting, biomaterials, bio-medical, and tissue engineering interaction, like biomarker detection, cell signaling pathways. Also, there are some challenges to translating ML and AI from practice to application.

Design requirements for various mechanical metamaterials, porous constructions, and lattice structures, used as tissue engineering scaffolds, lead to multi-objective optimizations, because of the intricate mechanical features of the biological tissues and structures they should mimic. Sometimes, traditional design and simulation methods for designing such tissue engineering scaffolds cannot be applied owing to geometrical difficulty, manufacturing defects, or large aspect ratios that result in numerical differences. AI and ML techniques are now discovering applications in tissue engineering, and they can demonstrate transformative resources for assisting designers in the field of regenerative medicine generally. As a source for predicting the mechanical properties of innovative scaffolds, 3D CNNs trained utilizing digital tomographies from CAD models have been validated as a powerful tool (Barrera and Franco-Martínez 2021). The presented AI-aided or ML-aided design strategy is generally assumed as a new methodology in the field of tissue engineering scaffolds, and of mechanical metamaterials (Barrera and Franco-Martínez 2021).

One of the pioneering large-scale AI-based materials innovation projects is the materials genome initiative, which has inspired other breakthroughs in new materials for advanced industrial applications (Qian, Siler, and Ozin 2015; Raccuglia et al. 2016; Barrera and Franco-Martínez 2021; Kisor Kumar Sahu et al. 2022). AI and ML have been utilized for the estimation of final properties and performance of materials from the chemical composition of the bulk materials under study. Occasionally, mechanical properties have been also forecasted with ML models. Moreover, the developing applications of ML/AI for the prediction, modeling, and control of mechanical properties is now creating an impact in the field of mechanical meta-materials, whose applications include medicine, healthcare, transport, energy, and space, to cite a few (Santos et al. 2009; Manuel Pegalajar Cuéllar et al. 2015; Lu et al. 2017; Winkler 2017; Gómez-Bombarelli 2018; Jose and Ramakrishna 2018; Naser 2018; Bonfanti et al. 2020; Lantada et al. 2020; Merayo, Rodriguez-Prieto, and Camacho 2020; Jiao and Alavi 2021; Lu, Ciucci, and Chen 2021; Gentili 2022). The DeepMind AI has analyzed and solved a 50-year-old biological problem. A 50-year-old biological challenge has been solved by DeepMind's protein-folding AI. AlphaFold can predict the shape of proteins upto a precision or resolution of an atom's width. By developing drugs and understanding diseases, the innovation will help scientists find cures (Figures 9.9 and 9.10). So, a long-standing and exceptionally difficult scientific problem regarding the structure and behavior of proteins has been successfully solved by new-found AI and ML systems.

DeepMind has designed and created an AI model which can quickly and precisely predict how proteins fold to get their 3D shapes. An AI company that achieved fame for designing computer systems that could beat humans at games has now made a massive advancement in biological science.

Prediction a protein's 3D building can assist in accelerating drug discovery by substituting "slow, expensive" structural biology experiments with quicker, lower computer simulations, among other advantages, in accordance with research by the NIH (National Institutes of Health). Also, it helps scientists gain a better understanding of the body, especially since misfolded or abnormal proteins can lead to diseases.

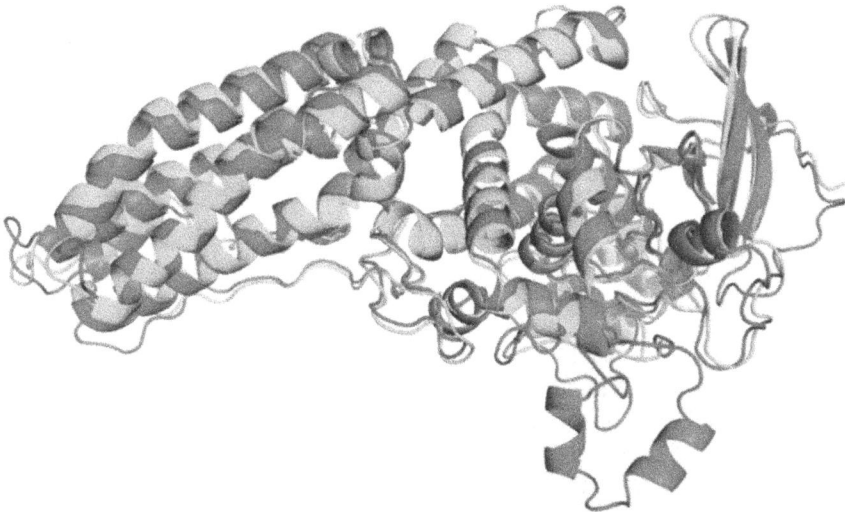

FIGURE 9.9   Protein folding (www.epistemologyontologyfoundationinstitute.org).

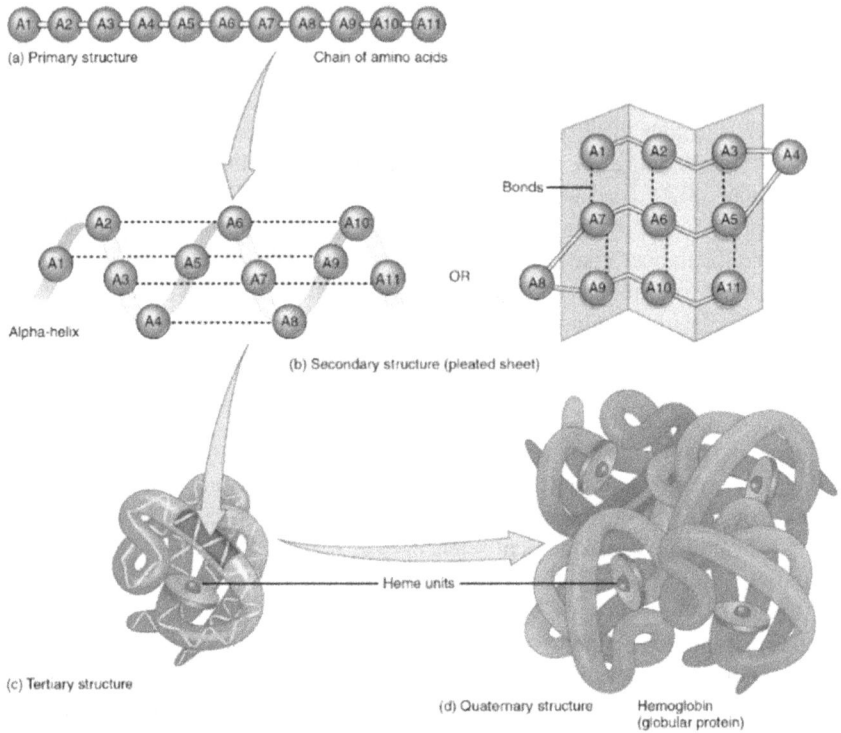

FIGURE 9.10 Developing in protein (www. techwoofafrica.com).

9.5 BRIEF ILLUSTRATIONS ABOUT THE AI/ML PROCESS

Here it is necessary to briefly explain the methodology of ML and AI. To get this aim, following is about illustrations to better understanding these techniques to predict some parameters and have a good decision with low error in the healthcare, medicine, and bio-systems.

9.5.1 General Information

Artificial intelligence with the abbreviation AI is a computer-based methodology for activities that typically need intelligence (transferring and translating data and information into knowledge) for efficient accomplishment. ML is a form of AI in which computer algorithms learn from data to form predictive models logically. ML is a part of AI focused on structure applications that learn from data and improve their accuracy over time without being programmed to do so. It should be mentioned that, in data science, an algorithm is a sequence of statistical processing steps.

ML is an application of AI that is based on the concept that we may provide machines with data and let them to understand and learn for themselves. ML uses neural networks

(NNs) to take data, and utilizes algorithms to explain sections of the problem, and produce an output. ML is a small part of the larger AI system-ML system that focuses in a specific way on computers that can learn and adjust based on what they know.

DL is a component of ML/AI, which basically senses the big NNs utilized to parse bigger datasets or more complicated problems. DL employs the same NNs and ML models, however on a much larger scale. This DL is crucial for larger datasets – DL is the approach through which we can get more knowledge, analyzing more data than has ever been possible before.

ML and DL concentrate on ensuring a program can go on to learn and expand based on what outputs it has come up with before. There are three different kinds of intelligence models and systems engaged in ML models and ML algorithms. SL focuses on providing an input and an output, and assisting the machine get there. SL helps an intelligent machine understand how their algorithms should get to the ultimate output. SL is more hands-on that other types of intelligent ML.

Unsupervised learning concentrates on giving a robot or intelligent machine the input, and then letting the algorithms do the rest. You give the robot the chance to take what you've given them and understand the output. Unsupervised learning has a higher risk of error than SL because you aren't saying what the response is. Unsupervised learning focuses on helping boost intelligence within a machine and its algorithms, permitting it to learn and develop as it figures out the output (Figures 9.11 and 9.12).

Reinforcement learning permits a machine to join goals while it is using its intelligence and algorithms to understand what it is doing well. Reinforcement learning focuses on helping a machine understand what it is doing appropriately as it gets toward the output. Reinforcement learning may or may not have an output, so it can be like both SL and unsupervised learning (www.wgu.edu). Reinforcement learning is occasionally classified as semi-supervised (www.wgu.edu). Based on the type of tasks, we can categorize ML models into the following types: classification models, regression models, clustering, dimensionality reduction, DL. So, the major difference between regression and classification algorithms is that regression algorithms are utilized to predict continuous values like price, salary, age, etc., and classification algorithms are employed to predict/classify discrete values such as male or female, true or false, spam or not spam, etc. Moreover, the important difference between clustering and classification is that clustering is an unsupervised learning technique that groups similar instances on the basis

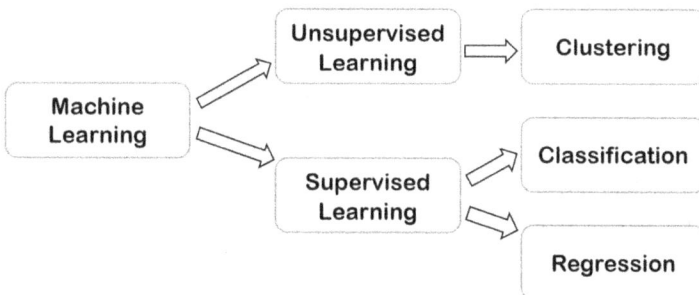

FIGURE 9.11   Simple chart of ML (www.dzone.com).

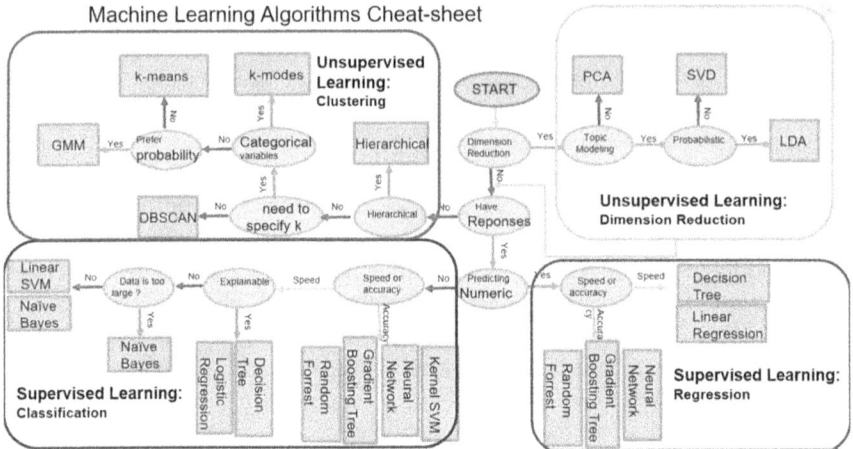

FIGURE 9.12   Comprehensive diagram for ML/AI (www.pinterest.com).

of features without labels (X) whereas classification is an SL technique that assigns pre-defined tags to instances on the basis of features with labels (X,Y).

In SL, the aim is to predict outcomes for new data. You know up front the type of results to expect. With an unsupervised learning algorithm, the objective is to get insights from large volumes of new data. ML itself determines what is different or interesting from the dataset.

In ML there are various models that normally fall into three different categories: (1) SL, (2) unsupervised Learning, and (3) reinforcement learning. For example, we present some famous models in ML/AI for regression aims:

1. Linear and nonlinear regression (polynomial and multiple LR)
2. Ridge regression
3. Lasso regression
4. K-neighbors regressor (KNN-R)
5. Decision tree regressor
6. Random forest regressor
7. Gradient boosting regressor (and XGBRegressor)
8. Adaboost regressor
9. Support vector regression (SVR)
10. Multilayer perceptron (MLP): artificial neural network (ANN) for regression
11. Robust regression
12. Elastic net
13. Stochastic gradient descent (SGD)
14. DL/NN (Keras, Pytorch)

Also, there are the same situations for classification purposes with minor changes. That is, the models can be LR, KNN, decision tree, random forest, SVM, XGBClassifier, ANN, NN (Keras, Pytorch). There are many classification models and algorithms for predicting some parameters in the supervised ML/AI models.

Start

↓

Identification of a Need

↓

Literature Review and Consulting with Physicians

↓

Definition of a Problem

↓

Data Collection

↓

Machine Learning / Data Science Researcher and Engineer

↓

Data Preprocessing / Cleaning / Normalizing / Scaling / Missing, Outliers, Drop, Filling Nan, Duplicates, Concatenating / Feature Selection / Ensemble / Splitting

↓

Verifying the Models: Random Forest, SVM, Logistic Regression, XGBClassifier

↓

Comparing the Accuracy: Confusion Matrix, AUC-ROC, F1- Score

↓

Selecting and Finalizing a Model with Excellent Accuracy

↓

End: Presentation
Conclusion Remarks

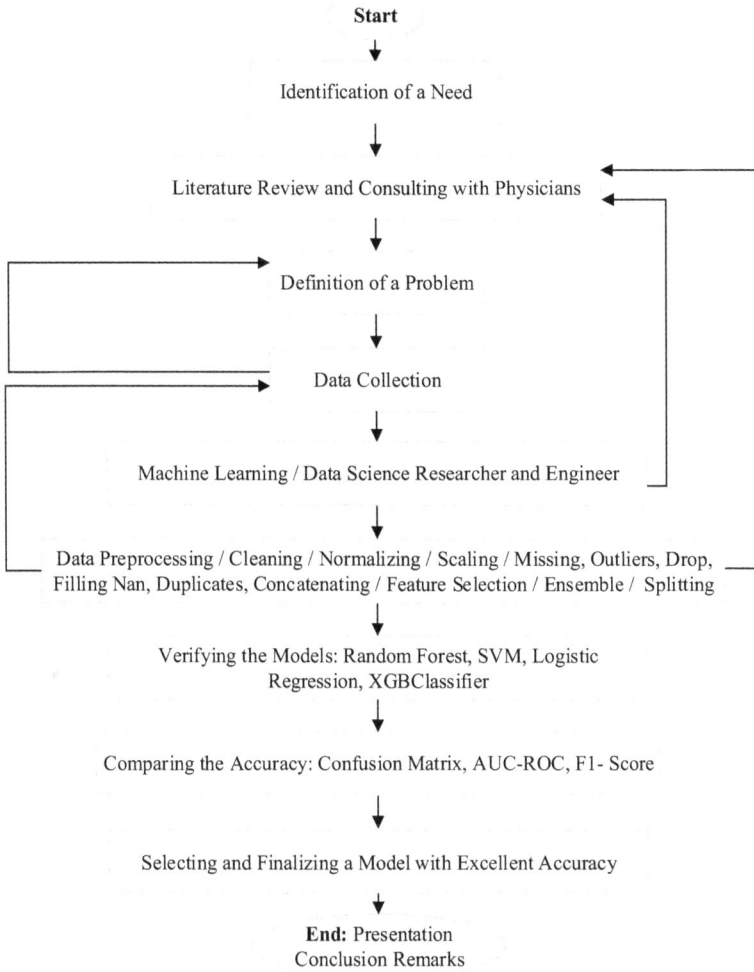

FIGURE 9.13 General road map and problem solution steps in a flowchart (algorithm): overall method.

With regard to classification models, some of the important algorithms include "Logistic Regression", "Decision Tree", "Random Forest", "Gradient-Boosted Tree", "XGBClassifier", "Multilayer Perceptron", "Artificial Neural Network/MLP", "Gaussian Naive Bayes", "K-Nearest Neighbors", and "Support Vector Machines". The general road map and problem solution steps are depicted in a flowchart (algorithm) as an overall methodology, which is shown in Figure 9.13.

Here, some models are illustrated.

9.5.2 Introduction to Some Models

An ML model is a mathematical description of the samples hidden in data. When the ML model is trained (built/fit) for the training data, it finds some governing structure within it. That governing structure is formalized into rules, which can be used under

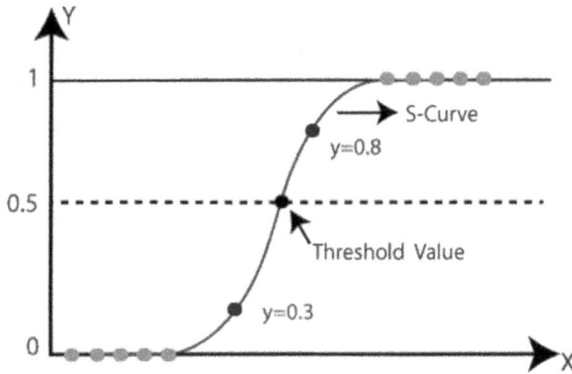

FIGURE 9.14   LR model (www.javatpoint.com).

different conditions for predictions. A model can infer a relationship within a set of data if it is trained on some training data, and then applied to new data.

9.5.2.1 Logistic Regression

The initial and classic model of LR is an instance of SL. LR is used to compute and predict the probability of a binary (0/1 or yes/no) event occurring. A simple example of LR can be employed in ML to obtain if a person is likely to be infected with a specific virus or not (see Figure 9.14).

Simply, the logistic function formulation is of the following sigmoid form,

$$Y(x) = \frac{1}{1 + e^{-X}} \tag{1}$$

Here, LR transforms its output utilizing the logistic sigmoid function to return a probability value generally. That is, LR is a special case of linear regression as it predicts the probabilities of the result employing the log function. We apply the activation function (sigmoid) to change the outcome into categorical and classified values.

9.5.2.2 Support Vector Machine

An SVM is a supervised ML algorithm which employs classification algorithms for two group classification problems. SVM utilizes a method and technique called the kernel trick to convert and transform a given data and so based on these transformations it finds a special optimum boundary between the possible outputs logically. Figure 9.15 shows components and terms of a SVM model schematically.

Regarding formulation of any hyperplane, it can be written as a set of points X satisfying the following formulation:

$$W^T X - b = 0 \tag{2}$$

Here, W is the normal vector to the hyperplane and the parameter $\frac{b}{W}$ determines the offset of the hyperplane from the origin along the normal vector W.

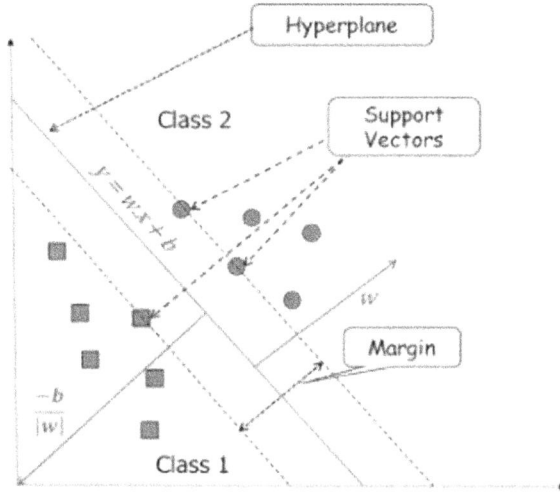

FIGURE 9.15   SVM model (Rani et al. 2022).

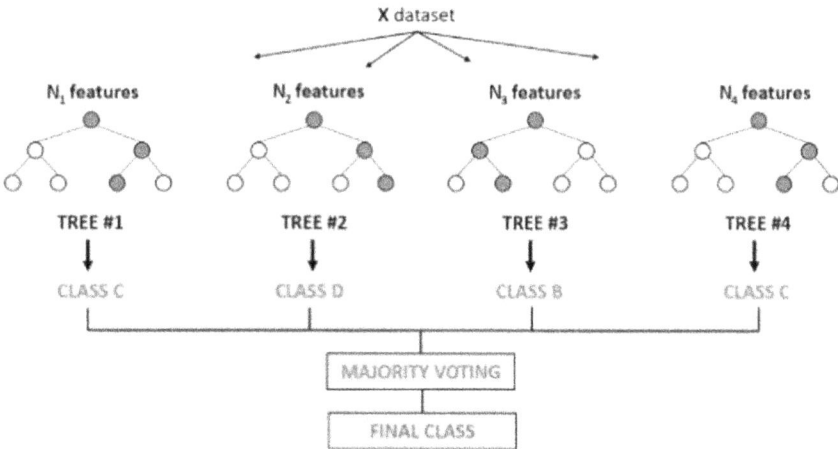

FIGURE 9.16   Random forest classifier model (www.freecodecamp.org).

9.5.2.3   Random Forest

The model of random forest is a supervised ML technique which is used extensively in classification and regression problems. It builds decision trees on dissimilar patterns and takes their majority vote for classification and average in case of regression. It should be mentioned that the model of random forests or random decision forests is an ensemble learning method for classification, regression, and other tasks that operates by constructing a multitude of decision trees at training time. For classification tasks, the output of the random forest is the class selected by most trees generally (Figure 9.16).

The solution steps for a random forest include choice random patterns from a given dataset and the construction of a decision tree for each pattern and sample and then getting a prediction result from each decision tree. Then, a vote is conducted for each predicted result. Finally, the result with the most votes is chosen as the final prediction.

The mathematical meanings behind the random forest (~ decision tree) consist of two important concepts: entropy and gain information. Entropy is a measure of the randomness of a system. Here, these two important formulations are presented as follows:

$$Entropy\ E(s) = -\sum P(X) * log_2(P(X)) \tag{3}$$

Here, P(X) is related to probability of X. Also, the mathematical formulation for entropy is as follows:

$$E(s) = -\sum_{i=1}^{c} - p_i\ log_2\ p_i \tag{4}$$

where c is the number of classes and p_i is basically the frequentist probability of an element/class "i" in our given dataset (p_i is a fraction of examples in a given class). Also, regarding gain information of Gain (S, A) we have,

$$Gain\ (S,A) = Entropy(S) - \sum_{\vartheta \in D_A} \frac{|S_\vartheta|}{|S|}\ Enropy(S_\vartheta) \tag{5}$$

We have the following expression for information gain from X on Y:

$$IG(Y,X) = E(Y) - E(Y \mid X) \tag{6}$$

Here, we can subtract the entropy of Y given X from the entropy of Y to compute the decrease of uncertainty about Y given an additional part of information X about Y. Generally, it is called information gain (IG). The larger the decrease in this uncertainty, the more information is obtained about Y from X logically.

Finally, the final vote and results are obtained by:

$$F = \frac{1}{B} \sum_{b=1}^{B} T_b(X) \tag{7}$$

That is, after training, predictions for unobserved samples X may be made by averaging the predictions from all the individual regression trees on the X field. The number of samples/trees, B, is a free parameter. Classically, several thousand trees are typically used, depending on the size and nature of the training set. An optimal number of trees B may be found utilizing cross-validation, or by detecting the out-of-bag error (bagging frequently: B times).

9.5.2.4 eXtreme Gradient Boosting (XGBoost)

eXtreme gradient boosting (XGB) is more precise than random forests and is a more influential and stronger model. It combines a random forest and gradient boosting (GB) to build a stronger model and set of outcomes. XGB (XGBoost) has smaller steps, predicting sequentially instead of individualistically. It uses the samples and patterns in residuals, strengthening the model. It means the predicted error is less than random forest predictions.

In the XGB model, GB refers to a class of ensemble ML algorithms which may can be employed for classification or regression predictive modeling problems. Also, ensembles are generally built from decision tree models (Figure 9.17).

The XGBoost model works as Newton-Raphson in a function field and space unlike GB, which works as a gradient descent in function space; a second-order Taylor estimation is applied in the loss function to build the connection to the Newton-Raphson method. So, a general unregularized XGBoost algorithm is as follows. In XGBoost, we describe the complexity as follows:

$$\omega(f) = \gamma T + \frac{1}{2} \lambda \sum_{j=1}^{T} w_j^2 \tag{8}$$

where w is the vector of scores on leaves, T is the number of leaves, and λ is a regularization term.

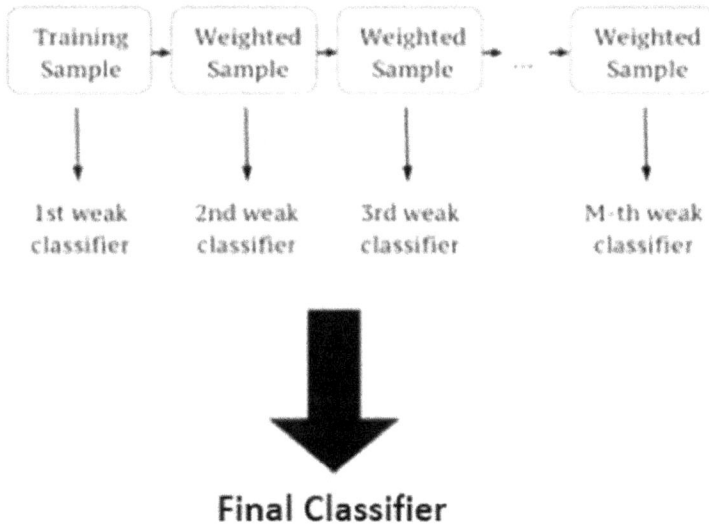

FIGURE 9.17 A simple model of XGB graphically (www.towardsdatascience.com).

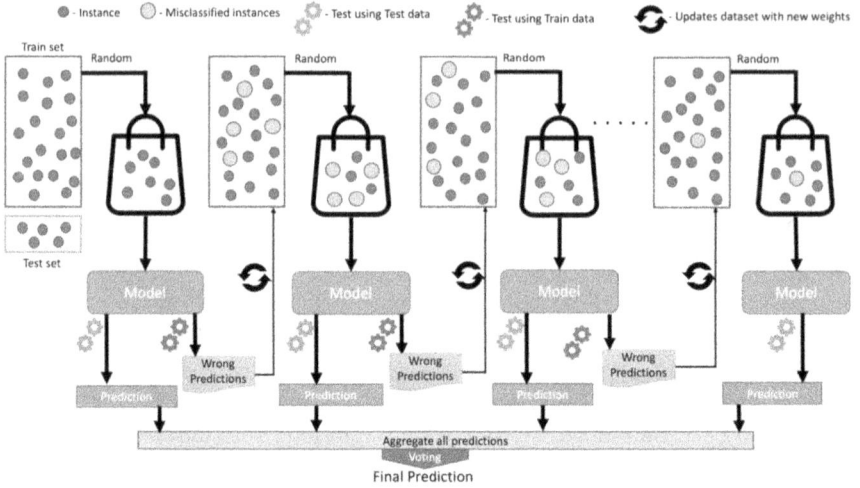

FIGURE 9.18   Internal working of boosting algorithm (www.dzone.com).

9.5.2.5   XGBoost Model for Regression

The XGBoost model for regression is called the XGBRegressor. So, we will build an XGBoost model for this regression problem and evaluate its performance on test data (unseen data/new instances) using the root mean squared error (RMSE) and the R-squared (R^2 coefficient of determination). This model is one of the strong algorithms for regression problems (see Figure 9.18). XGBoost utilizes second-order Taylor estimation for both classification and regression. The loss function containing output values can be approximated as follows:

$$L\left(y, p_i^0 + O_v\right) = L\left(y, p_i\right) + \left[\frac{d}{dp_i} L\left(y, p_i\right)\right] O_v + \frac{1}{2}\left[\frac{d^2}{dp_i^2} L\left(y, p_i\right)\right] O_v^2 \qquad (9)$$

Here, the first part is the loss function, the second part includes the first derivative of the loss function, and the third part includes the second derivative of the loss function.

9.5.2.6   Neural Networks

The new applications of NNs involve a different methodology in many instances. The NN method is promising for predicting analytical, numerical, and experimental trends and behaviors. Also, it has become progressively more popular in recent years. NNs may often solve the problems much faster compared to other methods with the additional ability to learn. For example, a graphical explanation of a three-layer feed forward network is shown in Figure 9.19.

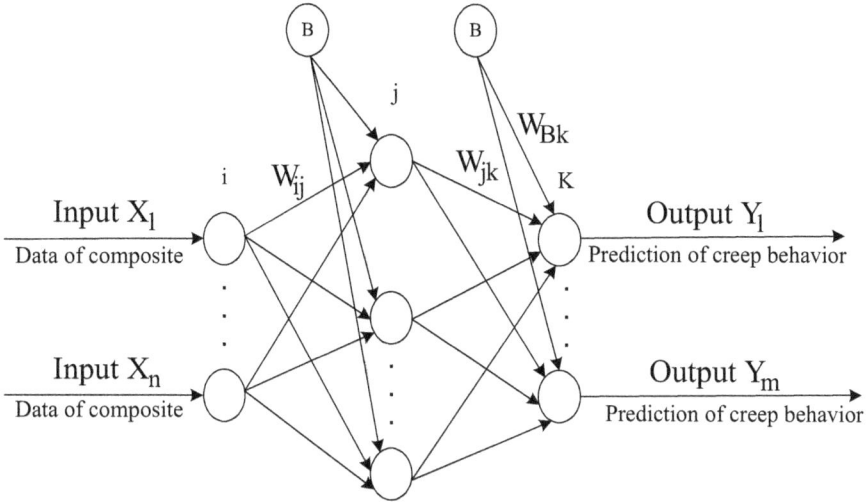

FIGURE 9.19 Schematically simple presenting the ANN structure and configuration.

Assuming that the network consists of n, p, and m neurons in the input, hidden, and output layers, respectively, then the net input (Z_j) to node j in the hidden layer is of the following form:

$$Z_j = \sum_{i=1}^{n} W_{ij} X_i + b_j, \quad j = 1, 2, ..., p \tag{10}$$

where X_j is the input of node j of the input layer, W_{ij} is the connection weight in relation with node i of the input layer and node j of the hidden layer, and b_j is the bias in relation with node j of the hidden layer. The output "h" from a neuron is achieved by transforming its input using a proper transfer function such as the following:

$$h_j = f\left(\sum_{i=1}^{n} W_{ij} X_i + b_j\right), \quad j = 1, 2, ..., p \tag{11}$$

Also, in the output layer, the net input Z_k to node k is of the following form:

$$Z_k = \sum_{j=1}^{p} W_{jk} h_j + b_k, \quad k = 1, 2, ..., m \tag{12}$$

The output " Y_k " of node k of the output layer is written as follows:

$$Y_k = g\left(\sum_{j=1}^{p} W_{jk} h_j + b_k\right), \quad k = 1, 2, ..., m \tag{13}$$

Furthermore, the error is computed utilizing Eq. (14) and is known as average squared error. Here, N indicates the total number of samples in the training set.

$$MSE = \frac{1}{N} \sum_{i=1}^{N} (e_i)^2 = \frac{1}{N} \sum_{i=1}^{N} (t_i - a_i)^2 \tag{14}$$

It should be stated that artificial neural networks (ANNs) are used as an interdisciplinary tool in numerous kinds of nonlinear problems. To create a NN for a particular problem, one requires a training algorithm. As NNs function based on samples, it is needed to make a set of practical examples representing the problem in the form of system inputs and outputs. Also, to improve an NN with appropriate and great performance, an adequate quantity of experimental data must be accessible. Throughout the training and testing sessions, the network architecture, learning algorithm, and the other related parameters of the NN should also be improved and optimized to the exact problem under analysis. The biggest disadvantage of an NN is its black box nature. Because it can estimate any function and study its structure, it doesn't provide any insights on the structure of the function being approximated. Also, vanishing gradient/zero gradient is another problem of NNs.

Once the NN is sufficiently optimum, and trained based on these data, it then becomes feasible to obtain suitable results when presented with any new input pattern it has never experienced before. Keras and Pytorch are open-source frameworks for DL that have achieved a reputation among data scientists. Keras is a high-level API capable of running on top of TensorFlow, CNTK, Theano, or MXNet (or as tf.contrib within TensorFlow). Pytorch and Keras are two DL libraries created by two big companies, Facebook and Google, targeted to facilitate DL applications like face recognition and self-driving cars, and so on. Some examples of these frameworks include TensorFlow, Pytorch, Caffe, Keras, and MXNet. If you're just beginning to discover DL, you should learn Pytorch first because of its reputation in the research community. But, if you're familiar with ML and DL and focused on getting a job in the industry as soon as possible, learn TensorFlow first.

9.5.2.7 Popular CNNs

CNN is a DL NN sketched for processing structured arrays of data like portrayals. CNNs are extremely efficient at picking up on design in the input image, like lines, gradients, circles, or even eyes and faces. Four CNN NNs that every ML engineer should know include the following:

1. LeNet-5 architecture is maybe the most broadly known CNN architecture. It was created by Yann LeCun in 1998 and was broadly used for written digits recognition (MNIST).
2. The AlexNet CNN architecture won the 2012 ImageNet ILSVRC challenge by a large margin.
3. VGG-16, as seen in the AlexNet architecture, CNNs were beginning to get deeper and deeper.
4. ResNet

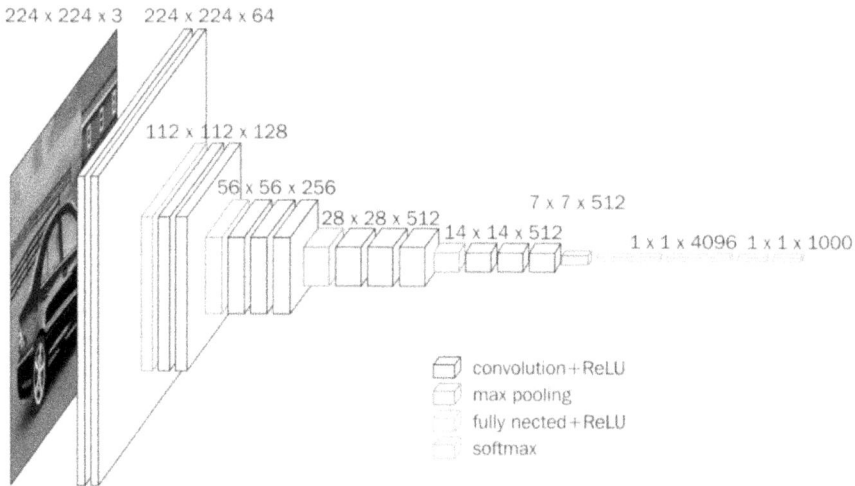

FIGURE 9.20 VGG-16 convolutional network for classification and detection (www.neurohive.io).

Figure 9.20 shows a VGG architecture that is a CNN model suggested by K. Simonyan and A. Zisserman from the University of Oxford in the paper "Very Deep Convolutional Networks for Large-Scale Image Recognition". The model achieves 92.7% top-5 test accuracy in ImageNet, which is a dataset of over 14 million images belonging to 1,000 classes. It was one of the famous models submitted to ILSVRC-2014. It makes the development over AlexNet by replacing large kernel-sized filters (11 and 5 in the first and second convolutional layer, respectively) with multiple 3×3 kernel-sized filters one after another. VGG-16 was trained for weeks and was using NVIDIA Titan Black GPUs. Minor disadvantages of CNNs are as the following:

1. A CNN is substantially slower due to an operation like Maxpool.
2. If the CNN has numerous layers, then the training process takes a lot of time if the computer doesn't have a great GPU.
3. A ConvNet needs a big dataset to handle and train the NN.

There are four main reasons why DL enjoys so much buzz now: data, computational power, the algorithm itself, and marketing (understanding the hype around DL). However, there are some disadvantages of NNs including black box, duration of development, amount of data, and computational expense. So, NNs are excellent for some problems and not so wonderful for others. DL is a little too hyped now and the expectations surpass what may be done with it, but then that doesn't mean it isn't helpful. We are living in an ML recovery stage and the technology is becoming more and more democratized, which permits more people to utilize it to build effective products. There are many problems and issues out there that may be solved with ML, and we are sure that we will soon achieve progress.

Overfitting happens when a model attempts to predict a trend in data that is extremely noisy. This is owing to an excessively complicated model with too many parameters. A model that is overfitted is inaccurate because the trend does not reflect the reality present in the data. Overfitting is an ML concept that arises when a model fits completely to train the dataset against its test and validation data. When this happens, the algorithm cannot operate precisely against unseen data, therefore contradicting its objective.

Note: There are some techniques to avoid overfitting such as given below:

1. Training with more data. With the increase in the training data, the important features to be extracted become important.
2. Removing outliers from the dataset and good handling/preprocessing the data.
3. Data augmentation.
4. Addition of noise to the input data.
5. Feature selection.
6. Cross-validation.
7. Simplification of data.
8. Regularization.
9. Ensembling.
10. Early stopping.
11. Adding dropout layers.
12. Using batch normalization.

9.5.3 ACCURACY

There are two key kinds of errors current in any ML model. They are reducible errors and irreducible errors. Accuracy is a measure of the degree of closeness of a measured or calculated value to its actual value. The percent error is the ratio of the error to the actual value multiplied by 100. The precision of a measurement is a measure of the reproducibility of a set of measurements. Here, we have tried to briefly explain three types of accuracy for classification problems (F1-score, area under the curve (AUC)-receiver operating characteristics (ROC), and confusion matrix (CM)). In this research, we have tried to use the F1-score for assessing the accuracy of the models. Also, we have the R2_Score for regression problems. If the value of the R2_Score is one, it means that the model is perfect and if its value is zero, it means that the model will perform badly on an unseen dataset (R2_Score = 1 − (RSS / TSS), where RSS = residual sum of squares & TSS = total sum of squares).

9.5.3.1 Criterion of F1-Score

The F1-score combines and balances precision and recall on the positive class whereas accuracy looks at appropriately classified observations that are both positive and negative. The F1-score is a generally utilized metric for classifying ML algorithms, but its definition is not widely understood, which can make it difficult to know what a good score is. The formula used for the F1-score is as follows:

$$F1 - Score = 2 * \frac{Precision * Recall}{Precision + Recall} \tag{15}$$

Precision is a quantity depicting how many of the positive predictions made are right (TP: True Positives):

$$Precision = \frac{TP}{TP + FP} = \frac{True\ Positives}{True\ Positives + False\ Positives} \tag{16}$$

Also, Recall is a quantity depicting how many of the positive cases the classifier properly predicted, over all the positive cases in the given dataset, and is defined as

$$Recall = \frac{TP}{TP + FN} = \frac{True\ Positives}{True\ Positives + False\ Negatives} \tag{17}$$

9.5.3.2 AUC-ROC

The AUC-ROC diagram (curve) is a performance measurement for graphically classifying problems at numerous threshold settings. ROC is a probability curve, and AUC presents the degree or measure of separability. It tells how much the model is capable of distinguishing between given classes. In ML, performance measurement is a vital task. So, when it comes to a classification problem, we can count on an AUC-ROC curve (Figure 9.21). When we require to check or visualize the performance of the multi-class classification problem, we apply the AUC-ROC curve. It is one of the most important evaluation metrics for checking any classification model's performance. It is also written as AUROC (Area Under the Receiver Operating Characteristics) (www.towardsdatascience.com).

An AUC of 0.5 indicates no discrimination (capability to diagnose patients with and without the disease or condition based on the test); the values between 0.7 and 0.8 are assumed acceptable, and between 0.8 and 0.9 are considered excellent, and more than 0.9 is considered outstanding and ideal logically.

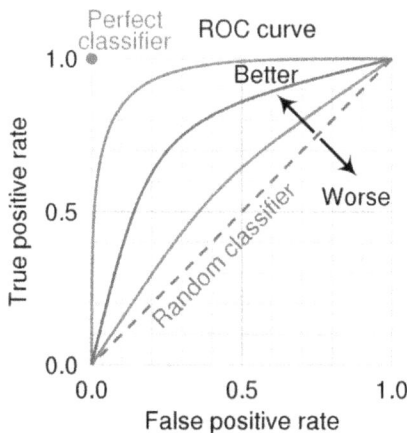

FIGURE 9.21   Presenting ROC-AUC curves and diagrams graphically.

9.5.3.3   Confusion Matrix

In general, a CM is a numerical table which is applied to express the performance of a classification algorithm. A CM visualizes and summarizes the performance of a classification algorithm. CMs are employed to imagine and visualize significant predictive analytics like recall, specificity, accuracy, and precision. CMs are beneficial because they give direct comparisons of values like True Positives (TP), False Positives (FP), True Negatives (TN), and False Negatives (FN) (Table 9.1).

CMs are extensively used because they provide a better overview of a model's performance than classification accuracy does. For example, in classification accuracy, there is no information about the number of misclassified instances.

Important Note: Because CM cannot process probability scores, all these class labels are binary. Here every class label is either 0 or 1 (0 represents negative and 1 represents positive labels). Therefore, the CM for a binary classification will be as given in Table 9.2.

TABLE 9.1
A simple sample of a CM schematically

Actual Values

		Positive (1)	Negative (0)
Predicted Values	Positive (1)	TP	FP
	Negative (0)	FN	TN

(www.towardsdatascience.com)

TABLE 9.2
Confusion matrix

		Actual Values	
		0 (negative)	1 (positive)
Predicted Values	0 (negative)	TN	FN
	1 (positive)	FP	TP
		N (total negative)	P (total positive)

(www.analyticsvidhya.com)

Here, N = total negative, P = total positive; we can see how a CM looks like for a binary classification model. TP means predicted value as well as the actual value both are positive, i.e., the model correctly predicts the class label to be positive. TN means the predicted value as well as the actual value both are negative, i.e., the model correctly predicts the class label to be negative. FP means the predicted value is positive, but the actual value is negative, i.e., the model falsely predicted these negative class labels to be positive. FN means the predicted value is negative, but the actual value is positive, i.e., the model falsely predicted the positive class labels to be negative.

9.5.4 PREPROCESSING THE DATA

Preprocessing the data is the heart of ML. Your ML instruments are as good as the quality of your data. Our data requires to go through a few steps before it can be used for making predictions. Here, a short description concerning how to preprocess data in Python step-by-step is provided:

1. Get the dataset and import all the crucial libraries
2. Load and import data in Pandas
3. Drop columns that aren't useful (feature selection)
4. Handle the data including removing outliers and duplications, dropping/ filling Not A Number (NAN) data
5. Drop rows with missing values or substitute with mean, median, or other techniques
6. Do feature scaling (standardization, normalization)
7. Create dummy variables and clean the dataset
8. Transform the data
9. Reduce the data and the dimensionality
10. Encode the categorical data
11. Balance the data
12. Take care of missing data and screening
13. Convert the data frame to NumPy
14. Divide the dataset into training data and test data (splitting the dataset)

Figure 9.22 shows graphically the preprocessing dataset in an ML model. This schematic can help to better understand the process. So, data preprocessing is necessary before its real use. In fact, data preprocessing is the idea of changing the raw data into a clean dataset. The dataset is preprocessed to sort missing values, noisy and outlier data, and other inconsistencies before executing it in the models.

9.6 ML AND AI IN AM

The use of AI and ML has growing considerably in AM owing to their unprecedented performance in data tasks like classification, regression, and clustering methods. For visualization of AI model in AM (Jin et al. 2021b), the location information is included to separate the condition that an infill raster is tied parallel together with

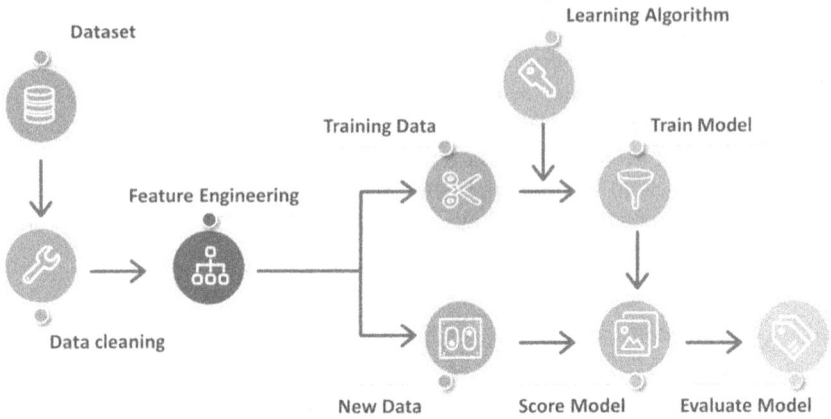

FIGURE 9.22 ML process diagram (www. medium.datadriveninvestor.com).

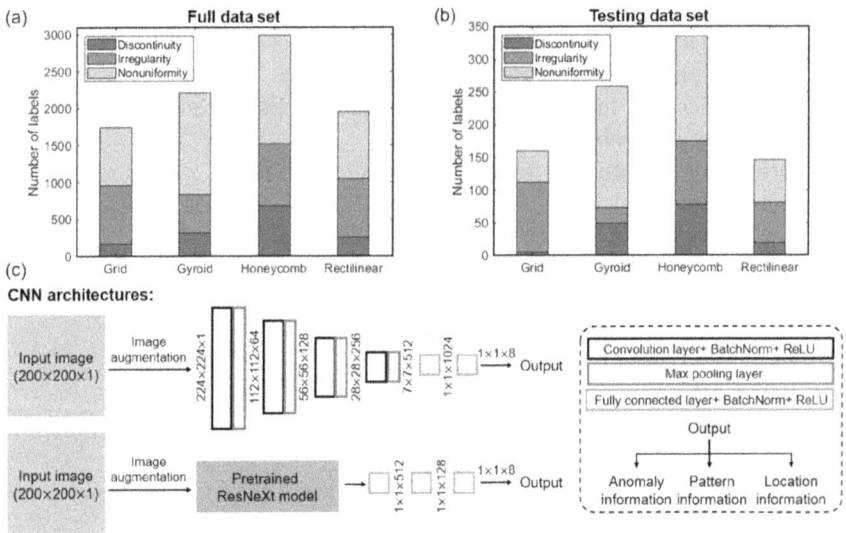

FIGURE 9.23 (a,b) Bar charts summarizing detailed anomaly information for both full and testing datasets. The number of labels for three types of anomalies under four infill patterns is counted. (c) Flow diagram showing the architectures of applied CNN models. The self-designed CNN architecture is illustrated at the top, and the pretrained model is shown at the bottom. Image size information is displayed in the vertical direction, and legends are marked in the bounding box with a dashed line (Jin et al. 2021b).

the perimeter from an irregularity anomaly. Figure 9.23(a and b) summarizes the thorough anomaly information on the full dataset and the testing dataset, presenting the number of labels for each anomaly under four infill patterns.

The architectures of the applied CNN models are shown in Figure 9.23c. However, given the rather small size of the bioprinting dataset, this model uses the final layer of a pretrained ResNeXt-50 as a static feature extractor that feeds into a small, optimized multilayer perceptron as shown in Figure 9.23c; despite ImageNet images being full-color and of natural scenes, the low-level pretrained feature detectors are still applicable (Jin et al. 2021b).

Recently the basic ideas of hierarchical constructions and their various features have been introduced to present the most powerful natural materials and compounds and their employment in synthetic made-up composites for tissue engineering, biomedical, and industrial applications. In these applications AI/ML models have been able to expand the characterization and design of natural and bio-inspired materials, optimizing the computational tools and overcoming the restrictions of old-style methods (Begg and Kamruzzaman 2005; Gobert et al. 2018; Gu, Chen, and Buehler 2018; Hanakata et al. 2019; Chen and Gu 2020; Jin, Zhang, and Gu 2020; Jin et al. 2021a; Milazzo et al. 2022).

One of the primary outcomes of the application of AI and ML for AM was presented in 1997 (Ikonen et al. 1997), and restricted amount of available published research studies is obtainable currently. One can make a few expectations to describe this situation:

1. Correctness and validity of dataset
2. AI/ML needs big data for precise performance
3. Lack of clean dataset
4. Costly experiments (material costs and tools) and hard-to-collect suitable data
5. Not many papers were found because of the former limits to standard AM terminology, and scientists frequently utilize dissimilar synonyms for the same process
6. Not all research has in the title or key words the terms "artificial intelligence and machine learning", "AI, ML", or its methods but in text it is explained to what extent ML methods are used

Consequently, scientists started evaluating diverse ML and AI techniques for designing, modeling, and simulation of AM processes. ML/AI models and their combinations, which are useful for simulation of PPBF AM processes, are illustrated in the literature (Shen et al. 2004; Wang et al. 2007, 2009; Li, Dong, and Zhang 2009; Munguía, Ciurana, and Riba 2009; Rong-Ji et al. 2009; Garg, Lam, and Savalani 2015; Garg and Lam 2015; Negi and Sharma 2016). The methods include ANN, genetic algorithm (GA), Ensemble-MGGP that consists of ANN, Bayesian classifier, and SVM algorithm, support vector regression (SVR).

For example, the best process parameters to manufacture parts with higher level of precision were determined. Rong-Ji et al. (2009) focused on such parameters of SLS as the layer thickness, hatch spacing, scanning speed, scanning mode, laser power, interval time, and work surrounding temperature. To get the optimal process parameters listed above, Rong-Ji et al. (2009) used a combination of GAs and back propagation (BP) NN algorithms. The results from BPNN were applied as input parameters for fitness function in GA. GA was employed as a method to obtain optimum process parameters based on minimum shrinkage ratio (Rong-Ji et al. 2009).

Moreover, some studies have designed an AI/ML model with just a small dataset with around ~100 samples. Garg and Lam (Garg, Lam, and Savalani 2015) made an effort to assess the environmental impact of 3D printing by application of ANN, genetic programming, and SVR methods. However, the polymer powder used by the SLS process decreases waste and saves fuel because it is biodegradable, which makes it environmentally sustainable. Their results (Garg, Lam, and Savalani 2015) showed that genetic programming performance is better than the other two; however, how open porosity is connected to the measurement of environmental impact is unclear. Available studies show that statistical analysis and AI/ML need a big dataset to be more precise. Also, mathematical simulation and modeling require deep knowledge on both process and material physics.

Some authors have developed an ANN model which can define the relation between the process parameters and SLS part density. Additionally, the effects of process parameters on part density can be analyzed quantitatively (Wang et al. 2009). An NN model for analyzing the density of parts prepared by SLS and also the effects of the process parameters, including layer thickness, laser power, hatch spacing, scanning speed, interval time, temperature of working environment, and scanning mode on the density was studied. Moreover, some interesting results have been extracted from this study (Wang et al. 2009). The NN model may be employed to analyze the effect of the SLS process parameters on density, quantitatively. The density of the SLS part is found to increase with the increase in laser power, the temperature of the work environment, and the interval time in a certain range, but it decreases with the increase of layer thickness, scanning speed, and hatch spacing.

While ML has been developed for many decades, ML applications in the AM field are only a few years old. These applications span processing parameter optimization, property prediction, defect detection, geometric deviation control, quality prediction, and assessment, etc. First, ML models learn the relevance of the relationship between the processing parameters and property by means of existing data to provide guidance for optimizing these processing parameters. Second, ML models predict the geometric deviation based on the designed geometry after training and offer guidance for geometric error compensation. Third, ML models are suitable at dealing with in situ images and acoustic emissions throughout printing and detecting defect formation in real time. But the obtainable data that may be extracted from the processing parameter-process-microstructure-property map have not been fully utilized. In this regard, exploiting more data acquisition methods, exploring more ML applications, and developing better algorithms will be the main research direction in this developing research field (Meng et al. 2020). A missing but useful functionality in SL in recent literature is active learning. In the AM field, labeling the output of each input data point is typically expensive in terms of the consumed time, cost, and human labor, because it requires conducting an experiment or a simulation at each input setting to make this observation. Active learning is a method that can alleviate this issue. The common procedure in ML models is acquiring enough input-output pairs first and then using them to train ML models without further query of labeling new data. On the contrary, the procedure in active learning is that the ML models can query interactively for labeling new data during training to maximize its performance. By this means, ML models may use fewer data points to achieve better performance. Therefore, active learning is strongly recommended if a dataset to be used to train the ML model has not been acquired. Another potential research field is the uncertainty quantification (UQ), which is critical for a robust design. The uncertainty in the AM field has been reviewed in Hu and Mahadevan (2017). In regression tasks, ML models like GP provide not only the mean value at a certain input as the prediction of its output, but also standard deviation, which represents the uncertainty at that point. Also, in classification tasks, ML models provide confidence on when they make a classification. These uncertainties are part of the epistemic uncertainty and have not been utilized in recent literature. In addition, a typical UQ procedure (Wang et al. 2019) may require hundreds of data points, which is impractical to obtain from experiments or simulations. In this regard, an ML-based surrogate model is very helpful in obtaining the required data and increasing the efficiency of the UQ procedure. Overall, UQ in ML applications in the AM field is a good research direction that has not been investigated in depth (Hu and Mahadevan 2017; Wang et al. 2019; Meng et al. 2020).

NDE techniques and AI technology have been presented and developed for efficient generation of digital material twins of Natural Fiber Reinforced Polymer Composites (NFRPCs), which will be valuable for designing next-generation bio-composites (Preethikaharshini et al. 2022).

9.7 SOLVED AND ANALYZED PRACTICES

We have presented five applied examples for more clarification of how we can predict AM parameters, healthcare, medicine, and biology problems (from beginning to the end as a historical orders). We have provided two healthcare and insurance examples for analyzing AI/ML because of their similarity with the solutions in our chapter topic.

9.7.1 EXAMPLE I

Preprocessing the dataset is required and necessary to model an AI/ML algorithm for prediction of parameters in AM of bio-composites. Present available methods to perform feature selection. There are usually three employed feature selection methods that are simple to make and yield excellent results, which are the following:

A. Univariate Selection
B. Feature Importance
C. Correlation Matrix with Heatmap

Let's take a closer look at each of these methods with an example (www.askpython.com). Datasets are available at the following address:
https://www.kaggle.com/iabhishekofficial/mobile-price-classification#train.csv

9.7.1.1   Univariate Selection

Statistics may be utilized in the selection of those features that carry a high relevance with the output. In the example below a statistical test for selecting the ten best features from the dataset is provided:

```
import pandas as pd
import numpy as np
from sklearn.feature_selection import SelectKBest
from sklearn.feature_selection import chi2
data = pd.read_csv("train.csv")
X = data.iloc[:,0:20]   #independent columns
y = data.iloc[:,-1]     #target column i.e price range
#apply SelectKBest class to extract top 10 best features
bestfeatures = SelectKBest(score_func=chi2, k=10)
fit = bestfeatures.fit(X,y)
dfscores = pd.DataFrame(fit.scores_)
dfcolumns = pd.DataFrame(X.columns)
#concat two dataframes for better visualization
featureScores = pd.concat([dfcolumns,dfscores],axis=1)
featureScores.columns = ['Specs','Score']   #naming the
dataframe columns
print(featureScores.nlargest(10,'Score'))
```

After running the code, the outputs provided in Table 9.3 appear.

9.7.1.2   Feature Importance

With this technique, you can understand the importance of every feature from your dataset with the use of the feature importance tool of the model. Feature Importance works by giving a relevancy score for every feature in your dataset; the higher the

TABLE 9.3

Output of above code for feature selection

Specs		Score
13	ram	931267.519053
11	px_height	17363.569536
0	battery_power	14129.866576
12	px_width	9810.586750
8	mobile_wt	95.972863
6	int_memory	89.839124
15	sc_w	16.480319
16	talk_time	13.236400
4	fc	10.135166
14	sc_h	9.614878

(www.askpython.com)

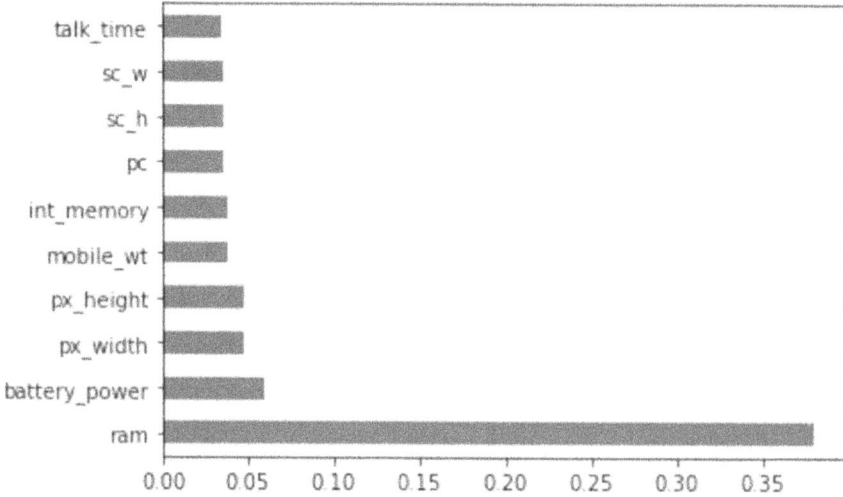

FIGURE 9.24 Output of above Python code for feature selection.

score given, the more relevant that feature for the training of your model. In the example below I will use the feature importance technique to select the top ten features from the dataset which will be more relevant in training the model.

```
import pandas as pd
import numpy as np
data = pd.read_csv("train.csv")
X = data.iloc[:,0:20]   #independent columns
y = data.iloc[:,-1]     #target column i.e price range
from sklearn.ensemble import ExtraTreesClassifier
import matplotlib.pyplot as plt
model = ExtraTreesClassifier()
model.fit(X,y)
print(model.feature_importances_) #use inbuilt class feature_
importances of tree based classifiers
#plot graph of feature importances for better visualization
feat_importances = pd.Series(model.feature_importances_,
index=X.columns)
feat_importances.nlargest(10).plot(kind='barh')
plt.show()
```

After running the code, the outputs given in Figure 9.24 appear.
   (www.askpython.com)

9.7.1.3 Correlation Matrix

With this technique, we can see how the features are correlated with each other and the target. The correlation matrix shows positive output if the feature is highly relevant and shows a negative output if the feature is less relevant to the data. A heatmap always makes it easy to see how much the data is correlated with each other and the target. In the example below I will create a heatmap of the correlated features to explain the correlation matrix technique.

```
import pandas as pd
import numpy as np
import seaborn as sns
data = pd.read_csv("train.csv")
X = data.iloc[:,0:20]   #independent columns
y = data.iloc[:,-1]     #target column i.e price range
#get correlations of each features in dataset
corrmat = data.corr()
top_corr_features = corrmat.index
plt.figure(figsize=(20,20))
#plot heat map
g=sns.heatmap(data[top_corr_features].corr(),annot=True,cmap="
RdYlGn")
```

After running the code, the outputs provided in Table 9.4 appear.

TABLE 9.4

Heatmap results of above Python code for feature selection

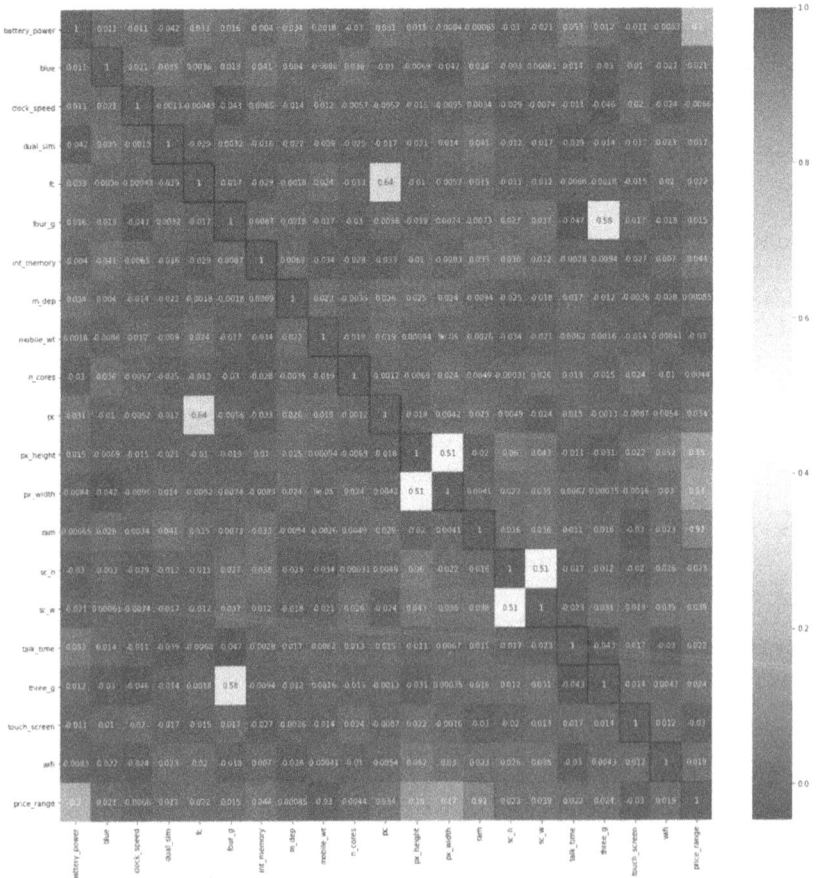

(www.askpython.com)

In this way, you can select the most relevant features from your dataset using the Feature Selection Techniques in Machine Learning with Python. All the above computations are according to the first 30 rows in Table 5.5. Also, the original file is available in the folder (www.askpython.com). Some samples of the mentioned dataset are presented in Table 9.5.

So, the researchers and engineers can use any of the above three methods to pre-process the data to start a suitable AI model for the prediction of parameters.

9.7.2 EXAMPLE II

Here, a simple numerical dataset is presented to build an AI predictive model which is similar to the AM dataset for predicting parameters related to the AM of bio-composites. Let us predict the health insurance cost (numerical dataset/Excel file). For analyzing this example, we must first check and screen the dataset. Datasets are available at the following address. This problem is a regression one. Here we present the first five rows (samples) https://github.com/AnuragMishra2311/Medical-Insurance-Charge-Predictor.

The shape of the Excel dataset is (1338, 7); it means we have 1338 samples/rows and 7 features/predictors/attributes including "age", "sex", "bmi", "children", "smoker", "region", "charges" (Table 9.6). Here, the body mass index (bmi) is a

---

TABLE 9.5

A section of samples

battery_po blue	clock_spee	dual_sim	fc	four_g	int_memor	m_dep	mobile_wt	n_cores	pc	px_height	px_width	ram	sc_h	sc_w	talk_time	three_g	touch_scre	wifi	price_range
842	0	2.2	0	1	0	7	0.6	188	2	2	20	756	2549	9	7	19	0	0	1	1
1021	1	0.5	1	0	1	53	0.7	136	3	6	905	1988	2631	17	3	7	1	1	0	2
563	1	0.5	1	2	1	41	0.9	145	5	6	1263	1716	2603	11	2	9	1	1	0	2
615	1	2.5	0	0	0	10	0.8	131	6	9	1216	1786	2769	16	8	11	1	0	0	2
1821	1	1.2	0	13	1	44	0.6	141	2	14	1208	1212	1411	8	2	15	1	1	0	1
1859	0	0.5	1	3	0	22	0.7	164	1	7	1004	1654	1067	17	1	10	1	0	0	1
1821	0	1.7	0	4	1	10	0.8	139	8	10	381	1018	3220	13	8	18	1	0	1	3
1954	0	0.5	1	0	0	24	0.8	187	4	0	512	1149	700	16	3	5	1	1	1	0
1445	1	0.5	0	0	0	53	0.7	174	7	14	386	836	1099	17	1	20	1	0	0	0
509	1	0.6	1	2	1	9	0.1	93	5	15	1137	1224	513	19	10	12	1	0	0	0
769	1	2.9	1	0	0	9	0.1	182	5	1	248	874	3946	5	2	7	0	0	0	3

(www.askpython.com)

TABLE 9.6

The first five rows (samples)

	age	sex	bmi	children	smoker	region	charges
0	19	female	27.900	0	yes	southwest	16884.92400
1	18	male	33.770	1	no	southeast	1725.55230
2	28	male	33.000	3	no	southeast	4449.46200
3	33	male	22.705	0	no	northwest	21984.47061
4	32	male	28.880	0	no	northwest	3866.85520

(www.askpython.com)

person's weight in kilograms (or pounds) divided by the square of height in meters (or feet). After preprocessing the dataset, it is time to incorporate this dataset in our model. So, here we introduce some Python code for getting this (https://github.com/AnuragMishra2311/Medical-Insurance-Charge-Predictor):

```python
import pandas as pd
import seaborn as sns
import matplotlib.pyplot as plt
from matplotlib import style
from sklearn.model_selection import train_test_split
from sklearn.linear_model import LinearRegression
df = pd.read_csv("insurance.csv")
df=dataset = dataset.dropna()
dataset.isnull().sum()
plt.figure(figsize=(18,8))
sns.heatmap(df.corr(),annot=True)
X=df.iloc[:,:-1].values
y=df.iloc[:,-1].values
from sklearn.model_selection import train_test_split
X_train,X_test,y_train,y_test=train_test_split(X,y,test_
   size = 0.2,random_state=0)
from sklearn.preprocessing import StandardScaler
sc = StandardScaler()
X_train=sc.fit_transform(X_train)
X_test=sc.transform(X_test)
```

9.7.2.1 Model I: Linear Regression

```python
from sklearn.linear_model import LinearRegression
from sklearn.metrics import
r2_score,mean_squared_error,mean_absolute_error
from sklearn.model_selection import cross_val_score
lin_reg=LinearRegression()
lin_reg.fit(X_train,y_train)
y_pred=lin_reg.predict(X_test)
print('Test Score(r2) : {}'.format(r2_score(y_pred,y_test))) #
#Avg score using cross val score
print('Cross Val Score : {}'.format(cross_val_score(lin_reg,X_
   train,y_train,cv=10).mean()))
```

9.7.2.2 Model II: XGBoost

```python
import xgboost
xgb=xgboost.XGBRegressor()
xgb.fit(X_train,y_train)
y_pred_xgb=xgb.predict(X_test)
print(r2_score(y_test,y_pred_xgb))
```

9.7.2.3   Model III: Random Forest

```
from sklearn.ensemble import RandomForestRegressor
forest=RandomForestRegressor()
# Without Hyperparameter tuning
forest.fit(X_train,y_train)
y_pred_forest=forest.predict(X_test)
print('Test Score : {} .format(r2_score(y_test,y_pred_forest)))
print('Cross Val Score : {}'.format(cross_val_score(forest,X_
   train,y_train,cv=10).mean()))
# No Overfitting
```

After running this short code, the following results are obtained:

For model I: Test Score(r2): 0.7238911072887935
Cross Val Score: 0.7273727994870052
For model II: r2_score: 0.8543091111065985
For model III: Test Score: 0.8766152602188266
Cross Val Score: 0.8072917702568052

Table 9.7 shows a heatmap for the correlations between the features. So, after learning the model, we can predict any features that we want. In addition to analyzing all features statistically, the model can predict for the data that was never seen by model. These regression models (predictive AI models) may be used for prediction of combinational percentages of basic material and optimum values of printer in designing bio-composite products via AI-/ML-based algorithms.

TABLE 9.7
Heatmap for showing the correlations between the features

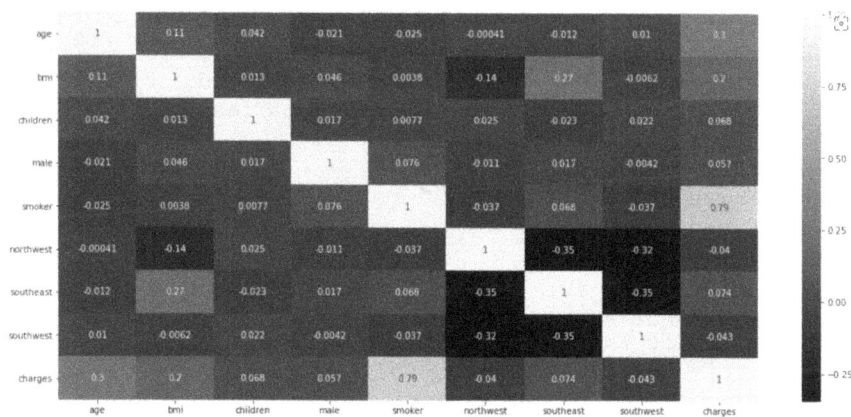

https://github.com/AnuragMishra2311/Medical-Insurance-Charge-Predictor

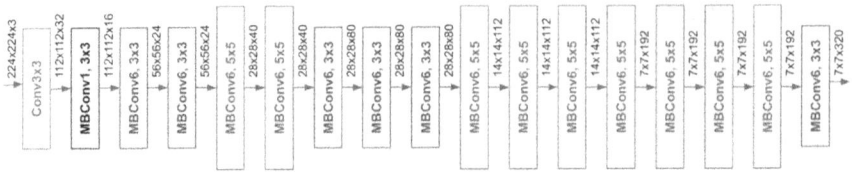

FIGURE 9.25 Efficient net architecture (www.opendl.in).

9.7.3 EXAMPLE III

Here an interesting example is introduced to predict cancer, which is similar to predicting the AM and 3D printing parameters for optimizing with the title of "Diagnosis and prediction of lung cancer along with the type of that". Here we can present a DL model for diagnosis of lung cancer with Python code (EfficientNetB3) (https://www.kaggle.com/code/vuppalaadithyasairam/78-test-acc-lung-cancer-efficientnetb3). This code is about the classification of chest CT images into one of the four cancerous conditions, namely, adenocarcinoma, large cell carcinoma, squamous cell carcinoma, and normal (Figure 9.26). Datasets are available at the following address. This problem is a classification. First EfficientNet architecture is presented and depicted as follows(Figure 9.25):

```
import numpy as np
import numpy as np
import pandas as pd
import matplotlib.pyplot as plt
import tensorflow as tf
from keras import Sequential
from tensorflow .keras.layers import*
from tensorflow .keras.models import*
from tensorflow .keras.preprocessing import image
train_set='../input/chest-ctscan-images/Data/train'
val_set='../input/chest-ctscan-images/Data/valid'
test_set='../input/chest-ctscan-images/Data/test'
train_datagen = image.ImageDataGenerator(
    rotation_range=15,
    shear_range=0.2,
    zoom_range=0.2,
    horizontal_flip=True,
    fill_mode='nearest',
    width_shift_range=0.1,
    height_shift_range=0.1
)
val_datagen= image.ImageDataGenerator(   rotation_range=15,
    shear_range=0.2,
    zoom_range=0.2,
    horizontal_flip=True,
    fill_mode='nearest',
    width_shift_range=0.1,
```

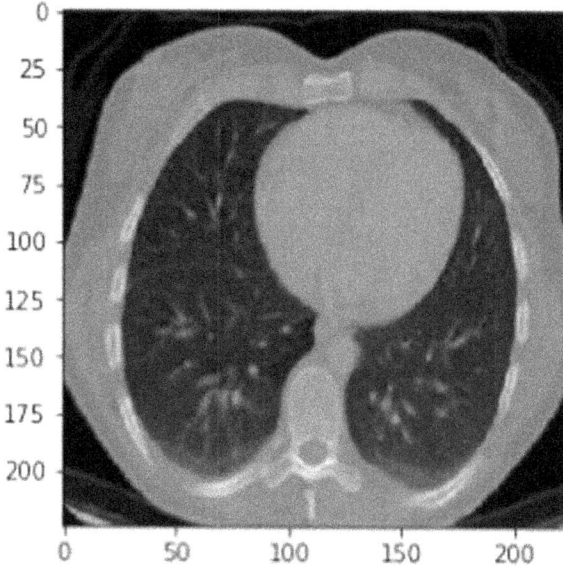

FIGURE 9.26 After learning, model prediction for this image is suspected to be adenocarcinoma (https://www.kaggle.com/code/vuppalaadithyasairam/78-test-acc-lung-cancer-efficientnetb3).

```
    height_shift_range=0.1)
test_datagen= image.ImageDataGenerator(  rotation_range=15,
    shear_range=0.2,
    zoom_range=0.2,
    horizontal_flip=True,
    fill_mode='nearest',
    width_shift_range=0.1,
    height_shift_range=0.1)
train_generator = train_datagen.flow_from_directory(
    train_set,
    target_size = (224,224),
    batch_size = 8,
    class_mode = 'categorical')
test_generator = test_datagen.flow_from_directory(
    val_set,
    target_size = (224,224),
    batch_size = 8,
    shuffle=True,
    class_mode = 'categorical')
validation_generator = test_datagen.flow_from_directory(
    test_set,
    target_size = (224,224),
    batch_size = 8,
    shuffle=True,
    class_mode = 'categorical')
base_model = tf.keras.applications.EfficientNetB2(weights='ima
  genet', input_shape=(224,224,3), include_top=False)
```

```python
for layer in base_model.layers:
    layer.trainable=False
model = Sequential()
model.add(base_model)
model.add(GaussianNoise(0.25))
model.add(GlobalAveragePooling2D())
model.add(Dense(256,activation='relu'))
model.add(BatchNormalization())
model.add(GaussianNoise(0.25))
model.add(Dropout(0.25))
model.add(Dense(4, activation='softmax'))
model.summary()
model.compile(loss='categorical_crossentropy',
        optimizer='adam',
        metrics=['accuracy','AUC','Precision','Recall'])
from tensorflow.keras.callbacks import ModelCheckpoint,
EarlyStopping
es=EarlyStopping(patience=3,monitor='val_loss')
filepath='best_model.h5'
checkpoint = ModelCheckpoint(filepath, monitor='val_accuracy',
verbose=1, save_best_only=True, mode='max')
history = model.fit(
        train_generator,
        epochs=18,
        validation_data=validation_generator,
        steps_per_epoch= 75,
        callbacks=checkpoint
        )
model.evaluate(train_generator)
model.evaluate(validation_generator)
model.evaluate(test_generator)
from keras.preprocessing import image
img = image.load_img('../input/chest-ctscan-images/Data/valid/
adenocarcinoma_left.lower.lobe_T2_N0_M0_Ib/000108 (7).
png',target_size=(224,224))
imag = image.img_to_array(img)
imaga = np.expand_dims(imag,axis=0)
ypred = model.predict(imaga)
print(ypred)
a=np.argmax(ypred,-1)
if a==0:
        op="Adenocarcinoma"
elif a==1:
        op="large cell carcinoma"
elif a==2:
        op="normal (void of cancer)"
else:
        op="squamous cell carcinoma"
plt.imshow(img)
print("THE UPLOADED IMAGE IS SUSPECTED AS: "+str(op))
```

After running this short code, the following results are obtained,

```
the dataset consists of about 1000 different lung CT images.
Found 613 images belonging to 4 classes.
Found 72 images belonging to 4 classes.
Found 315 images belonging to 4 classes.
training accuracy= 94.13%
training loss= 0.1919
training precision= 95.45%
training recall= 92.5%
training AUC= 0.9953
[[0.9059036 0.00122325 0.07508809 0.01778515]]
THE UPLOADED IMAGE IS SUSPECTED AS: Adenocarcinoma (0.9059036)
  according to maximum number and probability. That is,
  [Adenocarcinoma, Large cell carcinoma, Squamous cell
  carcinoma, Normal]~=[ 0.9059036, 0.00122325, 0.07508809,
  0.01778515]
```

9.7.4 EXAMPLE IV

Obtain the relationships among parameters of 3D printing pressure, nozzle moving speed, and printing distance by canonical correlation analysis (CCA) using AI/ML techniques as a mathematical meaningful formula (for AM of a bio-composite).

To understand which condition is the key parameter for 3D printing, CCA, which is a statistical method to study the correlations between two multivariate sets of variables, is used to examine the weight of these printing parameters. It may not only expose the correlations between multivariate variables, but also recognize the weight of every variable in the correlations, which is more thorough than traditional correlation analyses. The two multivariate sets of variables were line width and printing parameters (i.e., printing pressure, nozzle moving speed, and printing distance). Their relationships are obtained by CCA, as shown in the following equation (by regression multi linear regression model),

$$U = -0.192X_1 - 0.744X_2 + 1.177X_3 \tag{18}$$

where X_1 is the printing distance, X_2 is the moving speed, X_3 is the printing pressure, and U is the first typical variable (a new variable obtained after linear combination of each set of variables) of line width. In accordance with the formula, the total value of the correlation coefficient for printing pressure is the highest, proposing that the printing pressure had a higher impact on the line width than the printing distance and moving speed. It means that changing the printing pressure should be given priority in the process of optimizing the printing conditions among these three parameters. The signs of the coefficients indicated that the width of the printed line is positively correlated with the printing pressure, and negatively correlated with the printing distance and moving speed.

9.7.5 EXAMPLE V

Predictions of thermal fields in AM (predict the thermal areas in an AM using AI/ML).

First of all, we must collect and gather the dataset, and then preprocess the dataset for better learning the model. The provided files need code for AI/ML on data collected from FEM (Finite Element Method) simulations of AM models, including data collection, preprocessing, and ML. The requirements include libraries and software such as Python 3, Abaqus, Scikit-Learn, Numpy, Pandas, and Matplotlib, Seaborn. The provided code has the following folder structure, where the files are structured based on the stage of the pipeline (file structures).

- Abaqus
  - exp
- Figurer
- Machine_Learning
- Materials
- Preprocessing
  - feature_extraction
  - feature_improvement

The Abaqus folder contains the files to generate FEM models and to extract the relevant data from the models after the simulations have been completed. The methods are combined to create a script adapted to be run in Abaqus Python as the files in the Abaqus/exp folder. The Machine_Learning folder contains the code used for the ML process. The Materials folder contains the material information imported in the Abaqus scripts. The Preprocessing folder contains code for the feature engineering process of the project. The generated datasets are available at https://cutt. ly/QnqXV9Z   (https://github.com/kariln/Predictions-of-thermal-fields-in-additive-manufacturing). Some codes and predictions are presented briefly in Figure 9.27.

Here, some Python codes are available at https://github.com/kariln/Predictions-of-thermal-fields-in-additive-manufacturing/commit/351228afd076f560317639e9 f69e9305d3494f23#diff-5162d4c550c03a9cf9b75e7f580c92014321cde91280253a-1595759be2e0c878. The readers can refer to this address for more details.

9.8   CONCLUSION

This chapter shows that how to use ML and AI in AM, bio-composites, healthcare, medicine, and bio-systems. Also, it answers an important question, that is, which ML/AI models are proper for special problems and what algorithms are popular and famous in the ML/AI worlds with high accuracy and speed. Moreover, simple models were introduced in the chapter. This chapter tried to get a summary for beginners so that they can start using ML/AI techniques. Furthermore, the important notes in the text need to be considered. Some necessary steps were presented for doing and creating an ideal model from the beginning to the end like preprocessing data, feature selections, accuracy, and error, selecting suitable model (regression or classification), preventing overfitting, and other vital procedures to reach good outcomes. Finally, three examples were provided for more clarification on these problem solution steps by ML/AI with Python. Preprocessing data with an example, optimizing the insurance cost, diagnosis and prediction of lung cancer (design and build a model) were done and programmed. Also, some popular CNN

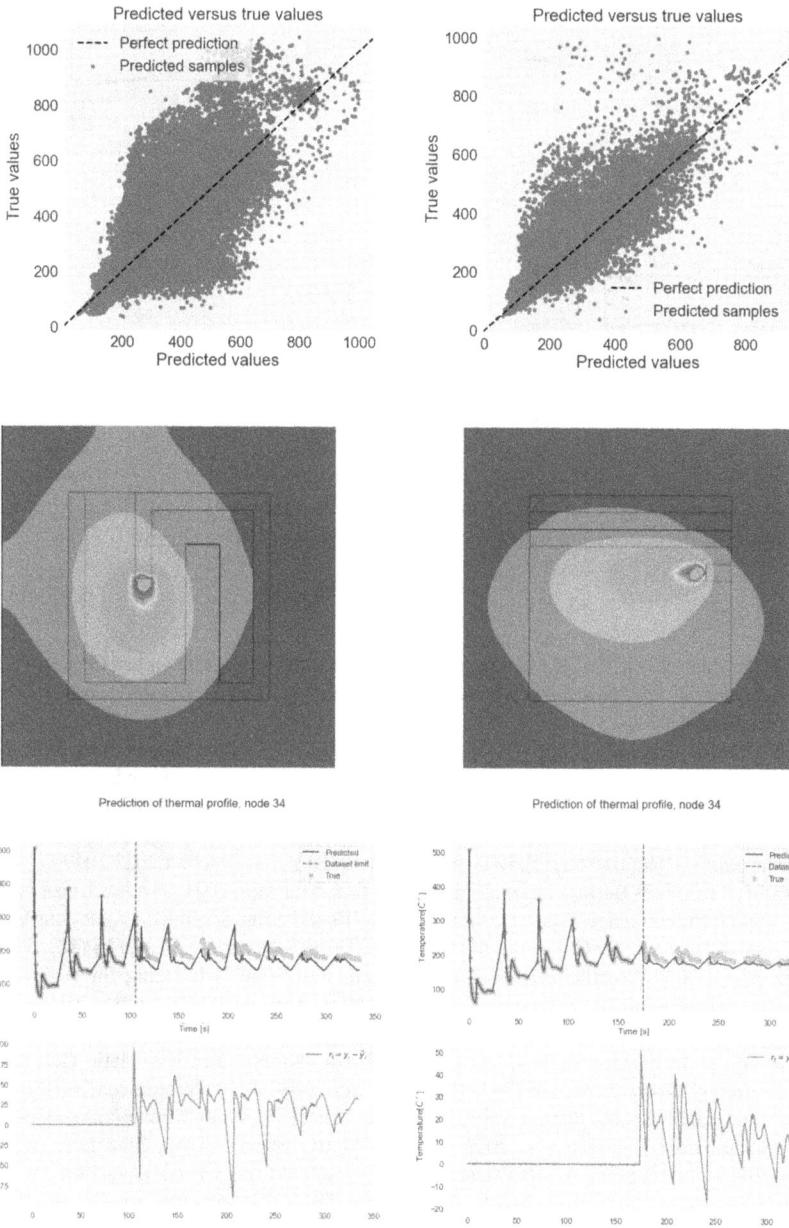

FIGURE 9.27 Prediction of thermal areas in AM using AI/ML (https://github.com/kariln/Predictions-of-thermal-fields-in-additive-manufacturing).

models were introduced to better understand image processing in healthcare and medicine like lung cancer prediction and diagnosis in patients. This chapter is an application and summary for entering the ML/AI worlds. It should be mentioned that this chapter has employed some codes from KAGGLE and GITHUB as shown

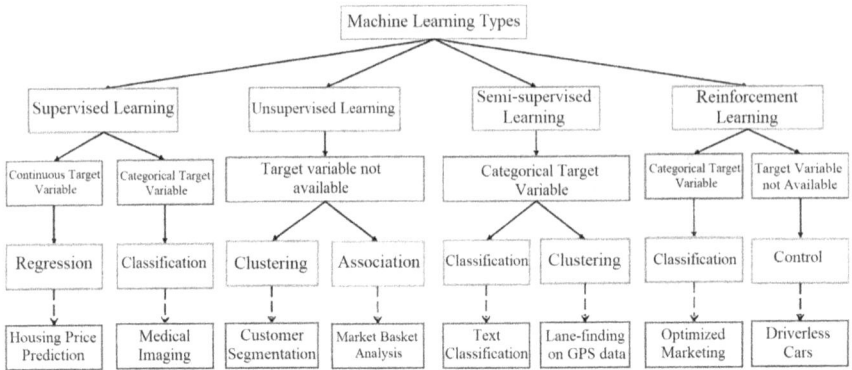

FIGURE 9.28   General chart of AI (www.analytixlabs.co.in).

and cited in the text. The author hopes that this chapter is a small and compact encyclopedia that provides useful context and guidance (see Figure 9.28).

REFERENCES

Alanazi, Abdullah. 2022. "Using Machine Learning for Healthcare Challenges and Opportunities." *Informatics in Medicine Unlocked* 30 (100924): 100924. https://doi.org/10.1016/j.imu.2022.100924.

Aquino, Yves Saint James, Wendy A. Rogers, Annette Braunack-Mayer, Helen Frazer, Khin Than Win, Nehmat Houssami, Christopher Degeling, Christopher Semsarian, and Stacy M. Carter. 2023. "Utopia versus Dystopia: Professional Perspectives on the Impact of Healthcare Artificial Intelligence on Clinical Roles and Skills." *International Journal of Medical Informatics* 169 (104903): 104903. https://doi.org/10.1016/j.ijmedinf.2022.104903.

Barrera, María Dolores Bermejillo, and Francisco Franco-Martínez. 2021. "Andrés Díaz Lantada, Artificial Intelligence Aided Design of Tissue Engineering Scaffolds Employing Virtual Tomography and 3D Convolutional Neural Networks." *Materials (Basel)* 14 (18): 5278.

Becker, Aliza. 2019. "Artificial Intelligence in Medicine: What Is It Doing for Us Today?" *Health Policy and Technology* 8 (2): 198–205. https://doi.org/10.1016/j.hlpt.2019.03.004.

Begg, R., and J. Kamruzzaman. 2005. "A Machine Learning Approach for Automated Recognition of Movement Patterns Using Basic, Kinetic and Kinematic Gait Data." *Journal of Biomechanics* 38 (3): 401–8. https://doi.org/10.1016/j.jbiomech.2004.05.002.

Boddu, Raja Sarath Kumar, Partha Karmakar, Ankan Bhaumik, Vinay Kumar Nassa, Vandana, and Sumanta Bhattacharya. 2022. "Analyzing the Impact of Machine Learning and Artificial Intelligence and Its Effect on Management of Lung Cancer Detection in Covid-19 Pandemic." *Materials Today: Proceedings* 56: 2213–16. https://doi.org/10.1016/j.matpr.2021.11.549.

Bonfanti, Silvia, Roberto Guerra, Francesc Font-Clos, Daniel Rayneau-Kirkhope, and Stefano Zapperi. 2020. "Automatic Design of Mechanical Metamaterial Actuators." *Nature Communications* 11 (1): 4162. https://doi.org/10.1038/s41467-020-17947-2.

Chen, Chun-Teh, and Grace X. Gu. 2020. "Generative Deep Neural Networks for Inverse Materials Design Using Backpropagation and Active Learning." *Advanced Science (Weinheim, Baden-Wurttemberg, Germany)* 7 (5): 1902607. https://doi.org/10.1002/advs.201902607.

Collins, Francis S., and Harold Varmus. 2015. "A New Initiative on Precision Medicine." *The New England Journal of Medicine* 372 (9): 793–95. https://doi.org/10.1056/NEJMp1500523.

Davidben-Israel, W., Steve Bradleyjacobs, Stefan Casha, Won Lang, A. Hyung, Madeleinede Ryu, and David W. Lotbiniere-Bassett. 2020. "The Impact of Machine Learning on Patient Care: A Systematic Review." *Artificial Intelligence in Medicine* 103.

Dias, Raquel, and Ali Torkamani. 2019. "Artificial Intelligence in Clinical and Genomic Diagnostics." *Genome Medicine* 11 (1). https://doi.org/10.1186/s13073-019-0689-8.

Ding, Yiming, Jae Ho Sohn, Michael G. Kawczynski, Hari Trivedi, Roy Harnish, Nathaniel W. Jenkins, Dmytro Lituiev, et al. 2019. "A Deep Learning Model to Predict a Diagnosis of Alzheimer Disease by Using 18F-FDG PET of the Brain." *Radiology* 290 (2): 456–64. https://doi.org/10.1148/radiol.2018180958.

Diprose, William, and Nicholas Buist. 2016. "Artificial Intelligence in Medicine: Humans Need Not Apply?" *The New Zealand Medical Journal* 129 (1434): 73–76. https://www.ncbi.nlm.nih.gov/pubmed/27349266.

Dong, Jingsi, Yingcai Geng, Dan Lu, Bingjie Li, Long Tian, Dan Lin, and Yonggang Zhang. 2020. "Clinical Trials for Artificial Intelligence in Cancer Diagnosis: A Cross-Sectional Study of Registered Trials in Clinicaltrials.gov." *Frontiers in Oncology* 10 (September): 1629. https://doi.org/10.3389/fonc.2020.01629.

Fukuchi, Reginaldo K., Bjoern M. Eskofier, Marcos Duarte, and Reed Ferber. 2011. "Support Vector Machines for Detecting Age-Related Changes in Running Kinematics." *Journal of Biomechanics* 44 (3): 540–42. https://doi.org/10.1016/j.jbiomech.2010.09.031.

Garg, A., and Jasmine Siu Lee Lam. 2015. "Measurement of Environmental Aspect of 3-D Printing Process Using Soft Computing Methods." *Measurement: Journal of the International Measurement Confederation* 75 (November): 210–17. https://doi.org/10.1016/j.measurement.2015.04.016.

Garg, A., Jasmine Siu Lee Lam, and M. M. Savalani. 2015. "A New Computational Intelligence Approach in Formulation of Functional Relationship of Open Porosity of the Additive Manufacturing Process." *The International Journal of Advanced Manufacturing Technology* 80 (1–4): 555–65. https://doi.org/10.1007/s00170-015-6989-2.

Gentili, Pier Luigi. 2022. "Photochromic and Luminescent Materials for the Development of Chemical Artificial Intelligence." *Dyes and Pigments: An International Journal* 205 (110547): 110547. https://doi.org/10.1016/j.dyepig.2022.110547.

Gobert, Christian, Edward W. Reutzel, Jan Petrich, Abdalla R. Nassar, and Shashi Phoha. 2018. "Application of Supervised Machine Learning for Defect Detection during Metallic Powder Bed Fusion Additive Manufacturing Using High Resolution Imaging." *Additive Manufacturing* 21 (May): 517–28. https://doi.org/10.1016/j.addma.2018.04.005.

Gómez-Bombarelli, Rafael. 2018. "Reaction: The near Future of Artificial Intelligence in Materials Discovery." *Chem* 4 (6): 1189–90. https://doi.org/10.1016/j.chempr.2018.05.021.

Gu, Grace X., Chun-Teh Chen, and Markus J. Buehler. 2018. "De Novo Composite Design Based on Machine Learning Algorithm." *Extreme Mechanics Letters* 18 (January): 19–28. https://doi.org/10.1016/j.eml.2017.10.001.

Hanakata, Paul Z., Ekin D. Cubuk, David K. Campbell, and Harold S. Park. 2019. "Erratum: Accelerated Search and Design of Stretchable Graphene Kirigami Using Machine Learning [Phys. Rev. Lett. 121, 255304 (2018)]." *Physical Review Letters* 123 (6): 069901. https://doi.org/10.1103/PhysRevLett.123.069901.

Hannun, Awni Y., Pranav Rajpurkar, Masoumeh Haghpanahi, Geoffrey H. Tison, Codie Bourn, Mintu P. Turakhia, and Andrew Y. Ng. 2019. "Cardiologist-Level Arrhythmia Detection and Classification in Ambulatory Electrocardiograms Using a Deep Neural Network." *Nature Medicine* 25 (1): 65–69. https://doi.org/10.1038/s41591-018-0268-3.

Hu, Z., and S. Mahadevan. 2017. "Uncertainty Quantification and Management in Additive Manufacturing: Current Status, Needs, and Opportunities." *The International Journal of Advanced Manufacturing Technology* 93.

Ikonen, I., W. E. Biles, A. Kumar, J. C. Wissel, and R. K. Ragade. 1997. "A Genetic Algorithm for Packing Three-Dimensional NonConvex Objects Having Cavities and Holes," ICGA, 591–98 .

Jaganathan, Kishore, Sofia Kyriazopoulou Panagiotopoulou, Jeremy F. McRae, Siavash Fazel Darbandi, David Knowles, Yang I. Li, Jack A. Kosmicki, et al. 2019. "Predicting Splicing from Primary Sequence with Deep Learning." *Cell* 176 (3): 535–48.e24. https://doi.org/10.1016/j.cell.2018.12.015.

Jiao, Pengcheng, and Amir H. Alavi. 2021. "Artificial Intelligence-Enabled Smart Mechanical Metamaterials: Advent and Future Trends." *International Materials Reviews* 66 (6): 365–93. https://doi.org/10.1080/09506608.2020.1815394.

Jin, Michael, Marc Schröder, and Victor E. Staartjes. 2023. "15-Artificial Intelligence and Machine Learning in Spine Surgery, Robotic and Navigated Spine Surgery." *Surgical Techniques and Advancements*, 213–29.

Jin, Zeqing, Zhizhou Zhang, and Grace X. Gu. 2020. "Automated Real-time Detection and Prediction of Interlayer Imperfections in Additive Manufacturing Processes Using Artificial Intelligence." *Advanced Intelligent Systems (Weinheim an Der Bergstrasse, Germany)* 2 (1): 1900130. https://doi.org/10.1002/aisy.201900130.

Jin, Zeqing, Zhizhou Zhang, Joshua Ott, and Grace X. Gu. 2021a. "Precise Localization and Semantic Segmentation Detection of Printing Conditions in Fused Filament Fabrication Technologies Using Machine Learning." *Additive Manufacturing* 37 (101696): 101696. https://doi.org/10.1016/j.addma.2020.101696.

Jin, Zeqing, Zhizhou Zhang, Xianlin Shao, and Grace X. Gu. 2021b. "Monitoring Anomalies in 3D Bioprinting with Deep Neural Networks." *ACS Biomaterials Science & Engineering*, no. acsbiomaterials.0c01761 (April). https://doi.org/10.1021/acsbiomaterials.0c01761.

Jose, Rajan, and Seeram Ramakrishna. 2018. "Materials 4.0: Materials Big Data Enabled Materials Discovery." *Applied Materials Today* 10 (March): 127–32. https://doi.org/10.1016/j.apmt.2017.12.015.

Kim, Hyo-Eun, Hak Hee Kim, Boo-Kyung Han, Ki Hwan Kim, Kyunghwa Han, Hyeonseob Nam, Eun Hye Lee, and Eun-Kyung Kim. 2020. "Changes in Cancer Detection and False-Positive Recall in Mammography Using Artificial Intelligence: A Retrospective, Multireader Study." *The Lancet. Digital Health* 2 (3): e138–48. https://doi.org/10.1016/S2589-7500(20)30003-0.

Kisor Kumar Sahu, Shibu, Abhilash M. Meher, M. K. Menon, V. Gangala, Saurabh Harsha Vardhan, Ashutosh Pandey, and Shreeja Kumar. 2022. "Artificial Intelligence and Machine Learning: New Age Tools for Augmenting Plastic Materials Designing, Processing, and Manufacturing, Encyclopedia of Materials: Plastics and Polymers" 3: 127–52.

Kosorok, Michael R., and Eric B. Laber. 2019. "Precision Medicine." *Annual Review of Statistics and Its Application* 6 (1): 263–86. https://doi.org/10.1146/annurev-statistics-030718-105251.

———, eds. n.d. *Annual Review of Statistics and Its Application Precision Medicine.*

Kumar, Law, and Singh Munish. 2023. "Artificial Intelligence Based Medical Decision Support System for Early and Accurate Breast Cancer Prediction." *Advances in Engineering Software* 175.

Langlotz, Curtis P. 2019. "Will Artificial Intelligence Replace Radiologists?" *Radiology. Artificial Intelligence* 1 (3): e190058. https://doi.org/10.1148/ryai.2019190058.

Lantada, Díaz, A. Franco-Martínez, F. Hengsbach, S. Rupp, F. Thelen, and R. Bade. 2020. "Artificial Intelligence Aided Design of Microtextured Surfaces: Application to Controlling Wettability." *Nanomaterials* 10.

Li, X., J. Dong, and Y. Zhang. 2009. "Information Engineering and Computer Science, 2009." In *ICIECS 2009. International Conference on, IEEE; 2009, Modeling and Applying of Rbf Neural Network Based on Fuzzy Clustering and Pseudo-Inverse Method*, 1–4.

Lin, Steven Y., Megan R. Mahoney, and Christine A. Sinsky. 2019. "Ten Ways Artificial Intelligence Will Transform Primary Care." *Journal of General Internal Medicine* 34 (8): 1626–30. https://doi.org/10.1007/s11606-019-05035-1.

Liu, Yun, Timo Kohlberger, Mohammad Norouzi, George E. Dahl, Jenny L. Smith, Arash Mohtashamian, Niels Olson, Lily H. Peng, Jason D. Hipp, and Martin C. Stumpe. 2019. "Artificial Intelligence-Based Breast Cancer Nodal Metastasis Detection: Insights into the Black Box for Pathologists." *Archives of Pathology & Laboratory Medicine* 143 (7): 859–68. https://doi.org/10.5858/arpa.2018-0147-OA.

Lu, Jingyan. 2016. "Will Medical Technology Deskill Doctors?" *International Education Studies* 9 (7): 130. https://doi.org/10.5539/ies.v9n7p130.

Lu, W., R. Xiao, J. Yang, H. Li, and W. Zhang. 2017. "Data-Mining Aided Materials Discovery and Optimization." *Journal of Materiomics* 3: 191–201.

Lu, Ziheng, Francesco Ciucci, and Chi Chen. 2021. "Editorial for the Special Issue 'Machine Learning and Artificial Intelligence for Energy Materials.'" *Materials Reports: Energy* 1 (3): 100056. https://doi.org/10.1016/j.matre.2021.100056.

Macrae, Carl. 2019. "Governing the Safety of Artificial Intelligence in Healthcare." *BMJ Quality & Safety* 28 (6): 495–98. https://doi.org/10.1136/bmjqs-2019-009484.

Manuel Pegalajar Cuéllar, Alejandro, Juan Manuel Lapresta-Fernández, Alfonso Herrera, María Salinas-Castillo, and Enrique Silviatitos-Padilla. 2015. "Luis Fermín Capitán-Vallvey, Thermochromic Sensor Design Based on Fe(II) Spin Crossover/Polymers Hybrid Materials and Artificial Neural Networks as a Tool in Modelling." *Sensors and Actuators B: Chemical* 208: 180–87.

Maxmen, J. S. n.d. "Long-Term Trends in Health Care: The Post-Physician Era Reconsidered BT – Indicators and Trends in Health and Health Care, Indicators and Trends in Health and Health Care." 109–15.

Meng, L., B. McWilliams, W. Jarosinski, H.Y. Park, Y.G. Jung, J. Lee, J. Zhang. 2020. "Machine Learning in Additive Manufacturing: A Review". *JOM* 72: 2363–2377.

Merayo, D., A. Rodriguez-Prieto, and A. M. Camacho. 2020. "Prediction of Physical and Mechanical Properties for Metallic Materials Selection Using Big Data and Artificial Neural Networks." *IEEE Access: Practical Innovations, Open Solutions* 8: 13444–56. https://doi.org/10.1109/access.2020.2965769.

Milazzo, Mario, Flavia Libonati, Shengfei Zhou, Kai Guo, and Markus J. Buehler. 2022. "Chapter 6 – Biomimicry for Natural and Synthetic Composites and Use of Machine Learning in Hierarchical Design, Biomimicry for Materials, Design and Habitats." *Innovations and Applications*, 141–82.

Munguía, J., J. Ciurana, and C. Riba. 2009. "Neural-Network-Based Model for Build-Time Estimation in Selective Laser Sintering." *Proceedings of the Institution of Mechanical Engineers, Part B: Journal of Engineering Manufacture* 223 (8): 995–1003. https://doi.org/10.1243/09544054jem1324.

Nam, Kyoung Hyup, Il Seo, Dong Hwan Kim, Jae Il Lee, Byung Kwan Choi, and In Ho Han. 2019. "Machine Learning Model to Predict Osteoporotic Spine with Hounsfield Units on Lumbar Computed Tomography." *Journal of Korean Neurosurgical Society* 62 (4): 442–49. https://doi.org/10.3340/jkns.2018.0178.

Naser, M. Z. 2018. "Deriving Temperature-Dependent Material Models for Structural Steel through Artificial Intelligence." *Construction and Building Materials* 191 (December): 56–68. https://doi.org/10.1016/j.conbuildmat.2018.09.186.

Negi, Sushant, and Rajesh Kumar Sharma. 2016. "Study on Shrinkage Behaviour of Laser Sintered PA 3200GF Specimens Using RSM and ANN." *Rapid Prototyping Journal* 22 (4): 645–59. https://doi.org/10.1108/rpj-08-2014-0090.

Panch, Trishan, Peter Szolovits, and Rifat Atun. 2018. "Artificial Intelligence, Machine Learning and Health Systems." *Journal of Global Health* 8 (2): 020303. https://doi.org/10.7189/jogh.08.020303.

Preethikaharshini, J., K. Naresh, G. Rajeshkumar, V. Arumugaprabu, Muhammad A. Khan, and K. A. Khan. 2022. "Review of Advanced Techniques for Manufacturing Biocomposites: Non-Destructive Evaluation and Artificial Intelligence-Assisted Modeling." *Journal of Materials Science* 57 (34): 16091–146. https://doi.org/10.1007/s10853-022-07558-1.

Qian, Chenxi, Todd Siler, and Geoffrey A. Ozin. 2015. "Exploring the Possibilities and Limitations of a Nanomaterials Genome." *Small* 11 (1): 64–69. https://doi.org/10.1002/smll.201402197.

Raccuglia, Paul, Katherine C. Elbert, Philip D. F. Adler, Casey Falk, Malia B. Wenny, Aurelio Mollo, Matthias Zeller, Sorelle A. Friedler, Joshua Schrier, and Alexander J. Norquist. 2016. "Machine-Learning-Assisted Materials Discovery Using Failed Experiments." *Nature* 533 (7601): 73–76. https://doi.org/10.1038/nature17439.

Rajkomar, Alvin, Jeffrey Dean, and Isaac Kohane. 2019. "Machine Learning in Medicine." *The New England Journal of Medicine* 380 (14): 1347–58. https://doi.org/10.1056/NEJMra1814259.

Rajkomar, Alvin, Eyal Oren, Kai Chen, Andrew M. Dai, Nissan Hajaj, Michaela Hardt, Peter J. Liu, et al. 2018. "Scalable and Accurate Deep Learning with Electronic Health Records." *Npj Digital Medicine* 1 (1). https://doi.org/10.1038/s41746-018-0029-1.

Rani, Alka, Nirmal Kumar, Jitendra Kumar, Jitendra Kumar, and Nishant K. Sinha. 2022. "Chapter 6 – Machine Learning for Soil Moisture Assessment, Deep Learning for Sustainable Agriculture." *Cognitive Data Science in Sustainable Computing*, Elsevier, 143–68.

Rong-Ji, Wang, Li Xin-hua, Wu Qing-ding, and Wang Lingling. 2009. "Optimizing Process Parameters for Selective Laser Sintering Based on Neural Network and Genetic Algorithm." *The International Journal of Advanced Manufacturing Technology* 42 (11–12): 1035–42. https://doi.org/10.1007/s00170-008-1669-0.

Rose, Sherri. 2018. "Machine Learning for Prediction in Electronic Health Data." *JAMA Network Open* 1 (4): e181404. https://doi.org/10.1001/jamanetworkopen.2018.1404.

Ross, Patrick, and Kathryn Spates. 2020. "Considering the Safety and Quality of Artificial Intelligence in Health Care." *Joint Commission Journal on Quality and Patient Safety* 46 (10): 596–99. https://doi.org/10.1016/j.jcjq.2020.08.002.

Rubinger, Luc, Aaron Gazendam, and Seper Ekhtiari. 2022. *Mohit Bhandari, Machine Learning and Artificial Intelligence in Research and Healthcare, Injury,* 2023;54 Suppl 3:S69–73. doi:10.1016/j.injury.2022.01.046.

Saheb, Tahereh, Tayebeh Saheb, and David O. Carpenter. 2021. "Mapping Research Strands of Ethics of Artificial Intelligence in Healthcare: A Bibliometric and Content Analysis." *Computers in Biology and Medicine* 135 (104660): 104660. https://doi.org/10.1016/j.compbiomed.2021.104660.

Santos, I., J. Nieves, Y. K. Penya, and P. Bringas. 2009. "Machine-Learning-Based Mechanical Properties Prediction in Foundry Production." In *Proceedings of the ICCAS-SICE 2009-ICROS-SICE International Joint Conference,* 4536–41. Fukuoka, Japan.

Sendak, M., S. Balu, and K. Schulman. 2017. "Barriers to Achieving Economies of Scale in Analysis of EHR Data: A Cautionary Tale." *Applied Clinical Informatics* 8 (3): 826–31.

Shen, Xianfeng, Jin Yao, Yang Wang, and Jialin Yang. 2004. "Density Prediction of Selective Laser Sintering Parts Based on Artificial Neural Network." In *Lecture Notes in Computer Science,* 832–40. Lecture Notes in Computer Science. Berlin, Heidelberg: Springer Berlin Heidelberg. https://doi.org/10.1007/978-3-540-28648-6_133.

Soenksen, Luis R., Timothy Kassis, Susan T. Conover, Berta Marti-Fuster, Judith S. Birkenfeld, Jason Tucker-Schwartz, Asif Naseem, et al. 2021. "Using Deep Learning for Dermatologist-Level Detection of Suspicious Pigmented Skin Lesions from Wide-Field Images." *Science Translational Medicine* 13 (581): eabb3652. https://doi.org/10.1126/scitranslmed.abb3652.

Topol, Eric J. 2019. "High-Performance Medicine: The Convergence of Human and Artificial Intelligence." *Nature Medicine* 25 (1): 44–56. https://doi.org/10.1038/s41591-018-0300-7.

Varga-Szemes, Akos, Brian E. Jacobs, and U. Joseph Schoepf. 2018. "The Power and Limitations of Machine Learning and Artificial Intelligence in Cardiac CT." *Journal of Cardiovascular Computed Tomography* 12 (3): 202–3. https://doi.org/10.1016/j.jcct.2018.05.007.

Wang, Rong Ji, Jianbing Li, Fenghua Wang, Xinhua Li, and Qingding Wu. 2009. "ANN Model for the Prediction of Density in Selective Laser Sintering." *International Journal of Manufacturing Research* 4 (3): 362. https://doi.org/10.1504/ijmr.2009.026579.

Wang, Rong-Ji, Lingling Wang, Lihua Zhao, and Zijian Liu. 2007. "Influence of Process Parameters on Part Shrinkage in SLS." *The International Journal of Advanced Manufacturing Technology* 33 (5–6): 498–504. https://doi.org/10.1007/s00170-006-0490-x.

Wang, Shuo, Hao Zhang, Zhen Liu, and Yuanning Liu. 2022. "A Novel Deep Learning Method to Predict Lung Cancer Long-Term Survival with Biological Knowledge Incorporated Gene Expression Images and Clinical Data." *Frontiers in Genetics* 13 (March): 800853. https://doi.org/10.3389/fgene.2022.800853.

Wang, Zhuo, Pengwei Liu, Yaohong Xiao, Xiangyang Cui, Zhen Hu, and Lei Chen. 2019. "A Data-Driven Approach for Process Optimization of Metallic Additive Manufacturing under Uncertainty." *Journal of Manufacturing Science and Engineering* 141 (8): 081004. https://doi.org/10.1115/1.4043798.

Winkler, David A. 2017. "Biomimetic Molecular Design Tools That Learn, Evolve, and Adapt." *Beilstein Journal of Organic Chemistry* 13 (June): 1288–1302. https://doi.org/10.3762/bjoc.13.125.

Yang, Zhou, Christina Silcox, Mark Sendak, Sherri Rose, David Rehkopf, Robert Phillips, Lars Peterson, et al. 2022. "Advancing Primary Care with Artificial Intelligence and Machine Learning." *Healthcare (Amsterdam, Netherlands)* 10 (1): 100594. https://doi.org/10.1016/j.hjdsi.2021.100594.

Zaharchuk, G., E. Gong, M. Wintermark, D. Rubin, and C. P. Langlotz. 2018. "Deep Learning in Neuroradiology." *AJNR. American Journal of Neuroradiology* 39 (10): 1776–84. https://doi.org/10.3174/ajnr.a5543.

Zeng, Tony, and Yang I. Li. 2022. "Predicting RNA Splicing from DNA Sequence Using Pangolin." *Genome Biology* 23 (1): 103. https://doi.org/10.1186/s13059-022-02664-4.

Zheng, R., M. Li, and Z. Liang. 2019. "SinNLRR: A Robust Subspace Clustering Method for Cell Type Detection by Nonnegative and Low Rank Representation." *Bioinformatics.*

10 Additive Manufacturing of Bio-Inspired Ceramic and Ceramic-Polymer Composite Lattice Structures

He R., and Zhang X.

10.1 INTRODUCTION

Cellular structures are required for a range of technological applications, owing to their comprehensive characteristics of superior mechanical properties and simultaneously for being as light as possible. Cellular structures are usually categorized into cellular ceramic structures (CCSs) (Maurath and Willenbacher 2017; Jana et al. 2018; Mei et al. 2019; He et al. 2020), cellular polymer structures (CPSs) (Kaur et al. 2017; Weeger et al. 2019; Bai et al. 2020a), and cellular metal structures (CMSs) (Yan et al. 2014; Huang et al. 2017; Dong 2019; Zheng et al. 2019) based on the materials. Extraordinary strength, stable chemical properties, and excellent lightweight makes CCSs the most promising candidate for structural components. Although the intrinsic brittleness of ceramics limits the manufacturing and promotion of CCSs, the advent of additive manufacturing technology significantly promotes the development of CCSs, making them stand out from other cellular structures (Mei et al. 2021; Wang et al. 2022; Zhang et al. 2022b).

However, the trade-off between the strength and toughness of ceramics is hardly achieved because of the strong chemical bond between the atoms in ceramics. Natural materials provide a perfect solution on a relatively larger scale to conquer the inferior toughness of ceramics (Munch et al. 2008; Zhang et al. 2022a). Natural materials, e.g. red abalone shell (Bouville et al. 2014; Huang et al. 2019; Wat et al. 2019), dactyl club of mantis shrimp (Patek and Caldwell 2005; Grunenfelder et al. 2014; Sun et al. 2020), and conch shell (Shin et al. 2016; Gu et al. 2017), are capable of withstanding multiple attacks, demonstrating superior strength and toughness at the same time. It is found that natural materials are composites that are constituted of stiff and tough constituents. Mimicking the organization type in natural composites is no doubt a reliable way to incorporate the high strength and damage tolerance simultaneously in ceramic-based materials.

 DOI: 10.1201/9781003362128-10

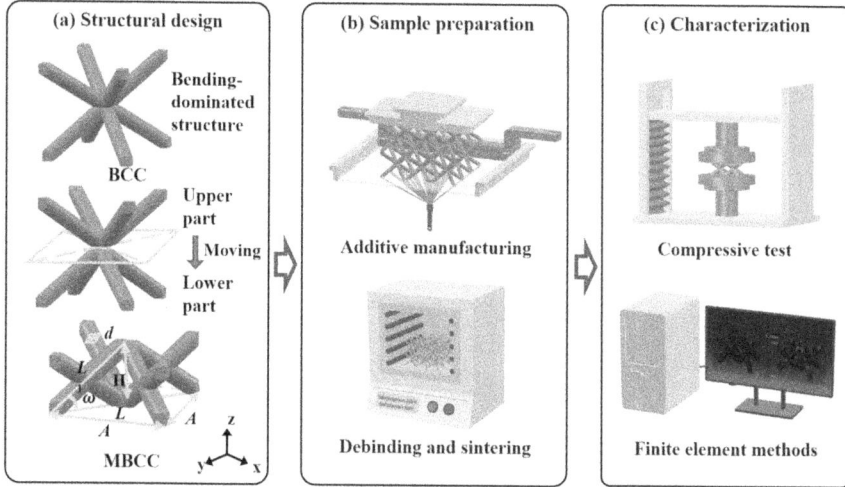

FIGURE 10.1 (a) Structural design, (b) sample preparation, and (c) characterization of additively manufactured CCSs.

In this chapter, we discuss the mechanical properties of additively manufactured CCSs and bio-inspired CCS/metal composite. CCSs with different relative densities and structural configurations were designed, as shown in Figure 10.1(a). Subsequently, CCSs were prepared by digital light processing (DLP) technology and sintering, as shown in Figure 10.1(b). The bio-composite was further fabricated by melting infiltration process. The compressive test and finite element methods involved in Figure 10.1(c) were adapted to examine the mechanical properties of CCSs and bio-composites.

10.2 MECHANICAL PROPERTIES OF CCSs

In the past decades, researches about CCSs have been successively deployed. It was proved that many factors would alter the response regularity of CCSs under compressive loading. In our previous work, we prepared Al_2O_3 CCSs by DLP technology and comprehensively investigated how structural parameters, containing relative density and structural configuration, influence the mechanical properties of CCSs.

10.2.1 Effect of Relative Density

Figure 10.2 shows the macro morphologies of CCSs with the same structural configuration, i.e., body-centered cubic (BCC) structure, but different relative densities. The relative densities of BCC CCSs were 20%, 30%, and 40%, respectively. The step-wise morphology could be observed on the surface of structures. Such step-wise morphology was ascribed to the layer-by-layer forming characteristic of DLP technology. It is worth noting that the dimensions of BCC CCSs were different from that of as-designed models. The as-designed 3D models of CCSs had a dimension of $30 \times 30 \times 30$ mm³. However, the final dimensions of BCC CCSs with 20%, 30%, and

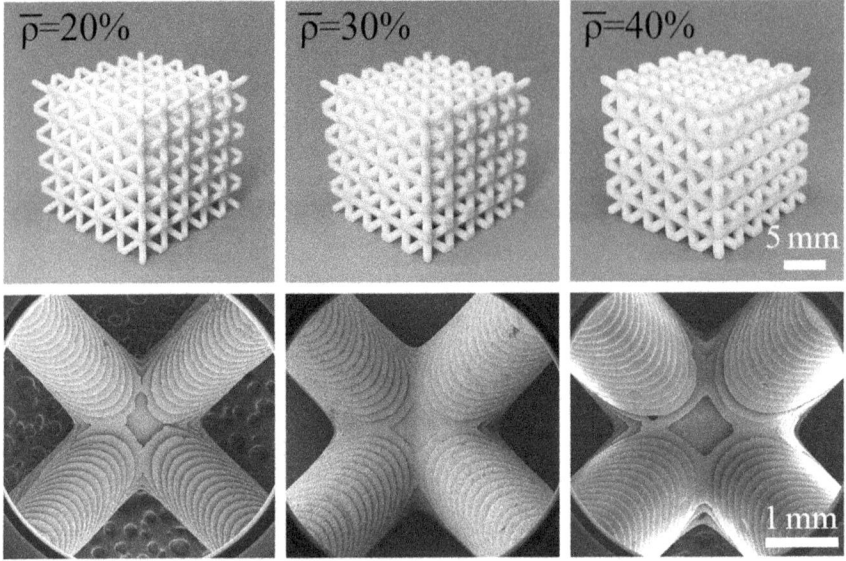

FIGURE 10.2   Macro and micro morphologies of BCC CCSs with different relative densities.

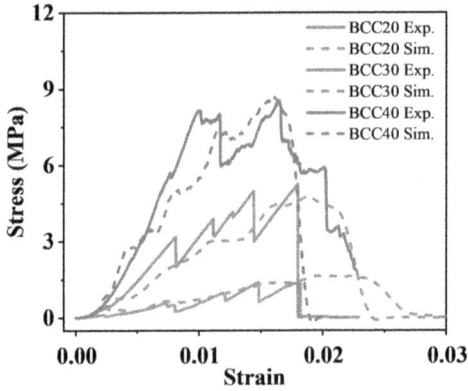

FIGURE 10.3   Experimental and simulating stress-strain curves of BCC CCSs with different relative densities.

40% relative densities were $23.81 \times 23.67 \times 22.37$ mm^3, $24.39 \times 24.31 \times 22.84$ mm^3, and $23.81 \times 23.70 \times 23.07$ mm^3, respectively. That is to say, there was approximately 20% contraction in the x, y, and z-axis of CCSs after sintering. The complete sintering shrinkage brought in low porosity, i.e., just 2%, of CCSs. Furthermore, the real relative densities of three types of BCC CCSs were 20.24%, 30.64%, and 38.95%, respectively, which testified that forming precision of DLP technology met the requirement of fabricating CCSs.

Figure 10.3 depicts the representative stress-strain curves of BCC CCSs with different relative densities. It was concluded that the stress CCSs suffer from increases during loading until reaching a maximum before decreasing dramatically. Besides, the compressive stress-strain curves of CCSs were significantly different from that of CMSs and CPSs (Maskery et al. 2018; Lei et al. 2019; Ma et al. 2020; Guerra et al. 2021). This is because ceramic is brittle with scarce deformation under loading. In contrast, polymers and metals are deformable and tough. Hence, only linear elastic stage was demonstrated on the compressive stress-strain curves of CCSs. The plateau stage and densification stage like that in the compression of CMSs and CPSs didn't exist (Tancogne-Dejean et al. 2016). Besides, it was found that the stress CCSs suffer from under compressive loading does not successively increase with strain. Sudden dropping happens during the loading process, which is deemed to be caused by the random distribution of manufacturing defects, e.g., pores, cracks, and interfaces, in a ceramic body. Furthermore, the compressive stress-strain curves of CCSs go upward with increasing relative density. It is extracted from curves that compressive strength of BCC CCSs with 20%, 30%, and 40% relative densities are 1.39 ± 0.22 MPa, 3.69 ± 0.63 MPa, and 6.72 ± 0.66 MPa, respectively. The Young's modulus of BCC CCSs is 0.18 ± 0.02 GPa, 0.28 ± 0.07 GPa, and 0.97 ± 0.26 GPa, respectively. The energy CCSs absorb during compression is 9.64 ± 0.98 kJ/m^3, 24.01 ± 2.97 kJ/m^3, 31.44 ± 4.97 kJ/m^3, respectively. It was concluded that the load-bearing capacity and stiffness of CCSs increased as the relative density increased because of the thicker struts. The energy the CCSs consumed in compression was decided based on the compressive strength and strain before CCSs were crushed. Due to the approximate equivalent strains, the varying tendency of energy absorbing abilities of CCSs with different relative densities demonstrated a similar tendency of compressive strength. That is to say, the higher the relative density, the more energy BCC CCSs consumed during compression. Besides, the stress-strain curves of CCSs extracted from simulation conformed with that obtained from testing, meaning simulation was reliable in predicting the mechanical properties of CCSs.

The failure mode of BCC CCSs under compressive loading was further analyzed. Figure 10.4 displays the morphologies of BCC CCSs after compression. It was found

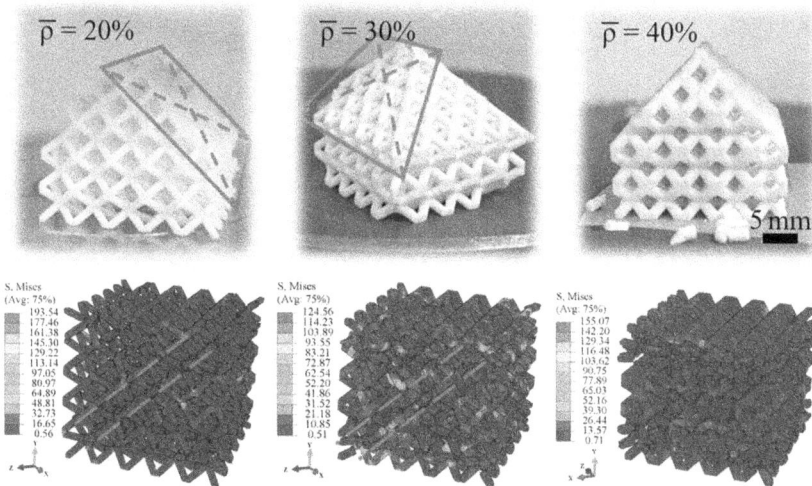

FIGURE 10.4 Final states of BCC CCSs with different relative densities.

that the collapsing behaviors of BCC CCSs with different relative densities were roughly the same, i.e., the struts in CCSs partially fractured at nodes located on a specific plane, liking the phenomena that dislocations slipping along close-packed directions in close-packed planes. Such failure mode of CCSs was markedly different from that of CMSs and CPSs under compressive loading, i.e., CMSs and CPSs were tightly compressed with considerable deformation due to their deformation ability and toughness because of the presence of metal and polymer (Bai et al. 2019; Liu et al. 2020; Rohbeck et al. 2020; Jing et al. 2021). Simulation results also demonstrated such failure mode. However, failure mode of CCSs depended on not only relative density, but also other structural configuration, which will be discussed in the next section (Zhang et al. 2022a; Zhang et al. 2022b).

10.2.2 EFFECT OF STRUCTURAL CONFIGURATION

CCSs with the same relative density, i.e., 30%, but different structural configurations were prepared to comparatively investigate the role of structural configuration in their mechanical properties. Three structural configurations, i.e., BCC, octet, and modified body-centered cubic (MBCC), were involved. Figure 10.5 displays the macro and micro morphologies of these CCSs. It was found that the step-wise morphology similarly appeared on the micro morphology of CCSs. Besides, the difference in the struts' diameters of three CCSs absolutely contributed to their structural configurations.

The typical stress-strain curves of CCSs with a relative density of 30% but different structural configurations were plotted in Figure 10.6. The compressive strength of BCC, octet, MBCC CCSs were 3.69 ± 0.63 MPa, 8.61 ± 0.59 MPa, and 14.38 ± 2.40 MPa, respectively. It was found that the compressive strength of CCSs with structural configuration of MBCC was the best. The mechanical properties of octet CCSs came second, and the mechanical properties of CCSs possessing a

FIGURE 10.5   Macro and micro morphologies of CCSs with a relative density of 30% but different structural configurations.

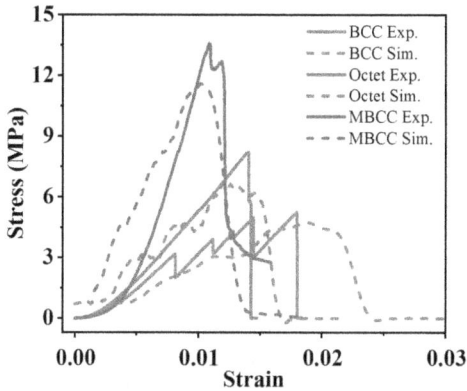

FIGURE 10.6 Experimental and simulating stress-strain curves of CCSs with a relative density of 30% but different structural configurations.

BCC structural configuration was the most insignificant. As known to all, BCC was a representative bending-dominated structure as its struts tended to bend in a compressive process. Octet, on the contrary, was a typical stretching-dominated structure whose struts suffered on stretching or compressive loading in compression. It was extensively proved that a stretching-dominated structure performed more outstandingly than a bending-dominated structure under compressive loading (Mei et al. 2019). Hence, the CCSs with a structural configuration of octet exhibited more excellent strength than that with a BCC structural configuration. Furthermore, MBCC CCSs displayed the most outstanding strength among three CCSs. This was on account of their average connectivity, i.e., 6.88. MBCC ought to be a bending-dominated structure (Li et al. 2020). That is to say, the mechanical properties of CCSs with the structural configuration of MBCC should considerably overlap with those of CCS possessing a BCC structural configuration. However, the fact was rather different. This is because MBCC originated from the typical BCC but possessed more nodes due to the intersected struts, which strengthened the stability of MBCC. Besides, the inclination angle of struts in MBCC was higher than that in BCC and octet, which simultaneously improved the strength of MBCC (Huang et al. 2017; Bai et al. 2020b). Hence, the MBCC CCSs demonstrated the most outstanding compressive load-bearing capacity. Likely, the Young's modulus of MBCC CCSs (2.04 ± 0.11 GPa) was also the highest among three CCSs. The Young's modulus of octet CCSs (0.58 ± 0.10 GPa) was the second-highest. And the BCC CCSs possessed the minimum stiffness (0.28 ± 0.07 GPa). Furthermore, the energy CCSs absorbed during compression depended on the compressive strength and strain. Although the outstanding compression and Young's modulus were usually accompanied by lower strain, the compressive strength prevailed. Hence, the energy absorbing abilities of MBCC CCSs (48.79 ± 13.03 kJ/m^3) and octet CCSs (68.35 ± 16.16 kJ/m^3) were more extraordinary than that of BCC CCSs (24.01 ± 2.97 kJ/m^3). The aforementioned results powerfully testified that appropriately adjusting the structural configurations of CCSs would facilitate tailoring their mechanical properties.

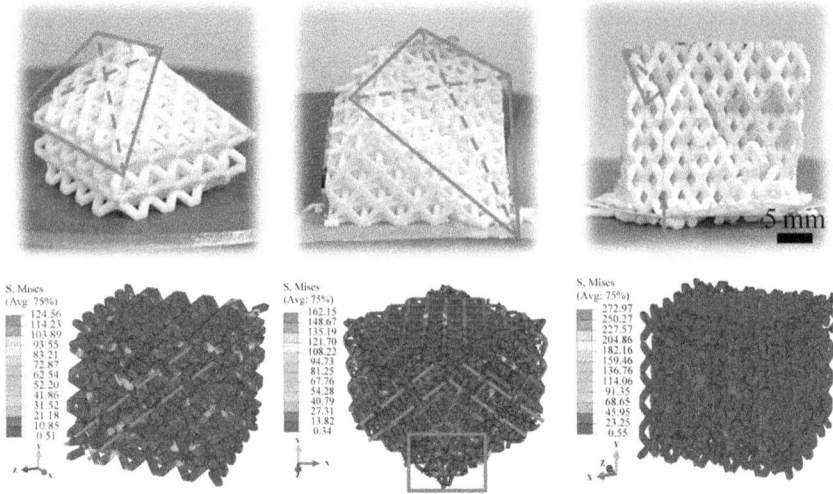

FIGURE 10.7   Final states of CCSs with a relative density of 30% but different structural configurations.

Figure 10.7 reveals the final states of BCC, octet, and MBCC CCSs after compression. It was observed that no matter what the structural configurations were, BCC, octet, and MBCC CCSs with the relative density of 30% exhibited similar final states, i.e., they fractured at nodes locating on a specific plane. During the initial test, CCSs cracked at some weak nodes because of the weakening of defects to the nodes' strength. As tests went on, the complete nodes at neighboring sites were forced to share more loading to make CCSs stable again. Hence, the neighboring nodes got damaged earlier because of the higher loading, making failure nodes concentrated on a specific plane.

However, coupling the effects of relative density and structural configuration, the failure mode of CCSs under compression was quite different. There was no doubt, as described in the last section, that all of BCC CCSs with 20%, 30%, and 40% relative densities fractured along a specific plane. However, something changed on octet and MBCC CCSs. As shown in Figure 10.8, the octet and MBCC CCSs with a relative density of 40% tended to fail as a whole, i.e., most nodes in CCSs were destroyed, rather than partially fracturing along a specific plane. The MBCC CCSs with a relative density of 30% exhibited a combined failure mode of partial fracture and integral fracture, which was more obvious on octet and BCC CCSs with a 30% relative density. That is to say, the proportion of fracturing along a specific plane in failure mode would decrease with the increase in relative density, and the proportion of integral fracture would increase at the same time. Such transformation would be accelerated or depressed by structural configurations.

10.3   MECHANICAL PROPERTIES OF BIO-COMPOSITES

The strong chemical bond among atoms of ceramics, on the one hand, endows ceramics with superior strength and Young's modulus, making them promising candidates

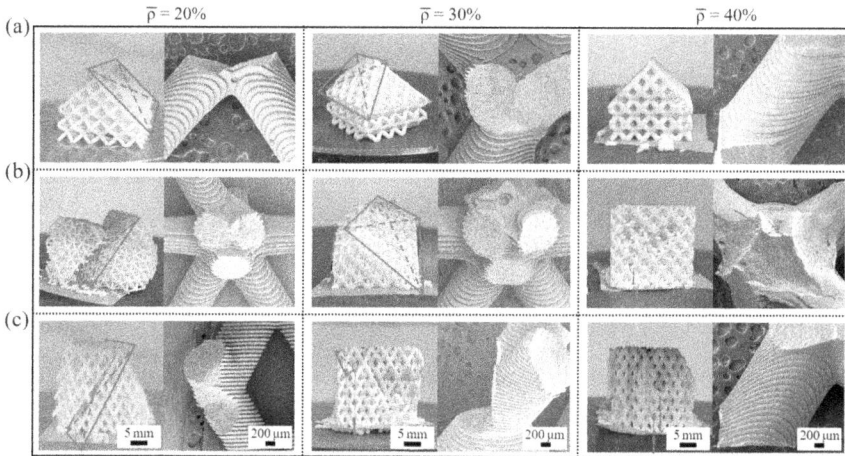

FIGURE 10.8 Final states of CCSs with structural configurations of (a) BCC, (b) octet, and (c) MBCC.

in engineering. On the other hand, strong chemical bonds weaken their toughness, making them too brittle to tolerate damage. Hence, improving the damage-tolerant ability of ceramic components is significant and urgent for their application.

10.3.1 BIOLOGICAL STRUCTURES IN NATURE MATERIALS

Nature materials have become more and more attractive in recent years due to the perfect combination of strong strength and excellent toughness, which can withstand frequent attacks from different prey and the cruel environment. It was revealed that a red abalone shell was composed of 95 vol.% stiff $CaCO_3$ tablets and 5 vol.% tough proteins. The stiff tablets were embodied in the soft protein in a brick-and-mortar form (Barthelat et al. 2007). Conch shell, another representative natural material, was also composed of stiff and tough components arranged in a cross-lamellar structure (Gu et al. 2017). It was concluded that these natural materials were composed of periodically arranged unit cells. Such unit cells were constituted with stiff and tough phases in a core-shell structure. That is to say, the stiff phase acted as the skeleton and the tough phase was the shell covering the skeleton.

Researchers duplicated the biological core-shell structures and prepared a series of bio-inspired composites, aiming to develop materials possessing excellent strength and toughness at the same time. Bouville et al. (2014) prepared the staggered $Al_2O_3/$ glass composite with a brick-and-mortar structure. With the comprehensive effect of pullout of inclusions, shear deformation of tough protein, and zigzag extension of cracks along the interface in the large zone, the strength and toughness of $Al_2O_3/$ glass composites reach as high as 470 MPa and 22 MPa $m^{1/2}$, respectively, far exceeding that of its constituents. Wat et al. (2019) achieved the trade-off and controllability between the strength and toughness of $Al_2O_3/$glass composite by duplicating the brick-and-mortar structure in a red abalone shell. In addition, the flexural strength and fracture work of Al_2O_3/epoxy composite with a cross-lamellar structure reached as high as 165 MPa and 8.2 kJ/m^2, approximately 2.5 and 2 times that of a natural conch shell. Besides, the Al_2O_3/epoxy composite could withstand not only quasi-static but also dynamic damage resistance. It was capable of stopping the drop of the impactor at a 3.61 m/s impacting velocity (Li et al. 2022). Plenty of remarkable achievements proved that duplicating the core-shell structure in natural structures would make the balance between the strength and deformation of ceramic-based composite come true.

10.3.2 Mechanical Properties of Bio-Inspired CCS/Metal Composite

We have discussed the mechanical properties of CCSs in the last section. It was found that once overloaded, CCSs would lose the load-bearing capacity suddenly and be catastrophically destroyed. Conquering the brittleness of CCSs was significant in prompting the application of CCSs. Inspired by the biological core-shell structure in natural materials, we proposed a CCS/metal composite to overcome the poor damage-tolerance ability of CCSs.

The CCSs with the structural configuration of gyroid, a kind of triply periodic minimal surfaces structures, was introduced here to explore the mechanical properties of bio-inspired CCS/metal composites. Gyroid CCSs made of Al_2O_3, as shown in Figure 10.9(a), with a 30% relative density, a $5 \times 5 \times 5$ mm^3 unit cell, and a $6 \times 6 \times 6$ array were prepared by DLP technology. The melt infiltration process was used to get melting AlSi10Mg liquid fill in the space in the Al_2O_3 CCSs in a graphite furnace, forming the bio-composite shown in Figure 10.9(b). It was found that the metal phase and the ceramic phase run through each other. And the interface of the ceramic phase and the metal phase was well combined. Besides, large scratches could be observed on the surface of metal, which was introduced in cutting the redundant metal forming in the melting and infiltration process (Lu et al. 2022).

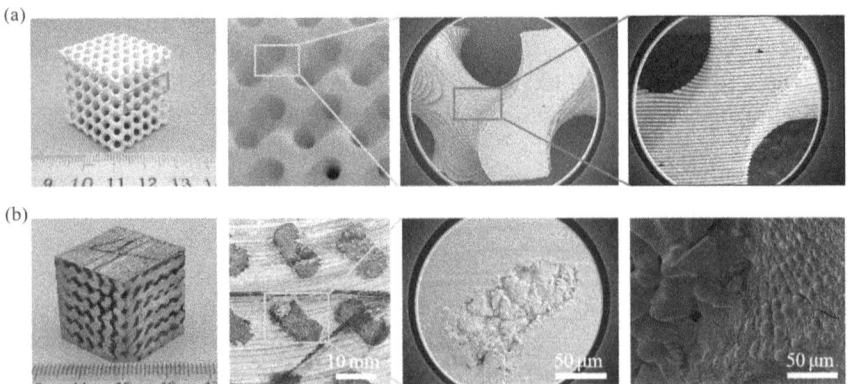

FIGURE 10.9 Macro and micro morphologies of (a) gyroid CCSs and (b) bio-inspired Al_2O_3 CCS/AlSi10Mg composites.

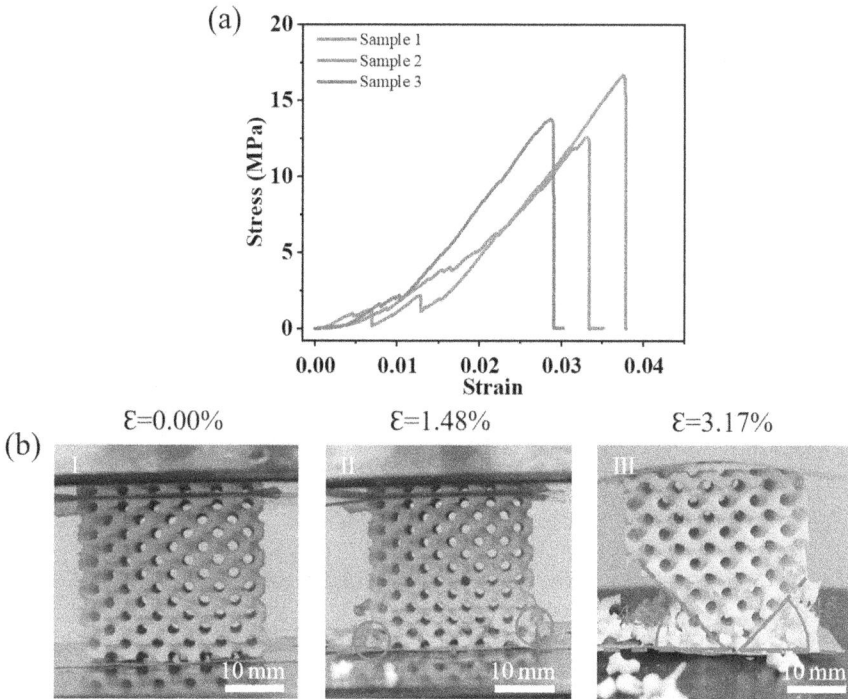

FIGURE 10.10 (a) Stress-strain curves and (b) failure process of CCSs with a relative density of 30% and a structural configuration of gyroid.

The three stress-strain curves of the gyroid CCSs are shown in Figure 10.10(a). The stress gyroid CCSs suffer from during compression increased with increasing deformation in a linear way, which was basically the same with other CCSs. The average compressive strength, Young's modulus, and energy absorption of gyroid CCSs were 14.35 ± 2.11 MPa, 0.62 ± 0.09 GPa, and 0.18 ± 0.03 MJ/m^3, respectively. Although the stiffness of gyroid CCSs was weak, its compressive strength was comparable to that of MBCC CCSs. This is because there was less stress concentrating on the surfaces of gyroid CCSs than that on struts' nodes (Shen et al. 2021). Furthermore, the energy gyroid CCSs absorbed during compression was even ~ four times of MBCC CCSs. The extraordinary mechanical properties proved that triply periodic minimal surface structures had unparalleled advantages with regard to mechanical properties. Besides, as shown in Figure 10.10(b), although gyroid CCSs were composed of connecting curves, they still demonstrated a similar final state as CCSs composed of intersected struts, i.e., with a fracture along a 45° plane.

Different from the stress-strain curves of gyroid CCSs, the stress-strain curves of the bio-composite, as shown in Figure 10.11(a), could be divided into two stages, including the elastic stage and densification stage. At the initial compression, the bio-composite underwent apparent elastic deformation dominated by the gyroid CCSs. Constrained by AlSi10Mg, the gyroid CCSs were capable of providing much higher load-bearing capacity. Hence, the compressive strength of the CCS/AlSi10Mg bio-composite at the end of the elastic stage reached as high as 84.90 ± 4.29 MPa, almost

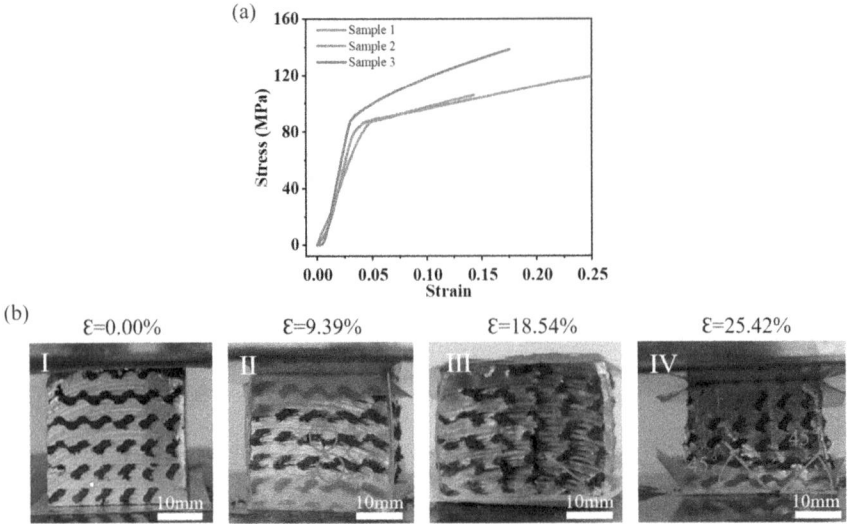

FIGURE 10.11 (a) Stress-strain curves and (b) failure process of bio-inspired Al_2O_3 CCS/ AlSi10Mg composite.

six times that of gyroid CCSs. It was significant that the deformation of bio-composite was significantly improved because of the constriction of AlSi10Mg to broken gyroid CCSs. As tests went on, the stress-strain curve became gentle. The metal in the CCS/ AlSi10Mg bio-composite was continuously squeezed outward, as shown in Figure 10.11(b). Meanwhile, cracks appeared on the surface, and the interface between the ceramic and metal was separated when the strain reached 9.39%. When the strain reached 18.54%, the lower part of the bio-composite gradually expanded outward, the densification of ceramics and metals was completed, and large cracks appeared on the surface of the bio-composite. When the strain increased to 25.42%, the bio-composite was pressed into a trapezoid. It is worth noting that the final cracking direction of the bio-composite was 45°, the same as that of gyroid CCSs. Benefits from the improved compressive strength and extended strain, the bio-composite consumed considerable energy during the compressive process. The energy absorbing ability of the bio-composite reached 7.55 ± 0.76 MJ/m^3 in the strain range of 0–0.1, approximately 43 times that of gyroid CCSs. Such results powerfully testified to the fact that bio-composites are highly suitable for structural components requiring load-bearing and energy absorbing simultaneously.

10.4 CONCLUSIONS

Cellular structures, especially CMSs and CPS, have been extensively investigated. Although CCSs possess outstanding strength and stiffness, poor deformation extremely restricts their fabrication and further investigation. The advent of additive manufacturing technology provides an efficient way to prepare complex and precise CCSs. It was found that structural parameters played a significant role in the mechanical properties and failure mode of CCSs. The higher the relative density, the more

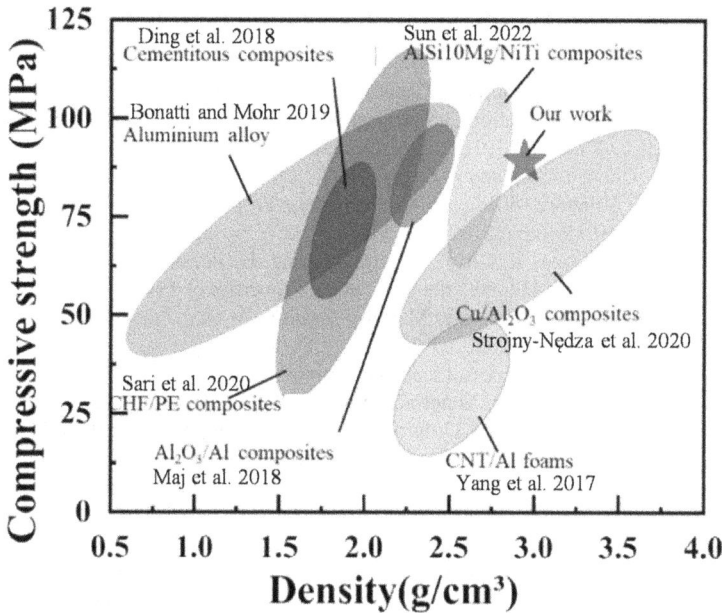

FIGURE 10.12   Ashby map of compressive strength and density of materials.

excellent the mechanical properties of CCSs. The mechanical properties were also closely related to the structural configurations. The failure mode of CCSs under compression contained a partial fracture along a specific plane and integral fracture. The proportions of the two modes were decided by the structural parameters. However, the poor damage tolerance of CCSs limits the application of CCSs.

The dedicated configuration mode of stiff and tough constituents in natural materials offers an elegant way to conquer the poor damage-tolerant ability of CCSs. Inspired by natural materials, a creative bio-inspired CCS/metal composite was designed and fabricated by DLP technology and melting infiltration process. The compressive strength and energy absorption of Al_2O_3 CCS/AlSi10Mg bio-composites were 6 and 47 times that of pristine gyroid CCSs. Furthermore, as seen in Figure 10.12, the bio-composite has a clear advantage with respect to mechanical properties. The experimental results reliably demonstrate that bio-composites have significant potential in not only compressive strength but also energy absorption, and can be reliably used as structural components.

ACKNOWLEDGMENT

This work was financially supported by the National Natural Science Foundation of China [52275310], the Open Project of State Key Laboratory of Explosion Science and Technology (QNKT22-15). The authors also want to sincerely thank the characterization at the Analysis & Testing Center, Beijing Institute of Technology.

REFERENCES

Bai, L., C. Gong, X. Chen, Y. Sun, L. Xin, H. Pu, Y. Peng, and J. Luo. 2020a. "Mechanical Properties and Energy Absorption Capabilities of Functionally Graded Lattice Structures: Experiments and Simulations." *International Journal of Mechanical Sciences* 182: 105735. http://doi.org/10.1016/j.ijmecsci.2020.105735.

Bai, L., C. Yi, X. Chen, Y. Sun, and J. Zhang. 2019. "Effective Design of the Graded Strut of BCC Lattice Structure for Improving Mechanical Properties." *Materials* 12 (13): 2192. http://doi.org/10.3390/ma12132192.

Bai, L., J. Zhang, Y. Xiong, X. Chen, Y. Sun, C. Gong, H. Pu, X. Wu, and J. Luo. 2020b. "Influence of Unit Cell Pose on the Mechanical Properties of Ti6Al4V Lattice Structures Manufactured by Selective Laser Melting." *Additive Manufacturing* 34: 101222. http://doi.org/10.1016/j.addma.2020.101222.

Barthelat, F., H. Tang, P. Zavattieri, C. Li, and H. Espinosa. 2007. "On the Mechanics of Mother-of-Pearl: A Key Feature in the Material Hierarchical Structure." *Journal of the Mechanics & Physics of Solids* 55 (2): 306–37. http://doi.org/10.1016/j.jmps.2006.07.007.

Bonatti, C., and D. Mohr. 2019. "Mechanical Performance of Additively-Manufactured Anisotropic and Isotropic Smooth Shell-Lattice Materials: Simulations & Experiments." *Journal of the Mechanics and Physics of Solids* 122, 1–26. http://doi.org/10.1016/j.jmps.2018.08.022.

Bouville, F., E. Maire, S. Meille, B. Van de Moortèle, A. J. Stevenson, and S. Deville. 2014. "Strong, Tough and Stiff Bioinspired Ceramics from Brittle Constituents." *Nature Materials* 13 (5): 508–14. http://doi.org/10.1038/nmat3915.

Ding, Y., J. Yu, K. Yu, and S. Xu. 2018. "Basic Mechanical Properties of Ultra-High Ductility Cementitious Composites: From 40 MPa to 120 MPa." *Composite Structures* 185: 634–45. http://doi.org/10.1016/j.compstruct.2017.11.034.

Dong, L. 2019. "Mechanical Response of Ti-6Al-4V Hierarchical Architected Metamaterials." *Acta Materialia* 175: 90–106. http://doi.org/10.1016/j.actamat.2019.06.004.

Grunenfelder, L. K., N. Suksangpanya, C. Salinas, G. Milliron, N. Yaraghi, S. Herrera, K. Evans-Lutterodt, S. R. Nutt, P. Zavattieri, and D. Kisailus. 2014. "Bio-Inspired Impact-Resistant Composites." *Acta Biomaterialia* 10 (9): 3997–4008. http://doi.org/10.1016/j.actbio.2014.03.022.

Gu, G. X., M. Takaffoli, and M. J. Buehler. 2017. "Hierarchically Enhanced Impact Resistance of Bioinspired Composites." *Advanced Materials* 29 (28): 1700060. http://doi.org/10.1002/adma.201700060.

Guerra Silva, R., M. J. Torres, J. Zahr Viñuela, and A. G. Zamora. 2021. "Manufacturing and Characterization of 3D Miniature Polymer Lattice Structures Using Fused Filament Fabrication." *Polymers* 13 (4): 635. http://doi.org/10.3390/polym13040635.

He, C., C. Ma, X. Li, L. Yan, F. Hou, J. Liu, and A. Guo. 2020. "Polymer-Derived SiOC Ceramic Lattice with Thick Struts Prepared by Digital Light Processing." *Additive Manufacturing* 35: 101366. http://doi.org/10.1016/j.addma.2020.101366.

Huang, J., S. Daryadel, and M. Minary-Jolandan. 2019. "Low-Cost Manufacturing of Metal-Ceramic Composites through Electrodeposition of Metal into Ceramic Scaffold." *ACS Applied Materials & Interfaces* 11 (4): 4364–72. http://doi.org/10.1021/acsami.8b18730.

Huang, Y., Y. Xue, X. Wang, and F. Han. 2017. "Mechanical Behavior of Three-Dimensional Pyramidal Aluminum Lattice Materials." *Materials Science & Engineering: A* 696: 520–8. http://doi.org/10.1016/j.msea.2017.04.053.

Jana, P., O. Santoliquido, A. Ortona, P. Colombo, and G. D. Sorarù. 2018. "Polymer-Derived SiCN Cellular Structures from Replica of 3D Printed Lattices." *Journal of the American Ceramic Society* 101 (7): 2732–8. http://doi.org/10.1111/jace.15533.

Jing, C., Y. Zhu, J. Wang, F. Wang, J. Lu, and C. Liu. 2021. "Investigation on Morphology and Mechanical Properties of Rod Units in Lattice Structures Fabricated by Selective Laser Melting." *Materials* 14 (14): 3994. http://doi.org/10.3390/ma14143994.

Kaur, M., T. G. Yun, S. M. Han, E. L. Thomas, and W. S. Kim. 2017. "3D Printed Stretching-Dominated Micro-Trusses." *Materials & Design* 134: 272–80. http://doi.org/10.1016/j.matdes.2017.08.061.

Lei, H., C. Li, J. Meng, H. Zhou, Y. Liu, X. Zhang, P. Wang, and D. Fang. 2019. "Evaluation of Compressive Properties of SLM-Fabricated Multi-Layer Lattice Structures by Experimental Test and µ-CT-Based Finite Element Analysis." *Materials & Design* 169, 107685. http://doi.org/10.1016/j.matdes.2019.107685.

Li, C., H. Lei, Z. Zhang, X. Zhang, H. Zhou, P. Wang, and D. Fang. 2020. "Architecture Design of Periodic Truss-Lattice Cells for Additive Manufacturing." *Additive Manufacturing* 34: 101172. http://doi.org/10.1016/j.addma.2020.101172.

Li, M., N. Zhao, M. Wang, X. Dai, and H. Bai. 2022. "Conch-Shell-Inspired Tough Ceramic." *Advanced Functional Materials* 32 (39): 2205309. http://doi.org/10.1002/adfm.202205309.

Liu, W., H. Song, and C. Huang. 2020. "Maximizing Mechanical Properties and Minimizing Support Material of Polyjet Fabricated 3D Lattice Structures." *Additive Manufacturing* 35: 101257. http://doi.org/10.1016/j.addma.2020.101257.

Lu, J., D. Wang, K. Zhang, S. Li, B. Zhang, X. Zhang, L. Zhang, W. Wang, Y. Li, and R. He. 2022. "Mechanical Properties of Al_2O_3 and Al_2O_3/Al with Gyroid Structure Obtained by Stereolithographic Additive Manufacturing and Melt Infiltration." *Ceramics International* 48 (16): 23051–60. http://doi.org/10.1016/j.ceramint.2022.04.283.

Ma, Z., D. Z. Zhang, F. Liu, J. Jiang, M. Zhao, and T. Zhang. 2020. "Lattice Structures of Cu-Cr-Zr Copper Alloy by Selective Laser Melting: Microstructures, Mechanical Properties and Energy Absorption." *Materials & Design* 187: 108406. http://doi.org/10.1016/j.matdes.2019.108406.

Maj, J., M. Basista, W. Węglewski, K. Bochenek, A. Strojny-Nędza, K. Naplocha, T. Panzner, M. Tatarková, and F. Fiori. 2018. "Effect of Microstructure on Mechanical Properties and Residual Stresses in Interpenetrating Aluminum-Alumina Composites Fabricated by Squeeze Casting." *Materials Science and Engineering: A* 715: 154–62. http://doi.org/10.1016/j.msea.2017.12.091.

Maskery, I., L. Sturm, A. O. Aremu, A. Panesar, C. B. Williams, C. J. Tuck, R. D. Wildman, I. A. Ashcroft, and R. J. M. Hague. 2018. "Insights into the Mechanical Properties of Several Triply Periodic Minimal Surface Lattice Structures Made by Polymer Additive Manufacturing." *Polymer* 152: 62–71. http://doi.org/10.1016/j.polymer.2017.11.049.

Maurath, J., and N. Willenbacher. 2017. "3D Printing of Open-Porous Cellular Ceramics with High Specific Strength." *Journal of the European Ceramic Society* 37 (15): 4833–42. http://doi.org/10.1016/j.jeurceramsoc.2017.06.001.

Mei, H., R. Zhao, Y. Xia, J. Du, X. Wang, and L. Cheng. 2019. "Ultrahigh Strength Printed Ceramic Lattices." *Journal of Alloys and Compounds* 797: 786–96. http://doi.org/10.1016/j.jallcom.2019.05.117.

Mei, H., W. Yang, X. Zhao, L. Yao, Y. Yao, C. Chen, and L. Cheng. 2021. "In-Situ Growth of SiC Nanowires@Carbon Nanotubes on 3D Printed Metamaterial Structures to Enhance Electromagnetic Wave Absorption." *Materials & Design* 197: 109271. http://doi.org/10.1016/j.matdes.2020.109271.

Munch, E., M. E. Launey, D. H. Alsem, E. Saiz, A. P. Tomsia, and R. O. Ritchie. 2008. "Tough, Bio-Inspired Hybrid Materials." *Science (American Association for the Advancement of Science)* 322 (5907): 1516–20. http://doi.org/10.1126/science.1164865.

Patek, S. N., and R. L. Caldwell. 2005. "Extreme Impact and Cavitation Forces of a Biological Hammer: Strike Forces of the Peacock Mantis Shrimp *Odontodactylus* Scyllarus." *Journal of Experimental Biology* 208 (19): 3655–64. http://doi.org/10.1242/jeb.01831.

Rohbeck, N., R. Ramachandramoorthy, D. Casari, P. Schürch, T. E. J. Edwards, L. Schilinsky, L. Philippe, J. Schwiedrzik, and J. Michler. 2020. "Effect of High Strain Rates and Temperature on the Micromechanical Properties of 3D-Printed Polymer Structures Made by Two-Photon Lithography." *Materials & Design* 195: 108977. http://doi.org/10.1016/j.matdes.2020.108977.

Sari, N. H., C. I. Pruncu, S. M. Sapuan, R. A. Ilyas, A. D. Catur, S. Suteja, Y. A. Sutaryono, and G. Pullen. 2020. "The Effect of Water Immersion and Fibre Content on Properties of Corn Husk Fibres Reinforced Thermoset Polyester Composite." *Polymer Testing* 91, 106751. http://doi.org/10.1016/j.polymertesting.2020.106751.

Shen, M., W. Qin, B. Xing, W. Zhao, S. Gao, Y. Sun, T. Jiao, and Z. Zhao. 2021. "Mechanical Properties of 3D Printed Ceramic Cellular Materials with Triply Periodic Minimal Surface Architectures." *Journal of the European Ceramic Society* 41 (2): 1481–9. http://doi.org/10.1016/j.jeurceramsoc.2020.09.062.

Shin, Y. A., S. Yin, X. Li, S. Lee, S. Moon, J. Jeong, M. Kwon, S. J. Yoo, Y. M. Kim, T. Zhang, H. Gao, and S. H. Oh. 2016. "Nanotwin-Governed Toughening Mechanism in Hierarchically Structured Biological Materials." *Nature Communications* 7: 10772. http://doi.org/10.1038/ncomms10772.

Strojny-Nędza, A., K. Pietrzak, F. Gili, and M. Chmielewski. 2020. "FGM Based on Copper-Alumina Composites for Brake Disc Applications." *Archives of Civil and Mechanical Engineering* 20 (3): 83. http://doi.org/10.1007/s43452-020-00079-1.

Sun, X., F. Jiang, D. Yuan, G. Wang, Y. Tong, and J. Wang. 2022. "High Damping Capacity of AlSi10Mg-NiTi Lattice Structure Interpenetrating Phase Composites Prepared by Additive Manufacturing and Pressureless Infiltration." *Journal of Alloys and Compounds* 905: 164075. http://doi.org/10.1016/j.jallcom.2022.164075.

Sun, Y., W. Tian, T. Zhang, P. Chen, and M. Li. 2020. "Strength and Toughness Enhancement in 3D Printing via Bioinspired Tool Path." *Materials & Design* 185: 108239. http://doi.org/10.1016/j.matdes.2019.108239.

Tancogne-Dejean, T., A. B. Spierings, and D. Mohr. 2016. "Additively-Manufactured Metallic Micro-Lattice Materials for High Specific Energy Absorption under Static and Dynamic Loading." *Acta Materialia* 116: 14–28. http://doi.org/10.1016/j.actamat.2016.05.054.

Wang, W., X. Bai, L. Zhang, S. Jing, C. Shen, and R. He. 2022. "Additive Manufacturing of Csf/SiC Composites with High Fiber Content by Direct Ink Writing and Liquid Silicon Infiltration." *Ceramics International* 48 (3): 3895–903. http://doi.org/10.1016/j.ceramint.2021.10.176.

Wat, A., J. I. Lee, C. W. Ryu, B. Gludovatz, J. Kim, A. P. Tomsia, T. Ishikawa, J. Schmitz, A. Meyer, M. Alfreider, D. Kiener, E. S. Park, and R. O. Ritchie. 2019. "Bioinspired Nacre-Like Alumina with a Bulk-Metallic Glass-Forming Alloy as a Compliant Phase." *Nature Communications* 10 (1): 961. http://doi.org/10.1038/s41467-019-08753-6.

Weeger, O., N. Boddeti, S. K. Yeung, S. Kaijima, and M. L. Dunn. 2019. "Digital Design and Nonlinear Simulation for Additive Manufacturing of Soft Lattice Structures." *Additive Manufacturing* 25: 39–49. http://doi.org/10.1016/j.addma.2018.11.003.

Yan, C., L. Hao, A. Hussein, P. Young, and D. Raymont. 2014. "Advanced Lightweight 316L Stainless Steel Cellular Lattice Structures Fabricated via Selective Laser Melting." *Materials & Design* 55: 533–41. http://doi.org/10.1016/j.matdes.2013.10.027.

Yang, K., X. Yang, E. Liu, C. Shi, L. Ma, C. He, Q. Li, J. Li, and N. Zhao. 2017. "Elevated Temperature Compressive Properties and Energy Absorption Response of In-Situ Grown CNT-Reinforced Al Composite Foams." *Materials Science and Engineering: A* 690, 294–302. http://doi.org/10.1016/j.msea.2017.03.004.

Zhang, M., N. Zhao, Q. Yu, Z. Liu, R. Qu, J. Zhang, S. Li, D. Ren, F. Berto, Z. Zhang, and R. O. Ritchie. 2022. "On the Damage Tolerance of 3-D Printed Mg-Ti Interpenetrating-Phase Composites with Bioinspired Architectures." *Nature Communications* 13 (1): 3247. http://doi.org/10.1038/s41467-022-30873-9.

Zhang, X., K. Zhang, B. Zhang, Y. Li, and R. He. 2022. "Additive Manufacturing, Quasi-Static and Dynamic Compressive Behaviours of Ceramic Lattice Structures." *Journal of the European Ceramic Society* 42 (15): 7102–12. http://doi.org/10.1016/j.jeurceramsoc.2022.08.018.

Zhang, X., K. Zhang, B. Zhang, Y. Li, and R. He. 2022a. "Quasi-Static and Dynamic Mechanical Properties of Additively Manufactured Al_2O_3 Ceramic Lattice Structures: Effects of Structural Configuration." *Virtual and Physical Prototyping* 17 (3), 528–42. http://doi.org/10.1080/17452759.2022.2048340.

Zhang, X., K. Zhang, L. Zhang, W. Wang, Y. Li, and R. He. 2022b. "Additive Manufacturing of Cellular Ceramic Structures: From Structure to Structure-Function Integration." *Materials & Design* 215: 110470. http://doi.org/10.1016/j.matdes.2022.110470.

Zheng, H., L. Liu, C. Deng, Z. Shi, and C. Ning. 2019. "Mechanical Properties of AM Ti6Al4V Porous Scaffolds with Various Cell Structures." *Rare Metals* 38 (6): 561–70. http://doi.org/10.1007/s12598-019-01231-4.

Index

acrylonitrile butadiene styrene (ABS) 7–10, 65–6

biodegradable 76–9

coconut fiber 62–74

extrusion process 31–3

fiber preparation 25–8
fused deposition modeling 13, 24, 31, 36, 44, 50, 98

lattice 142, 204–11

mechanical properties 34, 61, 66, 88–9, 205–14

poly lactic acid 6, 44, 48, 76–91

sandwich core 102–3
selective laser sintering 122–3

For Product Safety Concerns and Information please contact our EU
representative GPSR@taylorandfrancis.com
Taylor & Francis Verlag GmbH, Kaufingerstraße 24, 80331 München, Germany

www.ingramcontent.com/pod-product-compliance
Lightning Source LLC
Chambersburg PA
CBHW060403220326
41598CB00023B/3004